— THE —

Knife
Man

Portrait of John Hunter by Henry Bone after Sir Joshua Reynolds

— THE —

Knife
Man

The Extraordinary Life
and Times of John Hunter,
Father of Modern Surgery

Wendy
Moore

Broadway Books
New York

Broadway Books titles may be purchased for business or promotional use or for special sales. For information, please write to: Special Markets Department, Random House, Inc., 1745 Broadway, New York, NY 10019.

PRINTED IN THE UNITED STATES OF AMERICA

BROADWAY BOOKS and its logo, a letter B bisected on the diagonal, are trademarks of Random House, Inc.

Portrait of John Hunter on page ii: Reproduced by permission of the President and Council of the Royal College of Surgeons of England

Visit our website at www.broadwaybooks.com

First edition published 2005.

Book design by Donna Sinisgalli

Library of Congress Cataloging-in-Publication Data
Moore, Wendy.
The knife man / Wendy Moore.
p. cm.
1. Hunter, John, 1728–1793. 2. Surgeons—Great Britain—Biography.
I. Title.

RD27.35.H86M66 2005
617'.092—dc22
[B]
2005045723

ISBN 0-7679-1652-2

10 9 8 7 6 5 4 3 2 1

For Peter,

Sam,

and Susie

I have made candles of infants fat
The Sextons have been my slaves,
I have bottled babes unborn, and dried
Hearts and livers from rifled graves.

From "The Surgeon's Warning,"
Robert Southey, *Poems*, 1799

Contents

CHAPTER 1

The Coach Driver's Knee

⊙ᴍᴍᴖ

St. George's Hospital, Hyde Park Corner, London
December 1785
The patient faced an agonizing choice. Above the cries and moans of fellow sufferers on the fetid ward, he listened as the surgeon outlined the dilemma. If the large swelling at the back of his knee was left to continue growing, it would soon burst, leading to certain and painful death. If, on the other hand, the leg was amputated above the knee, there was a slim chance he would survive the crude operation—provided he did not die of shock on the operating table, or bleed to death soon after, or succumb to infection on the filthy ward days later—but he would be permanently disabled.

For the forty-five-year-old hackney coach driver, both options were unthinkable. Since he had first noticed the swelling in the hollow behind his knee three years ago, the lump had grown steadily, until it was the size of an orange.[1] It throbbed continuously and was now so painful, he could barely walk. Extended on the hospital bed before him, his leg and foot were hideously swollen, while his skin had turned an unsightly mottled brown. Once the coachman had gained admittance to St. George's, having persuaded the governors he was a deserving recipient of their charity, the surgeon on duty had lost no time in making a diagnosis. He had seen popliteal aneurysms at exactly the same spot on numerous occasions and knew the prognosis all too well.

It was a common-enough problem in the cabdriver's line of work. Aneurysms could happen to anyone, anywhere in the body, but they appeared to occur with particular frequency among coach drivers, and others

in equestrian occupations in Georgian London, in the popliteal artery behind the knee. The condition, in which a section of artery that has been injured or otherwise weakened begins to bulge to form a blood-filled sac, may
well have been triggered by the wearing of high leather riding boots, which
rubbed the back of the knee.[2] As the aneurysm swelled, it not only became
extremely painful but made walking exceedingly difficult. Whatever the
cause, the outcome was often an early death—if not from the condition itself, then from the treatment generally meted out. To lose his leg, even supposing the coach driver survived such a drastic procedure in an era long
before anesthesia or antiseptics, would mean never being able to work
again. But to carry on working, navigating his horse-drawn carriage over
London's rutted and congested roads, would be equally impossible if the
lump was left to grow. Either way, the cabbie feared destitution and the
workhouse.

But there was a third choice, the surgeon at his bedside now confided
on that early December day, for a coachman sufficiently willing or desperate. In his slow Scottish lilt, redolent of his humble farming origins, the
surgeon laid out his scheme for a daring new operation. Surrounded by the
poxed, maimed, and diseased bodies of London's poorest wretches, hud-
dled in their beds on the drafty ward, the cabbie resolved to put his life in
the hands of John Hunter.

Without a doubt, John Hunter's reputation was well known to the
coach driver long before he limped through the portal of St. George's, for
he was generally acknowledged as one of the best-skilled surgeons in
London, if not Europe, and was a favorite among the well-heeled and the
unshod alike. As well as working for no recompense patching up the poor
in St. George's, he was in constant demand from the fee-paying patients
who thronged each morning to his fashionable home in Leicester Square
or called him out for consultations in the elegant drawing rooms of their
West End villas. For all his blunt manners, coarse speech, and disdain for
fashion—he currently sported an unkempt beard and tied his tawny-
colored hair behind his head in preference to wearing the customary wig—
Hunter was firmly established in Georgian high society. He visited court as
surgeon extraordinary to George III, dined with the society artist Sir Joshua
Reynolds, and debated science with his close friend, the well-connected
naturalist Sir Joseph Banks.

Now aged fifty-seven, with seventeen years' service at St. George's under his belt, Hunter was renowned for his pioneering and controversial operations. Only two months before the coach driver's admission, he had skillfully cut away from the neck of a thirty-seven-year-old man a massive benign tumor weighing more than eight pounds and roughly the size of an extra head. The relieved patient had walked away with only a long, neat scar as souvenir of his ordeal.[3] Hunter was popular with the medical students, too. The coachman had watched the eager pupils trooping devotedly after their teacher on his ward rounds, for more students flocked to Hunter's side than to all the other surgeons at St. George's put together.[4] Aspiring young surgeons traveled not only from the far reaches of the British Isles but even from across the Atlantic to "walk the wards" at Hunter's side and hear their hero expound on his radical views in the private lectures he held at his home each winter.

But the cabbie would have heard darker stories, too, whispered on the wards, insinuated in newspapers, and muttered in coffeehouses and cockpits, for Hunter was as much feared and despised as admired in eighteenth-century London. Although his pupils idolized their master, and patients often had cause to thank the bluff but honest surgeon, Hunter's fiery temper and maverick views had earned him powerful enemies within the four walls of St. George's, and beyond. While aristocrats bowed to his medical advice, and denizens of the Royal Society—the engine room of eighteenth-century progress—hung on his every pronouncement, Hunter was isolated at St. George's. To his fellow surgeons, he was at best a laughingstock and at worst a reckless fool. And he had quarreled, too, with several of the city's other leading practitioners, not least his own elder brother.

To the students, the explanation for this was straightforward: Hunter was simply so far ahead of his contemporaries that he stood alone. But his rivals at St. George's had other opinions. They decried Hunter's novel approach and controversial methods, preferring to bleed, blister, and purge their patients to early graves—in strict accordance with classical teaching—than to question conventional modes of practice. They even encouraged Hunter's most vociferous enemy, a mediocre house surgeon named Jessé Foot, who worked in a neighboring hospital, and whom Hunter had upset by criticizing a surgical appliance the young upstart had invented. In Foot's jaundiced view, Hunter was "a very inferior, dangerous, and irregular practical surgeon" who was embroiled in "continual war" at St. George's.[5]

But there were stranger stories still about the rebellious surgeon.

Hunter was known to keep rare and exotic wild beasts—including a lion, a jackal, a dingo, and two leopards—at his country home in the tranquil village of Earls Court, a few miles west of London. In this rural retreat, the surgeon performed countless experiments on animals both dead and alive. Innumerable research papers, presented to his friends in the Royal Society, detailed his bizarre trials, such as grafting a cockerel's testicle into the belly of a hen—an early step toward transplanting body parts in humans—as well as the freezing of fish and rabbits' ears in a forlorn attempt to invent a scheme for human immortality. At this prototype research center, Hunter dissected great carcasses, including whales washed up on the banks of the Thames, apes sent back from explorations into unmapped territories, and elephants donated by Queen Charlotte. It was here, too, that he experimented on living animals, tying down squealing pigs, sheep, and dogs for lengthy dissections in order to explore how healthy organs function and to test ways to improve surgery. Although by now they had become inured to the sight of rare beasts grazing the lawns, Hunter's curious neighbors still gaped on occasions when the surgeon set out from Earls Court driving a cart pulled by three Asian buffalo, headed for the West End. Arriving at his Leicester Square town house, a drawbridge could be swiftly lowered—and just as swiftly raised—to allow mysterious cargoes to trundle in and out.[6]

The enterprising surgeon did not confine his zeal for research to the animal kingdom, however, for Hunter had built up his surgical expertise through an unrivaled knowledge of human anatomy. Since arriving in London almost four decades earlier, Hunter had dissected human bodies in unprecedented numbers. By his own admission, he had carved up "some thousands" during his lengthy career.[7] It was through this relentless first-hand exploration of the human body, rather than by reading the works of the ancient Greeks and Romans or passively watching over the shoulders of other practitioners, that Hunter had become such a skilled operator. Although other surgeons of the day had become adept at certain procedures through trial and error, many operations performed in London's charity hospitals were still risky gambles, due to ignorance of anatomy and physiology. Whenever Hunter cut, probed, sliced, and sawed, he knew precisely what lay beneath. He possessed a better knowledge than any other surgeon in town of the exact whereabouts, functions, and habits of every organ, muscle, blood vessel, and tissue, healthy or diseased, that he was likely to encounter.

Yet while many of Hunter's patients, rich and poor, were grateful for their surgeon's intimate knowledge of the human body, few gave their approval to the sinister extremes to which he went to obtain his research material. Like most surgeons and anatomists of the time, Hunter had no lawful source for the majority of the bodies he daily dissected. Like others, he was forced to adopt underhanded means to pursue his work, turning to London's criminal elements to supply his needs. Hunter, however, went further than any other anatomist of the day in his connections with the Georgian underworld. Desperate to expand his collection of animal and human specimens in the remarkable museum he had just established at his Leicester Square house, Hunter was notorious for paying above the norm for any kind of anatomical curiosity—whether rare human deformities or the results of a pioneering, but ultimately fatal, operation.

So as the tortured coach driver stared anxiously from his hospital bed into Hunter's pale blue eyes, he knew that volunteering to go under the renowned surgeon's knife in the operating theater at St. George's could very well mean going under his knife a second time—dead on the dissection table, with his mutilated leg destined for the anatomist's museum. Nonetheless, clinging to the slim prospect of recovery held out to him, he gave John Hunter his consent for the new operation.

For Hunter, the coach driver represented the perfect human guinea pig. Until now, he had tried to save patients with popliteal aneurysms from either a premature death or a life-threatening amputation by performing an operation that he knew was highly risky and exceedingly painful. This procedure, which had been attempted by surgeons across Europe with various refinements for some centuries, meant cutting straight into the back of the knee, tying the damaged artery above and below the aneurysm, and scraping out the blood-filled sac. Based on his knowledge of anatomy and research, Hunter believed that once prevented from taking its usual path down the popliteal artery, the blood supply would find alternative routes— or "collateral circulation"—through the smaller blood vessels in the area. Almost without exception, the technique had failed. More often than not, the already-weakened blood vessel burst and the patient bled to death, making Hunter the butt of his colleagues' scorn.

Given such outcomes, Percivall Pott, a respected surgeon at St. Bartholomew's Hospital, as well as Hunter's former teacher, insisted that amputation was the only viable remedy for popliteal aneurysms—even though he admitted the procedure was "terrible to bear" and "horrid to see."[8] The surrounding blood vessels in the leg were simply too few and too small, Pott argued, to foster collateral circulation once the main artery was tied. Conversely, William Bromfield, the recently retired senior surgeon at St. George's, who had once been Hunter's examiner for his surgeon's diploma, proclaimed that neither amputation nor "Hunter's operation" were of the slightest use in treating the condition. Mistakenly averring that a popliteal aneurysm indicated that a patient's entire arterial system was diseased, he insisted that there was simply no cure for the problem.[9] It was an opinion that effectively dealt a death sentence to every poor coachman who ventured into Bromfield's care.

Never one to bow to authority or defer to his elders, Hunter was not prepared simply to abandon hope and consign his patients to a lingering, agonizing death. His aversion to drastic surgery except as a last resort, and his firm belief in nature's healing powers, made him equally unwilling to perform amputations. Refusing to be defeated, he resolved to apply his singular approach to the problem. Unlike the vast majority of surgeons operating in London homes and hospitals, who wielded their lancets and saws in imitation of long-dead past masters while rarely considering any need for improvement, Hunter believed all surgery should be governed by scientific principles, which were based on reasoning, observation, and experimentation.

Techniques for most operations had changed little since medieval times, while treatment regimes still owed their basic principles largely to the theories of the ancient Greeks. Although medical students usually learned some rudimentary anatomy, this was considered a useful but not vital adjunct to on-the-job experience. And when patients died on the operating table as a result of ignorance and blundering, as they frequently did, few, if any, lessons were learned from the outcome.

Hunter, however, had enjoyed a distinctly different preparation for the job, having spent a full twelve years studying anatomy in his brother William's dissecting rooms and only a brief spell walking the wards, before embarking on his surgical career. As a consequence, he considered anatomy the foundation stone for all surgery. He believed that only by minutely studying the human body, in order to understand the whereabouts and functions of every living part, could surgeons possibly hope to improve their skills. Furthermore, his experience on the hospital wards had taught him principally, in ghastly and bloody detail, how primitive his profession really was.

So Hunter had set out systematically to question every established practice, develop hypotheses to advance better methods, and test by means of rigorous observation, investigation, and experiment whether these methods worked. Often he tried out his theories on animals before attempting new procedures on humans, while the results of the handiwork he performed in the operating theater were always carefully observed during the patient's subsequent recovery, or, if the person died, in autopsy investigations. The lessons he learned were diligently applied to modify his methods, resulting in a continuous research loop, which would still stand up to scientific scrutiny more than two centuries later. Instilling the same approach in his pupils, Hunter summed up his doctrine in characteristically concise style when his favorite protégé, Edward Jenner—who would later develop the smallpox vaccine—asked him for help in solving a problem. "I think your solution is just," responded his mentor, "but why think, why not trie the Expt."[10]

Accordingly, Hunter set out to apply his scientific doctrine to the coachman's problem. Despite the fact that his usual method of operating on popliteal aneurysms by tying the artery at the point of the damage—behind the knee—had so far failed, he remained convinced that the basic theory of collateral circulation held good. An understanding of collateral circulation had been grasped even by the Romans, who had tied small blood vessels when damaged and then relied on alternative routes for the blood supply, as far back as the first century A.D.[11] Yet Hunter's peers refused to believe that collateral circulation could provide an adequate blood supply to nourish an entire leg once the main artery was severed, or that it could develop sufficiently quickly before gangrene set in. Knowing his attempts to tie the artery behind the knee had failed, he asked his pupils, "Why not tie it up higher in the sound parts, where it is tied in amputation, and preserve the limb?"[12] Tying the artery in the patient's thigh, where the blood vessel was healthy, might have more chance of holding firm, he theorized, yet still allow the blood flow to create bypass routes around the diseased area lower down.

Applying his usual scientific method, first Hunter conducted experiments on animals to test out his novel hypothesis, choosing dogs as his substitute patients. In one trial, he instructed his assistant, Hunter's young brother-in-law, Everard Home, to cut open the leg of a live dog and expose the artery.[13] Under Hunter's directions, Home carefully peeled away layer after layer of artery walls, until the vessel was so thin that the blood could be seen as it pumped through. Home sewed up the dog's leg and six weeks later killed the animal, whereupon he discovered that the artery had fully recovered its normal thickness. The results clearly suggested that the thinness of the artery walls did not account for the failure of the operations, confirming Hunter's belief that the section of artery close to the aneurysm must be otherwise damaged or diseased. He was now convinced that tying the popliteal artery behind the knee, close to the aneurysm, was clinically doomed. The only chance of success was to tie the same vessel higher in the leg, where it is called the femoral artery.

Hunter even tested his theory on a stag from the royal herd in Richmond Park after gaining permission from George III.[14] Having caught the stag, Hunter cut into its neck and tied the carotid artery, which supplied blood to the antlers. Soon enough, the half-grown antlers grew cold to the touch, as he had expected. But a week or so later, he discovered that the antlers had not only regained their warmth but continued to grow. Ordering the deer to be slaughtered and sent to his dissecting room, he opened the beast and injected its blood vessels with a colored resin. Now he discovered that the stag's blood supply to the antlers had found a new route through smaller blood vessels, bypassing the tied artery. The experiment provided further proof to support his belief in collateral circulation. It was now abundantly clear that blood could find alternative routes, and quickly, too.

Despite his passion for experiment, Hunter never approached an operation lightly. No operation should be carried out unless absolutely necessary, he urged his students, nor should a surgeon operate unless he would undergo the same operation himself in similar circumstances. Outlining his plan for the proposed new popliteal aneurysm operation, he declared, "In this account, it may be supposed that I carry my notion too far; but it is to be understood that I only give my own feelings upon this subject, and I go no further in theory than I would perform in practice, if patients, being acquainted with the consequences of the disease, would submit to, or rather desire, the operation; nor do I go further than I now think I would have performed on myself were I in the same situation."[15] Having therefore acquainted the coach driver with the likely consequences of his condition if left unattended, and having advanced his opinion as to the most amenable remedy, Hunter left his patient in little doubt that the new operation was his best hope for the future.

Just days later, on December 12, the coachman was helped into the operating theater—situated rather inconveniently above the hospital's boardroom—to begin his ordeal. Although minor operations were still sometimes carried out in patients' beds in the wards—and some beds were equipped with straps and handcuffs for the purpose—Hunter desired an audience of students and fellow surgeons for his radical experiment. Covering his normal day clothes with an apron, stiff with the dried blood of past patients, he wore neither gloves nor mask. With no understanding of how infection transferred from a surgeon's begrimed fingers to a patient's open wounds—this was almost a century before Joseph Lister pioneered the use of antiseptics—he had no cause to wash his hands or sterilize his instruments. As his personal set of knives lay ready and junior surgeons craned their necks to see, Hunter checked to make sure that his patient still consented to the operation. Then, grabbed by muscular assistants, the cabbie was held tightly on the narrow wooden operating table, his leg firmly secured. With only a stiff shot of brandy, or at best some laudanum, a heady mixture of alcohol and opium, to dull the coming pain—it would be many decades before the introduction of effective anesthesia—the coach driver tensed for the first cut. Well aware the experimental operation was still very much a gamble, he knew that Hunter's whiskered face might be the last he would ever see.

As Hunter raised his knife to cut into the coachman's thigh, a young Italian surgeon, Paolo Assalini, was among the spellbound observers crowded in the theater to watch the spectacle. He noted every step in the pioneering operation.

> John Hunter, having placed the patient in a convenient position, made an incision about five inches long in the skin of the inner, lower part of the thigh with a small knife; he exposed the artery, then grasped it in his fingertips and separated it from the cellular membrane, the vein and the nerve for a distance of eight inches, making use only of the slender han-

dle of his knife. This done, he passed under the artery a grooved probe, and ran along this a silver needle threaded with a thick ligature which he tied on the lower part of the artery to block it off. The pulsation in the tumour ceased immediately but the blood thumped with such force against the ligature that to diminish the knocking and the danger of the thing bursting he made another ligature two inches away from the first one and tied it less tightly. Above this second one he made a third, tied even more loosely, and then a fourth. . . . Then, after pulling the threads to the outside, and separating them, he closed the lips of the wound.[16]

Swiftly, Hunter now bound the two sides of the wound together with sticking plaster and bandages—stitches were rarely used—leaving the ends of the threads hanging out to be removed later. He had to work quickly to keep blood loss to a minimum and to reduce the risk of traumatic shock, so the whole process would have taken less than five minutes.[17]

By the following day, the coachman's leg was already less swollen and the lump one-third smaller. By the fourth day, when the dressings were removed, the wound had sealed and the patient reported no pain. After two weeks, the silk threads fell away and the lump was significantly reduced. At the end of January 1786, six weeks after his operation, the coach driver walked unaided out of St. George's Hospital, and by July he was back in the driving seat of his hackney carriage, transporting passengers around London's chaotic streets.

As word of the dramatic new operation spread, it was soon copied in other London hospitals. Indeed, Hunter's operation rapidly became the standard procedure for aneurysms in various arteries, with his technique being adopted in France and Italy. The thrilled Assalini reported, "This operation, followed by such a happy outcome, excited the greatest wonder and awakened the attention of all the surgeons in Europe."[18]

As with so many pioneering medical advances, however, progress was not completely unhindered. After almost a year back at work, exposed to all the elements, the poor coach driver caught a fever—unconnected with either his aneurysm or the operation—and on April 1, 1787, he died. "He had not made any complaint of the limb on which the operation had been performed," Hunter's brother-in-law, Everard Home, later made plain, "from the time of his leaving the hospital." Seven days after his patient had died, and not without "some trouble and considerable expense," Hunter used his usual connections to obtain the cabbie's leg. Excitedly, he took the limb back to his attic dissecting room, anxious to see for himself the results of his great experiment. Once more, he wielded his knife and cut into the coach driver's leg. The aneurysm behind the knee was easy enough to find, being still "larger than a hen's egg," but the femoral artery above it had formed new routes through smaller blood vessels to bypass the diseased area and rejoin the artery farther down—exactly as Hunter had predicted. The leg was otherwise perfectly healthy; the operation had been a success.

Hunter went on to repeat his operation on at least four more occasions, and although one of these patients died soon after, his fourth patient, another coach driver, not only recovered completely—recuperating in the countryside at Hunter's expense—but went on to live for fifty more years.[19]

The operation that Hunter had performed on the first coach driver not only extended his reputation for daring and experimentation among Europe's surgeons and enhanced his prowess in the adoring eyes of his pupils, it would save countless lives, engender a whole new era of similar operations, and help to establish the foundations of scientific surgery. The procedure, which was a perfect example of the way Hunter applied scientific principles to every aspect of his work, would ensure his name lived on. The technique was henceforth called "Hunter's operation," and the site of his knife work was ever after known as "Hunter's canal."

These tributes alone might have been enough for any ambitious surgeon hoping to secure a place in the medical hall of fame, but they would never have satisfied the irrepressible John Hunter. He barely paused to acknowledge his advance, and never personally wrote about it, before plunging back into his daily round of ceaseless experiments. He had far too many other medical interests, from artificial insemination to electrical stimulation of the heart, and far too broad a vision, spanning botany, geology, early biology, and even the origins of life itself, ever to rest on his laurels.

Neither did the success of the remarkable operation do anything to stop the carping of Hunter's jealous colleagues at St. George's, or the scorn flowing from Jessé Foot's barbed pen, both of which continued regardless. But it was not Hunter's style to build bridges or attempt to resolve conflicts amicably, for he had courted controversy from the moment he dismounted from his horse, a brash, ill-educated, rough-mannered country boy fresh down from Scotland, in the middle of London's vibrant Covent Garden in 1748.

CHAPTER 2

The Dead Man's Arm

∽

Covent Garden, London

September 1748

The severed arm lay on the table in the dissecting room, turning putrid. The bloodless flesh looked grotesquely pale, but the nails on the fingers seemed as lifelike as ever. Keen to impress his brother William, ten years his senior, the twenty-year-old John Hunter picked up the knife before him to begin his first human dissection.[1] Virtually strangers, the two brothers had not cast eyes on each other since William had left the family farm to make his future in London when John was twelve years old. The intervening years, at least as far as William's critical eye now assessed, had done little to improve his younger brother.

A shrewd and ambitious social climber, William had moved in ever more rarified circles since leaving Glasgow University and arriving in his "darling London" eight years previously.[2] Now thirty, he had already visited both Paris and Leiden to further his anatomy studies, obtained his surgeon's diploma in order to embark on a surgical career, and begun to build a fashionable practice attending the deliveries of the babies of the well-to-do. Although he had only just secured his first hospital post—as "surgeon-man-midwife" at London's new Middlesex Hospital—and had but a few years' private practice to his name, William was steadily accruing both wealth and esteem. Already he had pretensions to leave his lowly surgical colleagues behind and join the elevated ranks of the physicians, who enjoyed the highest status in London's rigid medical hierarchy.

Yet for all William's self-assurance, climbing the greasy pole of Georgian English society was no easy ascent for a Scot of relatively humble beginnings and with limited connections. The Jacobite uprising in 1745 and the British army's bloody victory at Culloden the following year were still fresh in English memories. No matter that most Lowland and urban Scots had no sympathy with the revolt, anti-Scots prejudice was rife. Accordingly, William had smoothed his Scottish burr and anglicized his country manners in order to blend in with polite London society. Elegantly attired in brocade and lace, and sporting a full powdered wig, he dined with fellow Scottish intellectuals, including the novelist Tobias Smollett, the physician John Pringle, and the painter Allan Ramsay, who had similarly gravitated to the metropolis in a collective migration of talent from universities in Glasgow and Edinburgh. But astute William had managed to penetrate English social circles, too, forging connections with influential men, such as the magistrate and writer Henry Fielding and the diarist Horace Walpole. Between coffeehouses and theaters, dissecting rooms and salons, William skillfully bridged both the divergent Scots and English cultures and the contrasting worlds of science and the arts.

Not only was William a sophisticated socialite, he also possessed a smart business sense—vital in the risky world of Georgian enterprise, where commercial know-how mattered as much as knowing the right people. The anatomy school he had established in Covent Garden two years earlier had proved both a popular and a profitable venture. Aspiring young surgeons, attracted to London from throughout the British Isles and overseas to walk the wards of the capital's grim charity hospitals, had flocked to the school for expert tuition in dissection. Indeed, so successful had his classes become that William now desperately needed help to run the business. With classes beginning in just two weeks' time, he was anxious that his younger brother should prove a competent and biddable assistant—especially when it came to taking over the more sinister side of his enterprise.

The prospects did not appear promising. An awkward, uncultured, and largely uneducated country lad, the slight youth with his shock of red hair had failed utterly to distinguish himself to date. Eschewing all scholarly pursuits, he had so far frittered away his days on the family farm, displaying no obvious aim or ambition. After briefly toying with enlisting in the army, he had been prompted by an apparent sudden impulse to offer his assistance to William and make the arduous two-week journey on horseback from Lanarkshire down to London. Now he found himself shrinking in the full glare of his clever elder brother's attention, presented with the biggest challenge of his uneventful life so far: to dissect a rotting limb and tease out the muscles under William's rigorous scrutiny.

Surrounded by skeletons and skulls, bones hanging from rafters, and organs pickled in jars, he steeled himself against the sight and stench of human remains as he picked up the knife William proffered. Making the first cut through the leathery skin of the man's arm, he was careful not to flinch or show his distaste to watchful William. He knew that this was a crucial test, the most important moment in his life so far. At last, he was being given a chance to prove his worth.

Born in deep midwinter, in the early hours of February 14, 1728, John Hunter was the tenth child in a farming family struggling to make ends meet in the rugged countryside south of Glasgow.[3] With three children, including an earlier John, having already died, and their aging father, also named John, worried over finances for his large family as well as his own failing health, the arrival of another mouth to feed was not particularly welcome. John Hunter, Sr., who was sixty-five by the time of his youngest son's birth, plainly neglected his latest offspring in favor of his studious older boys, James and William, then thirteen and ten. It was left to John's mother, Agnes, and his four older sisters—Janet, Agnes, Dorothy, and Isobel—to bestow attention on the baby of the family.

Reputedly descended from Norman ancestors, the Hunters of Hunterston in Ayrshire, John Hunter, Sr., had married Agnes Paul, the daughter of a prosperous Glaswegian family, in 1707, the year of English and Scottish union. He was forty-four and she was twenty-two.[4] They baptized their first child, the original John, in 1708 and buried him fourteen years later. Their second child, Elizabeth, died at just one year, and a second son, Andrew, survived only to the age of three. Trading as a grain merchant in the market village of East Kilbride, John Hunter, Sr., had also acquired some farmland, and after the birth of four more children, the family moved from their village home to a thatched stone house at Long Calderwood, a mile to the northeast, making this farmstead the center of their small world.[5]

Although they were relatively well off by the standards of most local people, who lived from hand to mouth in pitiful conditions, it was nevertheless a comfortless existence for a large family, squeezed as they were into a smoky two-bedroom cottage.[6] A single peat fire provided the only source of heat and the children slept in box beds, which were pulled out of the walls like giant drawers each night. Winters were long and severe, and the Clyde Valley soil poorly drained, so it was a relentless battle to eke out an existence for a growing family. Just a year after John Hunter was born, his father had to sell a parcel of land in order to send fourteen-year-old James to train as a lawyer in Edinburgh and William to Glasgow University to study theology for a career in the Scottish kirk.

Ignored by his father, indulged by his mother, and pampered by his older sisters, young "Jock" or "Johnny," as he was variously nicknamed, grew up a headstrong boy. While his mother found her wayward son difficult to control, his father appeared to neither know nor care how his youngest child spent his days. So although Johnny was sent daily to the village school where his brothers had distinguished themselves, whenever he could, he skipped lessons to go rambling through the woods and fields, stalking wild animals in the undergrowth. For all his mother's encouragement, he detested lessons and books with a vengeance, as he later recalled:

> When I was a boy it was a little reading and writing, a great deal of spelling and figures; geography which never got beyond the dullest statistics, and a little philosophy and chemistry as dry as sawdust, and as valuable for deading [deadening] purposes. I wanted to know about the clouds and the grasses, why the leaves change colour in the autumn. I watched the ants, bees, birds, tadpoles, and caddis worms. I pestered people with questions about what nobody knew or cared anything about.[7]

According to his sister Dorothy, young Johnny "would do nothing but what he liked and neither liked to be taught reading nor writing nor any kind of learning."[8] Although he was "by no means considered as a stupid boy," he made such little progress with his lessons that at thirteen he abandoned formal education.

At a time when Scotland prided itself on its kirk-run school system and the Scots boasted literacy levels well in excess of their English neighbors, John's obstinacy toward learning was a heartfelt family disappointment. His hatred of books was so intense that he probably suffered some form of dyslexia, for not only did he struggle with schooling throughout childhood, being unable to read or write until his teens, but he also retained his distaste for all things academic throughout his life. Certainly he could dash off a letter in a well-formed script as well as anyone—although his spelling was erratic even by eighteenth-century standards and his grammar equally idiosyncratic—but whenever possible, in later life, he dictated his thoughts to assistants and even had books read out loud to him.[9] He almost reveled in his disdain for the printed word. Although he would build up a respectable library, he liked to declare that he "totally rejected books," preferring to take up "the volume of the animal body."[10]

Not surprisingly, John Hunter's aversion to literary pursuits would provide powerful ammunition for later enemies, particularly the snobbish Jessé Foot—although he only dared publish his malevolent biography the year after Hunter's death. Mistakenly attempting to equate Hunter's literacy difficulties with a lack of intelligence, Foot claimed that the surgeon "was incapable of putting six lines together grammatically into English."[11] As a result, alleged Foot, Hunter had not personally written a single word of any of his publications, employing instead his friend Smollett—among others—as a ghostwriter. Foot even went so far as to blame this "want of the polish of education" for Hunter's notoriously colorful language. Hunter certainly loved to damn and curse in an age when polite conversation was considered the mark of a gentleman. His pupils would later feel moved to excuse his bad language, while nineteenth-century editors of his works deemed it necessary to excise his oaths. And he admitted enlisting some editorial help with his writings. But the reams of written material Hunter produced, in scientific papers to the Royal Society, voluminous published works, and innumerable letters, easily demonstrated his ability to put pen to paper. Indeed, Foot wanted to have it both ways: He also charged Hunter with concealing his reading prowess in order to plagiarize rivals without suspicion.

Hunter's early distrust of the written word would make him forever skeptical of classical teaching and the slavish repetition of ancient beliefs. He always preferred to believe the evidence of his own eyes rather than the recorded views of others. From his earliest years, roaming the countryside of Lanarkshire when he should have been poring over Latin primers at a school desk, he sought the solutions to the questions that puzzled him through painstaking observation and experiment.

There was ample opportunity on and around the family farm, with its steady stream of births and deaths among the horses, cattle, and fowl. The inquisitive Johnny would have witnessed local farmers' attempts at crude veterinary care when their animals fell ill and may have helped to nurse sickly youngsters back to life. He studied the behavior of wildlife in the nearby hills and valleys and would remember some of these observations all his life. Possibly he was encouraged by his older sisters, since his closest sister in age, Isobel, was said to have enjoyed petting the foals and calves born on the farm.

Many of the questions that were evidently sparked by these earliest observations—such as pinpointing the exact moment when an embryo begins to form in a hen's egg, determining what distinguishes a living, breathing creature from a lifeless carcass, and even puzzling over the origins of the human species itself—would occupy him all his life. As he would later comment, "I love to be puzzled, for then I am sure I shall learn something valuable."[12] Equally, his childish awe at the powers of nature would never leave him. "No chemist on earth can make out of the earth a piece of sugar but a vegetable can do it," he later wrote, and he marveled at reproduction: 'This production of Animals out of themselves excites wonder, admiration, and curiosity."[13]

But at the age of thirteen, leaving the village school in the same year that his father died, the adolescent John Hunter was left with no more material purpose in life than to help his mother and sisters run the farm. Both of his older brothers had since changed their earlier plans and thrown in their lot with medicine, yet Johnny felt no urge to follow suit. William left Glasgow University after five years without taking his degree in theology, having fallen under the spell of one of the founding figures of the Scottish Enlightenment, the philosopher Francis Hutcheson. Imbibing Hutcheson's liberally minded approach to religion was manifestly at odds with William's intended career in the kirk. His plans for the church abandoned, William was taken under the wing of the family physician, the kindly William Cullen, who would become a towering force for scientific progress on the Scottish medical scene. Cullen trained William in basic medicine, sent him to Edinburgh University to study anatomy, and in 1740 packed him off to London to learn the latest approach to childbirth, all with a view to setting up partnership together on William's return. Eldest brother James, meanwhile, had followed William to London in a joint bid to make their names in medicine.

A passion for medical care did little to help the family. Agnes fell ill and died at just twenty-five years old. Her sister Isobel followed her to the grave at only seventeen, and finally James, the eldest son and brightest star in the family's firmament, returned home with the classic signs of consumption. Seventeen-year-old John stood helplessly at his brother's bedside as he coughed up blood and died a slow, painful death in the prime of life. His premature demise left William head of the family, with more necessity than ever to make his London enterprises a success, while John languished at home in an increasingly empty house.

So six of John Hunter's nine siblings, as well as his father, had been carried to the little graveyard encircling the East Kilbride parish church. Although the loss of those closest to him through his teenage years left a deep mental scar, such a death toll was by no means uncommon by Georgian standards. Life expectancy as an average in England in the middle of the eighteenth century was just thirty-seven years—and would have been roughly similar in Scotland—although this stark figure masks a disproportionately high death rate in childhood.[14] Surviving infancy was life's biggest battle; having reached adulthood, there was a fair chance of attaining a ripe old age. In London, where child death rates were highest, almost half of the babies born between 1750 and 1769 never reached their second birthdays. Burials far exceeded baptisms in the capital throughout most of the century.[15] In spite of the city's thriving funeral business—guaranteeing ample research material for anatomists—London's population continued to mushroom, since impoverished families migrated from the countryside in their search for work faster than their predecessors expired. Contagious diseases ran rampant in the city, thanks to appalling poverty and malnutrition, overcrowding in slum dwellings, and filthy living conditions. While the carcasses of dead animals lay rotting in the streets, open sewers carried human excrement and animal offal down central gutters. But prospects of health were little better in the countryside, where poor harvests and disease epidemics could equally lead to early death.

Although Georgian families no longer needed to fear the medieval scourges of plague and leprosy, there were plenty of other menaces to worry about. A child born into eighteenth-century Britain required a tough constitution and no small amount of luck to evade or defeat the innumerable infectious diseases—diphtheria, measles, mumps, scarlet fever, influenza, and consumption (tuberculosis), to name but a few—lying in wait in the first years of life. Where disease failed to snare the youngster, malnutrition, neglect, and misguided child rearing often succeeded. Babies were force-fed with sugared and alcohol-laced pap from germ-ridden cups, soothed with weak beer or straight gin, and swaddled in blankets that were rarely changed. And throughout adolescence, there were many more epidemics to battle, from the typhus fever that reigned in jails, hospitals, and ships to the dreaded smallpox virus, which accounted for 10 percent of all deaths, killing a fifth of those it struck and hideously disfiguring many more.

Surviving this obstacle course conferred lifelong immunity against a raft of diseases, but negotiating adulthood was still precarious. Life hung on a slender string. With no modern antibiotics to fight bacteria, a healthy adult could fall suddenly sick with a contagious fever and be dead within days. For women, childbirth was an additionally hazardous experience. Many died during agonizing deliveries or succumbed to sepsis days later, especially in the "lying-in" hospitals that sprang up from the middle of the century and provided new breeding grounds for bacteria. Even for those adults who reached middle age, there were still innumerable ailments and complaints—from gout to bladder stones, venereal disease to toothache—that could cause pain and discomfort, with little hope of cure or relief.

With so many risks to life and limb, it is little wonder that the Georgians were fanatical in their pursuit of health. Letters, diaries, novels, and newspapers reveal an almost unhealthy preoccupation with the work-ings of the human body. Whether conversing with family or friend, busi-ness associate or acquaintance, the Georgians reveled in reporting intimate details of their bodily functions and personal ailments, while everyone had a favorite homespun remedy to recommend. Among the least palatable were toxic potions, violent purges that acted as laxatives, emetics that in-duced vomiting, and the copious letting of blood.

Those who could afford the time and money were obsessed with the latest fads in treatment, to the point of hypochondria. And there was no lack of variety; when it came to health care, the Georgians were spoiled by the number of choices. With no effective regulation or policing, and no con-sensus on diagnosis or therapy, a patient was equally likely to buy a noxious potion from a traveling quack, an herbal concoction stewed by the village wise woman, or a traditional remedy prescribed by a learned physician—and sometimes all three.

It should have been a golden age for medicine. In the previous century, the philosopher Francis Bacon had confidently predicted that scientific progress would bring nature firmly under control. In 1628, the English physician William Harvey had discredited prevailing medical beliefs when he proved that the heart, not the liver, propelled the blood around the body in a continuous circulation. Further anatomical discoveries served to reinforce the notion promulgated by the French scientist René Descartes, the driving force of the scientific revolution, that the body was more like a machine than a mysterious receptacle for the soul. A new spirit of optimism was in the air as the Royal Society, founded in 1660, and, in France, the Académie Royale des Sciences, ushered in an era of natural philosophy, in which scientific experimentation was heralded as the stepping-stone to a modern society.[16]

So much for the theory; the reality was rather more down-to-earth. The promise of the scientific revolution did little, if anything, to render medicine safer or more effective. Health care still waited for its enlightenment, and the eighteenth century became a free-for-all for mountebanks, cranks, and meddling medics. While learned physicians, flourishing doctorates from Oxford, Cambridge, and universities across Europe, paid lip service to scientific progress, in practice they clung like leeches to their classical doctrines. Apart from a few notable exceptions, most held firm to the teachings of the Greek father of medicine, Hippocrates, who taught in the fifth century B.C. that all illness was due to an imbalance of the four "humors" of the body: blood, phlegm, black bile, and choler, or yellow bile. An excess of one humor or a deficiency in another was diagnosed from patients' eating, sleeping, and toiletary habits, inspecting their urine and stools, and occasionally by feeling the pulse, but rarely if ever by examining the site of the complaint. The remedies prescribed were designed to restore the "humoral balance," whether through potions and pills made up by apothecaries, clysters—enemas—injected by syringe, or copious bloodletting.

Remedies were rarely anything less than drastic; that way, Georgian patients could at least feel they were getting value for their money. Patients happily submitted to being dosed with toxic elixirs, blistered by heated glass cups applied to their backs, and bled to the point of unconsciousness. Despite Harvey's demonstration that the theory behind bloodletting—to extract blood from a specific site—was nonsense, since blood circulated through the entire body, phlebotomy only increased in popularity as a cure-all for every ill. If not recommended by physicians, it was often demanded by patients, being administered for almost every condition, from ailments in young children to prolonged childbirth. As much as thirty ounces might be removed at one sitting, often disastrously weakening the body's natural defenses against disease.

Yet while physicians prescribed bloodletting in every circumstance, their taboo on touching the human body meant they never personally slit open a vein to let the blood flow. In the strict medical hierarchy that prevailed, all messy and distasteful jobs that involved touching or cutting flesh were left to surgeons, or to barbers and barber-surgeons, who offered bloodletting as a sideline to cutting hair and shaving beards. Far from being interlopers in the field of surgery, barbers were the first surgeons. The earliest organized medical care, in medieval times, had been centered on monasteries. But the church frowned on its devotees spilling blood, and so barbers—who were frequent visitors to the brethren in order to keep tonsures and beards in trim—assisted the monks in their medical work by excising warts, removing abscesses, and letting blood. The familiar red-and-white-striped poles outside barbershops are leftover reminders of their erstwhile professions. Originally, they signified the bandaged and bloodied stick gripped by patients during minor surgical procedures.

Understandably distrustful of organized medicine, patients turned to alternative healers—the so-called quacks—who were no less likely to prove effective. In the month before John Hunter arrived in London, *Gentleman's Magazine* listed more than two hundred nostrums available in apothecary shops, ranging from "horseballs" for coughs to powders for piles.[17] In the same issue, readers were enthralled by reports of Bridget Bostock, a folk healer in Cheshire, who reputedly cured every ill with the simple remedy of applying her spit. Ailing Georgians also sought relief in electric shocks and hypnotism, and even by inhaling the breath of young women. And like many infants of his day, in 1712 the toddling Samuel Johnson was lifted up to receive the "royal touch" from Queen Anne in the forlorn hope of curing his scrofula, a form of tuberculosis.[18] He wore around his neck all his life the gold "touch piece" the monarch had given him. Yet these remedies were no crankier than the ingredients authorized by the country's most elite medical men in the directory of medicines, the *Pharmacopoeia Londinensis*, published by the Royal College of Physicians in 1746.[19] Although the updated formulary had jettisoned unicorn's horn and moss from human skulls, it wholeheartedly endorsed oyster shells, crabs' eyes, and ground wood lice among its "speedy, safe and pleasant Cures."

Interventions were not only radical and unpleasant; they were often downright harmful. Indeed, of all the pills, potions, and plasters in the practitioner's medicine bag, there was only one truly effective ingredient: cinchona, or Peruvian bark. Originally a Native American remedy brought back from the New World by Jesuit missionaries in the seventeenth century, it contained quinine, a valuable therapy for malaria, which was still prevalent in the British Isles in the eighteenth century. Another innovation, inoculation against smallpox, where a minute amount of infectious pus taken from an afflicted person was used, had similarly been adopted from a folk remedy popular in Turkey. Initially opposed by the snooty physicians when it was introduced to Britain in 1721 by Lady Mary Wortley Montagu, the wife of the British ambassador in Constantinople, inoculation saved countless children from certain death—although it could also prove fatal in overzealous hands. In addition, the ubiquitous prescription of opium at least provided some welcome respite from the miseries of sickness. Beyond these few aids, which owed little to learned medicine, the rest was largely useless, or worse.

It was this maelstrom of muddled, ignorant, and incompetent medical practice that the twenty-year-old John Hunter decided to enter after losing six siblings to the ravages of disease. His mother, ever concerned for her favorite son's future, had already attempted to find him suitable employment earlier in 1748. That summer, he had gone to live briefly in Glasgow with his eldest sister, Janet, and her new husband, a wealthy but idle timber merchant.[20] Although Johnny was already noted for the dexterity and neatness of his hands, it was in his brother-in-law's timber yard, watching carpenters at work, that he learned to handle knives and saws with expert precision. Naturally enough, this short interlude, lasting at most a few months, would later provide Hunter's enemies with an excuse to dismiss the surgeon as an "ignorant carpenter." But before the summer was over, the timber yard had closed, its owner bankrupt, and John Hunter was back home, kicking his heels once more.

Quickly dismissing an impulse to join the army, he finally took the plunge and wrote to William, asking to join him in London, in a move that would not only transform John Hunter's life but would alter the course of medicine forever. Desperate for a new assistant, especially one who would be bound by family loyalty to keep the murky secrets of the dissecting room, William immediately assented. So, early in September 1748, the farmer's son left his Scottish homeland for the first time, riding the four hundred miles to London with a family friend, Thomas Hamilton, to arrive in Covent Garden just two weeks before the anatomy school opened for the autumn term.

T*he contrast between* rural Lanarkshire and bustling, chaotic London could not have been more startling. With a population of around 675,000, almost forty times the size of Glasgow, London was the biggest city in Europe.[21] As Hunter approached from the northeast, through the pleasant villages of Tottenham, Islington, and Pentonville, the rough, rutted road became increasingly busy, while houses, shops, and taverns wrestled for space along the way. As he neared the city, the narrow, towering tenements, which housed whole families in single cellars and attic rooms, almost blocked out the sky. Negotiating the congested streets, where stagecoaches and private carriages battled for passage with farm carts and livestock, seemed hopelessly confusing; the sounds of horses' hooves, creaking wheels, and complaining cattle were deafening. Mud, animal dung, refuse, and human waste splashed pedestrians as they walked the pavements and tried to dodge the swinging shop signs, speeding bearers of sedan chairs, and downpours of foul water from upper-story windows. By late afternoon, oil lamps lighted the smoky streets and candles illuminated shop windows displaying silk clothing and exquisite jewelry, their luxury forming a pantomime backdrop to the squalor of ragged children begging in the gutters.

In Hatton Garden, located in Holborn, where John initially joined William in lodgings, the affluent residents still enjoyed the pastoral view north toward fields and market gardens. But in Covent Garden, where William had rented rooms for his anatomy school, the once-fashionable square was now a shambling market edged by taverns, gin houses, and brothels. London's gentry had forsaken the elegant piazza, designed in the previous century by Inigo Jones, in a concerted shift toward the cleaner air and pleasant squares of the West End. In their stead, the apartments beneath the arcades had been taken over by writers, artists, actors, and pimps. By day, hawkers yelled out their wares to passersby. By night, the neighborhood was stalked by press gangs, pickpockets, prostitutes, and armed robbers. And the view north from here, toward the desperately poor neighborhood of St. Giles, would provide Hogarth with inspiration for his famous *Gin Lane,* with its grotesque images of lives destroyed by the ubiquitous spirit.[22]

Well used by now to such extremes of wealth and poverty, William had been settled in London for five years. Having first lodged as a pupil in midwifery with William Smellie, a fellow Scot, aspiring William had soon abandoned his down-to-earth teacher—and dissolved his partnership plans with William Cullen—in favor of a more promising alliance with another compatriot, the physician James Douglas. Inveigling himself a position with the family, he had been appointed tutor to James Douglas, Jr., and even become engaged to Douglas's daughter, Martha. But when first Douglas and then Martha died within a year of each other, William's plans for a rewarding future—both financially and emotionally—were scuppered. He would never again become romantically involved. Determined, nonetheless, to make his way in London, he established himself as a fashionable male midwife and forged ahead with plans to launch an anatomy school.

Amid the squalor, sexual scandal, and violent crime that characterized Covent Garden, William's anatomy school was fast gaining its own notoriety. The first of its kind in Britain, offering young medical students daily hands-on tuition in human dissection, the school was a daring, even dangerous, venture. The study of anatomy still languished in its infancy in mideighteenth-century England, with inevitable ignorance about the workings of the human body among medical practitioners as a consequence. While would-be physicians spent years at universities poring over ancient texts and aspiring surgeons blithely copied the errors of their tutors during long apprenticeships, the study of the human body remained a backwater of medical education.

Some thirty years before William's enterprise, William Cheselden, the most famous surgeon of his day, had invited medical students to private anatomy demonstrations at his home. The young surgeons crowded around the celebrated Cheselden as he demonstrated parts of the body on a corpse dragged from the gallows, cutting up the cadaver on his dining room table. But Cheselden's anatomy dinners soon attracted the ire of the Company of Barber-Surgeons. He was disciplined in 1715, on the grounds that he "did frequently procure the Dead bodies of Malefactors from the place of execution and dissected the same at his own house."[23] The rebuke was less in response to Cheselden's domestic arrangements than to his timing: The demonstrations clashed with the company's own lectures.

Since its foundation by royal charter in 1540, the Company of Barber-Surgeons had secured the right to the bodies of four criminals—the number later increased to six—hanged in London each year. The corpses were dissected in public demonstrations staged several times a year, in an arrangement designed as much as the ultimate deterrent for felons as for educational purposes. By 1631, however, the company was also encountering some difficulties reconciling its anatomy duties with its catering facilities. Minutes noted that "the bodies have been a great annoyance to the tables, dresser boards and utensils in the upper kitchen by reason of the blood, filth and entrails of these anatomies."[24] A custom-built anatomy theater was finally opened in 1638. But when the barbers and surgeons split in 1745, the resulting new Company of Surgeons was temporarily homeless and therefore devoid of an anatomy theater in which to stage any demonstrations. This sorry situation would continue for almost a decade—until a new theater was built in 1753—seriously curtailing anatomy teaching in London.

Although the Company of Barber-Surgeons had sought to ban rival anatomy lectures, private classes had, in fact, flourished unchecked. Several practitioners, including James Douglas and Percivall Pott, ran lectures at their hospitals and in their own homes during the first half of the century. But these courses were short, sporadic, and generally limited in scope. There was no dedicated anatomy school in London—indeed, there was no medical school at all, in contrast to the flourishing centers in major cities in Scotland, Ireland, and on the Continent. Moreover, the difficulties associated with obtaining human bodies allowed few opportunities for students even to witness, let alone practice, dissection. Students in crowded anatomy lessons were lucky to catch a glimpse inside a single body as it decayed over several weeks. When bodies were impossible to obtain, an animal corpse was substituted. William Hunter described a typical class under Professor Alexander Monro at Edinburgh University: "There I learned a good deal by my ears; but almost nothing by my eyes; and therefore, hardly anything to the purpose. The defect was, that the professor was obliged to demonstrate all the parts of the body, except the bones, nerves and vessels, upon one dead body. There was a foetus for the nerves and blood-vessels; and the operations of surgery were explained, to very little purpose indeed, upon a dog."[25] He was equally unimpressed by his studies in London, where the course he attended fielded only two corpses over thirty-nine lectures.

William's experience of anatomy lessons in England and Scotland had convinced him that such courses were hopelessly inadequate as a preparation for a career in medicine. But he had also witnessed anatomy teaching in Paris and Leiden, where more liberal laws provided an abundant supply of bodies for research purposes. His entrepreneurial nose had immediately sniffed a gap in the market for a dedicated school offering comprehensive studies in anatomy in London. With the new Company of Surgeons in disarray, unable even to police private anatomy classes, let alone provide its own, William seized his opportunity. Setting up his lecture theater and dissecting room in a rented Covent Garden apartment, he imported to London the Continental approach, advertising his first course in the *London Evening Post* in September 1746 with this inducement: "Gentlemen may have the opportunity of learning the Art of Dissecting during the whole winter session in the same manner as at Paris."[26]

When it opened its doors on October 13, 1746, the new school was a revolutionary venture. With lectures running from 5:00 P.M. until 7:30 P.M. every evening except Sundays, each course spanned more than seventy lessons. The courses were repeated in two terms, lasting almost four months each, from October to early January and from late January until April, in a step-by-step program providing a complete introduction to the subject. Perched on his podium, William held the students' attention as he imparted up-to-the-minute knowledge on the human body in his eloquent style. As the pupils took notes, bottles of pickled organs and dried specimens of muscle and bone were passed around for them to handle and inspect.

But the school offered more than just lively lectures and interesting visual aids; each student was also guaranteed hands-on experience. Adopting the "Paris manner," as William so discreetly put it, meant quite simply that every student was guaranteed a corpse of his own on which to practice dissection and surgical methods. For the first time, not only were pupils able to learn the basics of anatomy from an expert teacher; they could see with their own eyes precisely how the different organs, tissues, muscles, and bones interconnected. With their own hands, they could unravel the coils of the gut, trace the branches of the delicate air passageways in the lungs, and feel the weight of a human heart. Their waistcoats and breeches covered with aprons, they could spend hours delving inside a human carcass, up to their elbows in blood and guts. The school was an overnight success. Fired by a thirst for knowledge, eager to explore and discover for themselves, young students flocked to enroll from across Britain, Europe, and even the American colonies.

At last William's finances were on a secure footing. Now determined to capitalize on the achievements of two years' success, William—the self-made man—was keen to recruit young John—the self-taught youth—to his cause. With no time to lose before the students clamored once more at his doors, he put his brother immediately to work, setting him the critical test of dissecting an arm. Stooped over the bench, John Hunter mustered all the skills he had learned in carpentry, applied all his inborn curiosity for nature, and meticulously picked the fat from the various muscles of the limb with a dexterity unknown in a novice. The neatness and delicacy of his workmanship exceeded even William's exacting standards. But still unconvinced, William set his brother an even harder task: to dissect a second arm, in which the arteries had been injected with colored wax, separating the vivid blood vessels from the fleshy muscles. Once again, John's performance surpassed all his brother's expectations. William even went so far as to predict that his young brother had the makings of an excellent anatomist who should never want for employment.[27]

This was high praise indeed from the fastidious elder brother. As far as William was concerned, it was plain that his stubborn, rough-edged, untutored brother—the wild child of the family—could be successfully molded into the competent, compliant workhorse he needed to help run the school. William was certain that under his careful tutoring and guidance, his young brother could become a skilled assistant and perhaps even an able surgeon, leaving William free to pursue his own goal of becoming a respectable physician and a wealthy gentleman. Encouraging and ambitious for his new ward, William was like a true father figure—everything John's own father had not been—or at least it appeared that way so long as John remained firmly in his shadow.

So on the first day of the autumn term, John Hunter took his place in the lecture theater as his brother rose to greet the new class. After detailing the purpose of anatomy, the history of the subject, and the program for the busy next three months, William outlined to the impressionable youngsters gathered in the room the stark choice that now lay before them. If they chose to idle away their days and skimp on their studies, then they would forever struggle to find work, want for money, and tread "a low path in life." But if they worked hard, studied diligently, and pursued perfection, then, he assured them, they would be respected, courted, and satisfied with their lot. In the eighteenth-century can-do society, an ambitious young man with a will to work, a few connections, and a mind to adopt the correct manners could truly take the world by storm, insisted William. Rising to his finale, he urged, "and I firmly believe, that it is in your power not only to chuse, but to have, which rank you please in the world."[28]

Nobody listened more attentively than his own young brother. Almost overnight, the carefree youth became an industrious young man; the aimless boy found his lifelong passion; the diffident Jock of the Scottish Lowlands was transformed into the popular, fun-loving Jack who frequented the taverns, theaters, and coffeehouses of Covent Garden. His days were fully employed: preparing specimens at first light every morning for the students to peruse, demonstrating anatomy skills and the art of preserving parts in the dissecting room all day, and listening beside his fellow pupils at William's lectures every evening. But even then there was little rest. One more task fell to Jack Hunter, but this job could only take place under cover of darkness. While William built up his bank balance and his reputation, rapidly becoming London's most sought-after accoucheur, John was put in charge of the seamier side of his business: procuring the school's essential teaching material.

The Stout Man's Muscles

When two-year-old John Race died in early December 1747, his grieving parents laid his body in its tiny box in the frozen ground of the paupers' graveyard at Whitechapel, located in east London. But the infant was not to be allowed to rest in peace. Shortly after midnight on Sunday, December 6, shovels broke soundlessly through the freshly dug soil, rough hands wrenched the coffin from its shallow grave, and John's limp corpse was dragged from its box, stripped of its shroud, and shoved into a sack.

For once, the burial ground was under watch. James Thomas and Charles Pritty were caught in the act of exhuming the infant's body and promptly brought to justice. Ten days later, the two men stood before a judge and jury, charged that they "unlawfully did dig up" the dead body of John Race and carried it away "to the evil example of all others."[1] The jury hearing the case lost no time in finding the pair guilty. Thomas and Pritty were each fined one shilling—a relatively paltry sum—but their six months' confinement in the fearful Newgate Prison provided ample opportunity for remorse.

Placed in charge of procuring dead bodies for his brother's anatomy school at the tender age of twenty, less than one year after the Thomas and Pritty court case, John Hunter faced a formidable challenge. He had been left in no doubt that the school required an abundant supply of human corpses in order to continue its success. William had made clear his conviction that regular practical dissection was the surest way by which any aspiring medical man could properly learn his trade. Indeed, if future surgeons were denied the chance to practice their primitive operations on dead bodies, to make their blunders on unfeeling flesh, the only alternative was to practice on live patients, with inevitably fatal results. So, as William explained to the new students at the start of their autumn term in 1748, it was therefore necessary "for giving a complete course of Anatomy, to provide a number of *fresh subjects.*"[2]

William's promise that each of his pupils would enjoy firsthand access to a corpse was the very raison d'être of his teaching. It was precisely what set his school apart from those of his rivals. The handful of other anatomists offering their skimpy courses in London presented little threat; William knew his main competition came from back home in Scotland. It was there, at the highly respected medical school of Edinburgh University, that his own former tutor, Professor Alexander Monro, was attracting students from across Europe with rising success.

For Monro, however, working in a small city like Edinburgh presented considerable difficulties with regard to obtaining the corpses he needed for anatomical exploration. Accordingly, he had become adept at eking out just two bodies for his entire course.[3] For William, the vast, overcrowded, and disease-ridden urban sprawl of London guaranteed a plentiful supply of teaching material—so long as underhanded means were employed—as well as providing the anonymity to cover his procurement methods. But as the popularity of his lectures grew, so his requirement for bodies increased. And since a dead body rarely lasted more than a week before decomposing beyond use, even in winter, the school needed a steady stream of cadavers on an almost daily basis. These needed to be fresh bodies, laid out on the dissecting bench within a day or two of death. William informed his class why this was necessary: "The dead body cannot be too fresh for dissection. Every hour that it is kept, it is losing something of its fitness for anatomical demonstrations."[4] In addition, he needed bodies to order, ready to show particular parts of the anatomy in men, women, children, and fetuses, both healthy and diseased.

With his genteel manners, expensive finery, and influential friends, William had no desire to go ferreting out corpses himself in the dead of night. As his polished midwifery skills and charming bedside manner endeared him to the wives and daughters of London's nobility, so he grew more and more inclined to distance himself from the darker side of his business. To date, therefore, the job of obtaining dissection material had been undertaken by his assistant, a former pupil named John Symons.[5] But with Symons leaving to set up practice on his own, the task now fell to young John. With his ungainly country ways, plain dress, and the vulgar oaths he seemed unable or unwilling to suppress, John would have no trouble mixing in the seedy taverns of the London underworld, while William could continue supping claret with his distinguished associates.

Completely dependent on William for his board and lodging, his job, and even his pocket money, John Hunter was in no position to decline his allotted role. In any case, he was anxious to excel at the only work for which he had ever felt any interest. Bodies were needed and he was the man to make sure they were delivered. But accomplishing the task in accordance with William's demanding requirements was an organizational and logistical challenge of staggering proportions. Obtaining human cadavers with which to teach medical students the rudiments of anatomy was no easy matter in Georgian Britain. Not one was available legally to private anatomists like William.

Since it was almost unheard of in the mid–eighteenth century for anyone to donate his or her body for dissection in the cause of aiding research into medical care, there were few opportunities for lawful postmortems. Accordingly, surgeons in Britain had fought for legal rights to bodies for dissection ever since the vogue for anatomy had filtered across the English Channel from Renaissance Europe in the fifteenth century. In Scotland, the Guild of Surgeons and Barbers was granted in 1506 the right annually to "ane condampnit man after he be deid to mak anatomea of."[6] Within a few years, the Company of Barber-Surgeons in London was given similar privileges. On the allotted dates, spectators gathered at Surgeons' Hall to witness the demonstrations and listen to lectures that spanned several days, with a first-rate banquet to follow. The diarist Samuel Pepys was among the spectators in 1663, and he recorded in his journal, "I did touch the dead body with my bare hand; it felt cold, but methought it was a very unpleasant sight."[7]

Almost a century later, when John Hunter was charged with the job of obtaining bodies for William's school, legal provision remained unchanged. Although anatomists on the Continent were guaranteed plentiful supplies of bodies in order to conduct their essential work, throughout the English-speaking world, in Britain, Ireland, and the American colonies, there was no legal source of corpses for anatomical training or research beyond the narrow allocation to the elite medieval guilds.[8] Consequently, any teacher who wished to offer his students a practical foundation in anatomy, any surgeon who wanted to hone his knife skills, or any anatomist who desired to enhance understanding of the human body was forced to break the law.

There was, at least, no shortage of bodies. The first, and most obvious, source for an enterprising young man charged with obtaining corpses was the gallows. With hanging the penalty for nearly two hundred crimes, from filching a watch to cold-blooded murder, as many as fifty men, women, and even children were executed in a single year. And so, on October 28, 1748, the first hanging day after William's school opened its doors for business, John Hunter would have been standing in the shadow of Tyburn Tree, the notorious triple-legged gibbet on London's western edge, as nine men and one woman were hauled up in front of the jubilant crowds.[9] Condemned for assorted acts of burglary, smuggling, horse stealing, and highway robbery, the ten prisoners had been drawn in an open cart through packed streets on the three-mile journey from Newgate Prison to Tyburn. Stopping for the convicted to imbibe quantities of beer at taverns along the route, the manic parade had become an uproarious melee by the time it arrived at the gallows. As the nooses were positioned around the necks of the felons, spectators crushed into a rickety stadium for a better view. But for Sarah Kenningham and her nine fellow condemned, there was no last-minute reprieve, and as the cart in which they stood lurched away from the gibbet, the frenzied onlookers roared with approval.

Now the theater really began as waiting surgeons and their accomplices tusseled with relatives of the convicted to wrest control of the bodies. The minute the jerking bodies swung above the crowd, John Hunter and his rival combatants rushed forward. Friends and family of the executed leapt up to grab the dangling corpses' feet in an effort to shorten the agonizingly slow death by strangulation that usually ensued and to claim the bodies for a decent burial. At the same time, the beadles of the Company of Surgeons battled to seize their legitimate booty for the dissecting table, while fighting off private anatomists and their agents to the prize. The crowd loved it: It was all part of the festival atmosphere on the public holidays known as "Paddington fair day." The novelist Samuel Richardson was less enthusiastic when he described a typical scene: "As soon as the poor creatures were half-dead, I was much surprised before such a number of peace-officers, to see the populace fall to hauling and pulling the carcasses with so much earnestness, as to occasion several warm rencounters, and broken heads. These were the friends of the persons executed . . . and some persons sent by private surgeons to obtain bodies for dissection. The contests between these were fierce and bloody, and frightful to look at."[10]

Although the right of the Company of Surgeons to six bodies annually was endorsed by royal authority, it was by no means accepted by the Georgian populace. Hanging days were a furious free-for-all, especially when some of the period's most notorious villains went to their deaths. When the infamous burglar Jack Sheppard was hanged in 1724, the surgeons battled all day with his friends and well-wishers, who were anxious to give their hero a respectful burial. The mob prevailed and Sheppard was laid to rest.[11] But when Jonathan Wild, the self-styled "Thief-Taker General," who governed organized crime in London, went to the gallows the following year, the surgeons had their revenge. After the body was buried by his devotees, it was secretly dug up and smuggled away to Surgeons' Hall. His skeleton remains at the Royal College of Surgeons to this day.[12]

Popular feeling ran high. The notion of being dissected after death, or consigning the body of a loved one to that fate, provoked intense horror. Desecrating a dead body offended deeply held religious convictions.[13] Most God-fearing Georgians were convinced that if their bodies were mutilated by anatomists and their remains scattered far afield, they would never be resurrected whole on Judgment Day. Typically heartfelt emotions were expressed by Vincent Davis, a butcher at Smithfield market, who was sentenced to hang in 1725 for murdering his wife, when he declared, "I have killed the best wife in the world, and I am certain of being hanged, but for God's sake, don't let me be anatomised!"[14]

But there was a more immediate reason why condemned Georgians feared the anatomist's knife. With the hangman's art far from scientific, criminals executed at Tyburn usually died from slow asphyxiation rather than from a swiftly broken neck. Consequently, it was not uncommon for such convicts to regain consciousness after being cut down from the scaffold, occasionally on a dissecting table. Cases of people reviving after being hanged were well known; the phenomenon would evoke particular interest for John Hunter.

Although of slender build and, at five two, slightly less than average height for the time, the twenty-year-old Hunter was broad-shouldered and strong, his muscles developed by years of farm labor, as well as fiercely determined. He could reasonably hold his own with any of the roughnecks and villains, official or otherwise, clustered beneath the gibbet. Certainly he managed to obtain corpses from Tyburn, whether by bribery, skulduggery, or brute force, on a number of occasions. Although ordinarily careful not to divulge the sources of his dissection "subjects," at one point he casually referred to an instance when two bodies at once were whisked away from Tyburn: "In the spring of 1753 there was an execution of eight men, two of whom I knew had at that time very severe gonorrheas. Their bodies being procured for this particular purpose, we were very accurate in our examination, but found no ulceration."[15]

The bodies of those who were hanged were particularly valuable to William's school. Since the executed felons died suddenly, their corpses were generally free of the diseases that proved fatal to most people dying a natural death. The pupils could therefore study the human anatomy in its normal, healthy state. Yet Tyburn could never supply the school with sufficient bodies to satisfy the pupils' needs, nor meet William's demands for the corpses of women who were pregnant—a pet interest, given his growing obstetrics business—since the law prohibited pregnant women from being hanged. Neither could the gibbet provide the bodies of fetuses, babies, or young children, which were equally necessary to the pupils' tuition.

Ingenious alternatives were required. Anatomists took to loitering at Newgate in the days before execution, bartering to buy the bodies of condemned prisoners in advance of their deaths. One desperate felon, William Signal, had been persuaded in 1752 to sell his body to a surgeon solely so that he could buy decent clothes for his day of execution. Another common ruse was to bribe unscrupulous undertakers to sell bodies prior to burial. Many a bereaved relative followed a coffin packed with stones in a funeral procession through Georgian London. Some anatomists even took to dissecting the bodies of their own relatives. William Harvey conducted autopsies on his father and sister when they died—although he insisted that upon his own death, his body should be wrapped in lead to thwart any fellow enthusiasts.[16] But for the serious anatomist determined to procure a regular and reliable supply of bodies in mid-eighteenth-century London, there was really only one viable source: the grave.

Grave robbing had occurred on occasion in places throughout the British Isles since at least the seventeenth century; John Hunter would help transform it into an industry. When first he arrived in Covent Garden in 1748, a raw, fresh-faced youth, body snatching was in its infancy. During the span of his career, with his encouragement and support, the trade would grow exponentially, spawning warring gangs of professional grave robbers who stalked the city's churchyards night after night. By the end of the eighteenth century, John and William Hunter's disciples would have exported the practice to Ireland and the United States, stoking up mass protests in both countries. Eventually, the practice would reach such a scale—as London burial grounds had to be policed by armed vigilantes to thwart the thieves—that Parliament was forced in 1828 to investigate.[17] Even then, it was only the discovery of murders committed by Burke and Hare in Scotland, and by similar but lesser-known assassins in London, which ultimately compelled the government to legislate and end more than a century of body snatching by introducing the Anatomy Act of 1832.

Driven initially by his brother's demands, but soon compelled by his own zeal to extend knowledge of human physiology, John Hunter would dissect more bodies, and therefore require more bodies stolen from graves, than any other anatomist of the eighteenth century. He would develop closer relationships with the grave robbers than any other surgeon, even— in all probability—joining their nighttime expeditions, and he would ultimately pull off the most dramatic body-snatching ruse of all time.

In its earliest days, body snatching had been carried out principally by a handful of surgeons and their students, who occasionally visited graveyards by night to seek out dissection material for lessons the next day. Instances had been reported in Scotland, where the medical schools in Edinburgh and Glasgow competed both for students and for corpses, as far back as 1711, and medical students in Dublin and Oxford organized similar nocturnal raids. John Bellers, a prominent Quaker, noted the situation existing in 1714: "It was not easy for the students to get a body to dissect at Oxford, the mob being so mutinous to prevent their having one."[18]

So in October 1748, with William clamoring for more bodies for his pupils, John Hunter almost certainly set out himself under cover of darkness from the Covent Garden school, armed with a shovel and crowbar, to scour local burial grounds for freshly dug graves. Most likely, he commandeered parties of students, probably bolstered by several rounds of ale in a tavern beforehand, to help in his grisly undercover work. Although he generally took care to conceal the source of his anatomy material, one indiscreet entry in his casebooks makes plain that he obtained bodies from London graveyards: "Sept 1758. In this Autumn we got a stout Man for the Muscles from St George's ground."[19] His ambivalent "we" suggests he may well have led the crew of daredevils who raided the burial ground, which was probably St. George's churchyard in Bloomsbury, and was not just the recipient of the stout man's corpse. Well-built men were always in demand: With their skin stripped off, they could be used to display the muscles to best advantage. But such nocturnal outings were precarious, as well as unpleasant in the extreme. So John quickly sought to recruit willing accomplices to his cause.

Professional grave robbers had begun supplying anatomists with corpses for dissection in the early 1700s.[20] Concentrated mainly in London and Edinburgh, but also working in provincial English towns and later the American colonies, the body snatchers developed unique methods of unearthing their quarry with speed, stealth, and efficiency. Strongly despised and universally feared, the rival packs of ruthless men were nicknamed the "Resurrectionists," after their ability to raise the dead from their graves, or the "Sack 'Em Up Men," after their method of transporting their wares. Initially, the perpetrators were grave diggers and sextons bribed by surgeons to exhume the corpses they had just buried. One grave digger who had been thus seduced was caught in 1736 lifting bodies from St. Dunstan's churchyard in Stepney, located in east London, where he worked, and selling them to a surgeon for dissection. Sentenced to be publicly whipped, he received hundreds of lashes from the angry crowds that gathered in the churchyard to dole out his punishment.[21]

So not only were the grave robbers willing to supply bodies to Scottish and English anatomists; they could also prove to be convenient scapegoats for outraged public opinion. By the time Pritty and Thomas were caught stealing John Race's corpse in 1747, the growing business in dead bodies was beginning to attract all manner of lawless desperadoes and petty crooks, who were greedy for rich pickings from the eager surgeons. Before long, with the Hunter brothers' backing, body snatching would become a carefully orchestrated, highly managed, and increasingly lucrative industry. Rising public vigilance and widespread condemnation would only prompt the Resurrectionists to develop ever more sophisticated techniques.

Working on moonless nights throughout the winter, when the anatomy classes ran, the teams used wooden shovels to dig silently through the freshly turned soil of a new grave. By the discreet glow of covered lanterns, they dug a narrow shaft at the head of a grave down to the top end of a coffin, exposing about one-third of the lid. Using crowbars or grappling hooks, they pulled on the coffin lid, snapping the cheap thin wood under the weight of the remaining soil, to reveal their trophy. It was now but an easy matter to sling a rope around the arms and drag the body from its dank hole. Quickly, the gang stripped the corpse of its funeral shroud, working on the principle that while there was no law against stealing a body, since it had no owner, taking clothes, a coffin, or even a wedding ring could be punishable by hanging. In practice, the courts still found ways to punish miscreants—as Thomas and Pritty had found, to their misfortune. Next, the naked and usually limp body—rigor mortis would normally have passed within a day or two of death—was bundled into a sack or hamper, ready for delivery before dawn to the basement steps of a waiting anatomist.[22]

At the peak of their reign of terror, before the Anatomy Act ended the practice, professional gangs would unearth a body from a shallow grave in fifteen minutes flat. A single team could procure as many as ten bodies in a night's work or three hundred in one year. Paupers' graves were always the most popular, since the poor were conveniently buried in mass graves left uncovered until full or in shallow graves in public burial grounds. Richer folk took care to protect their dead in locked vaults or lead coffins.

Grave diggers, church sextons, and night watchmen appointed to guard the graves were often in the pay of the body snatchers. Fearful of discovery, the crooks took meticulous care to disguise the violation they caused by carefully replacing soil, flowers, and mementos in their original positions. The crimes were often so stealthy that a gang would sometimes burrow into a suitable-looking new grave, only to find it empty, their rivals having gotten there before them. In such cases, when one team had "trespassed" on the burial ground considered the territory of another, revenge was swift; there was no honor among body thieves. The crooks gave one another up to the authorities, raised a mob against their rivals, or left a trail of guilt, propping up coffins in their graves and leaving shrouds strewn on the ground. Ultimately, the reckless men became so adept at their trade, and the demand from anatomists so intense, that London's churchyards were honeycombed with empty tombs. On several occasions when thefts were suspected, horrified relatives would frantically dig up grave after grave, only to find every corpse was gone.

Under the laws of the black market, rising demand meant rising prices. An adult body in the mid–eighteenth century could be bought by surgeons for about one guinea, but the price had doubled by the 1780s and then leapt to as high as sixteen guineas over the next twenty to thirty years. The bodies of children, known as "smalls," were priced by the inch, while a rare medical condition always commanded a premium. At a time when a merchant seaman earned just over one pound for a month's hard labor, the body snatchers' nightly toil was a lucrative business, even if they were idle throughout the summer, when bodies would not keep and anatomy classes were therefore forced to close.[23] Often surgeons connived with their suppliers, directing them to the grave of an unlucky patient whose condition they wished to investigate or whose diseased organ they wished to obtain. There was nowhere to hide, as Sir Astley Cooper, one of John Hunter's most fervent devotees, would make clear when he looked the members of the 1828 parliamentary select committee in the eye and chillingly told them, "There is no person, let his situation in life be what it may, whom, if I were disposed to dissect, I could not obtain."[24]

Certainly the law seemed largely indifferent to the antics of the grave robbers, as well as to the receivers of their stolen goods. While the Resurrectionists were sometimes caught and prosecuted, police officers often looked the other way when they encountered the thieves at work. If the crooks were prosecuted, the courts usually dealt out relatively lenient penalties, and even then their surgeon masters often paid their bail and supported their families while they were imprisoned. Only a handful of surgeons was ever brought to justice. John Hunter would never face legal action over his relentless trade. Likewise, government and Georgian high society turned a blind eye to the buoyant market in bodies. Those in power knew that the army and navy were desperately short of skilled surgeons, they appreciated the need for anatomists to practice their art, and, in any case, the desecrated graves were rarely those of their relatives. But the practice was not without its risks. Body snatchers were often stoned, shot at, and, in Ireland, even killed by outraged members of the public.

Although the grave robbers were understandably discreet about their craft, a rare firsthand account of a gang in action in the early nineteenth century describes the methods they employed. In *The Diary of a Resurrectionist,* an anonymous grave robber outlined a typical January night: "Got up at 3 in the morning, the whole party went to Guys and St Thomas, got 3 adults, 1 from Guy's and 2 from St Thomas's, took them to St Thomas, came home and met again, took one of the above to Guy's, settled for the Horse £24."[25]

Hospital patients, like those stolen from Guy's and St. Thomas', provided an easy source of bodies. After burial in the hospital graveyards, they were often unearthed and returned to the very surgeons who had caused their demise, now curious to examine their handiwork. Certainly John Hunter would have obtained corpses from hospital burial grounds, as an entry in his casebooks recounts: "In 1759, I stripped the bones of an old Woman that died in St George's Hospital." The elderly woman was not, however, a patient whose family had consented to a postmortem—which is evident from his adjoiner: "I knew nothing of her history."[26]

From the day he started work in William's anatomy school, John Hunter embarked on a long and fruitful relationship with the grave robbers that would plunge him deep into London's criminal underworld. As procurer in chief of corpses for the school, he immediately sought out willing grave robbers and recruited more for the job. It was his role to give the villains their orders for the night, haggle over prices, and keep a candle burning in the dissecting room while waiting for the cumbrous sacks to arrive. But far from finding the job distasteful, Jack Hunter—as he now liked to be known—soon acquired a reputation for relishing his task. His youthful liking for a drink, lack of social airs, and colorful language evidently endeared him to his sinister suppliers. One early biographer, writing within living memory of the anatomist, called him "a great favourite" with the body snatchers.[27]

Plainly, he was good at his job. William was a demanding taskmaster when it came to obtaining subjects for dissection; he required not just random men, women, and children but also fetuses, babies, and pregnant women at various stages of gestation, both for his own research purposes and for lectures and students' practice. His young brother supplied them all. His records reveal: "In the winter, 1749, a child was brought into the room used for dissection in Covent-garden."[28] The lack of an accompanying medical history betrays the obvious source of the unnamed child. And on page after page, the vagueness of the descriptions of the bodies detailed in his records points clearly to their origins. "There was an Old Man dissected at our house" reads one entry, while another notes, "A Young Boy about four years was dissected." Later, Hunter would publish sketches of teeth and jaws he had dissected during his years in Covent Garden, taken from fetuses, babies, and children. Betraying similar uncertainty as to their ages, he would refer only to a fetus "about seven or eight months old," a child "of five or six years of age," and "a youth about eleven or twelve years old."[29]

Pregnant women were a recurrent requirement throughout the time he spent with William. Partly due to the difficulty in obtaining the corpses of pregnant women, the structure of the pregnant womb had been only rarely investigated by anatomists to date. Through the underground contacts he had established, John Hunter learned the whereabouts and acquired the bodies of women who had died at different stages of pregnancy, their bloated corpses being dragged from their coffins and shoved into sacks as usual. In William's later treatise on the pregnant womb, *The Anatomy of the Human Gravid Uterus,* an advertisement boasted that William had "had more opportunities of examining this subject than any other anatomist."[30] It was John who had made most of those opportunities possible.

In all, during the twelve winters John Hunter would spend with his brother, as he would later admit, he had "been present at the dissection of more than two thousand human bodies."[31] The intense industry involved was later captured dramatically in a watercolor by Thomas Rowlandson entitled *The Dissecting Room,* which shows three naked corpses in the process of being cut open by a huddle of more than a dozen excited anatomists.[32] Prominent among the dissectors are William and John Hunter. Given the scale of their operation, both in dissecting and acquiring human corpses, it was little wonder that William made sure to swear his pupils at the start of every term to an oath of silence: "In a country where liberty disposes the people to licentiousness and outrage, and where Anatomists are not legally supplied with dead bodies, particular care should be taken, to avoid giving offence to the populace, or to the prejudices of our neighbours. Therefore it is to be hoped, that you will be upon your guard; and, out of doors, speak with caution of what may be passing here, especially with respect to dead bodies."[33]

Out half the night scavenging in graveyards or waiting in the dissecting room for a muffled knock on the door, John Hunter soon learned to survive on minimal rest. He would manage on only four or five hours' sleep a night for most of his life. But his days were fully employed, too. Up at sunrise to catch the best light for dissection, he was kept busy preparing bodies, helping students with their practical dissection, and making specimens until dusk, before attending William's lectures for close to two hours every evening. Though school lessons had left him cold, he was in awe of his con-

fident elder brother's eloquent lecturing style, which, he declared, "probably no man could excell."[34] Indeed, William's lectures would eventually become so acclaimed that society intellectuals and aristocratic gentlemen, as
well as keen medical students, clamored to listen. In 1777, the economist
Adam Smith would attend a complete course, along with his friend Edward
Gibbon, the historian; at the end of each lecture, Gibbon would make a
point of thanking William for his instruction.[35]

Yet if William still held his little brother's attention in the lecture theater, it was not long before pupil began to outshine master in the dissecting room. After just six months of instruction, by the end of the spring
course in 1749, William declared his protégé sufficiently accomplished to
take over all the dissecting work at the school. With his deft fingers and
sharp powers of observation, John excelled at preserving the body parts he
dissected as specimens—or "preparations," as they were known—which
were passed around at lectures for the pupils' perusal. Spending long hours
hunched over the dissecting bench, he pickled organs and tissues in spirits, dried bones and muscles before varnishing, and injected intricate systems of arteries and veins with vividly colored wax. Some survive to this
day, despite the handling of numerous students, their colors as bright as
ever. For the moment, William found this arrangement both fortuitous and
convenient. It allowed him to devote more time to his advancing obstetrics
career and society engagements, while his brother enhanced the reputation
of the anatomy school and filled the shelves with stunning preparations. It
would not be long, however, before John's exceptional prowess provoked
jealousy and recrimination.

What little time John Hunter had left each day was occupied principally in demonstrating anatomy skills to the students. Working shoulder-to-
shoulder with other young enthusiasts in the dissecting room by day and
debating anatomical questions of the moment into the night, he would
forge lifelong friendships. Compared to William's sometimes austere and
forbidding demeanor, John's easygoing, personable manner won him many
devoted companions. Although there would be no shortage of people ready
to testify to his outspokenness and quick temper—he never suffered
fools—later pupils would assert that his manners were "extremely companionable" and that he possessed "a very considerable humour."[36]

Despite the overwhelming demands on his time, the young Jack
Hunter still found time to sample the many social pleasures of Georgian

London, jeering with the audiences in local theaters and carousing in neighborhood drinking venues.[37] Even his venomous first biographer, Jessé Foot, conceded that, in his youth at least, Hunter was "a companionable man: he associated in company, drank his bottle, told his story and laughed with others."[38]

The *first warm* rays of spring sunshine brought the tireless John Hunter's six months' labor in the dissection room to a necessary end, for even the most hardened of anatomists could not stomach cutting up rotting corpses in summer heat. With warm days ahead and the coming harvest to consider, he was entitled to a well-deserved break back home on the farm. His eldest sister, Janet, had died that spring at the age of thirty-six. His grieving mother was herself now sixty-four, in failing health, and sorely in need of comfort from her favorite son. But William had other plans for his young brother. Having honed his knife skills on dead bodies throughout the winter, John Hunter was now dispatched to practice on the living.

CHAPTER 4

The Pregnant Woman's Womb

∽ഇഇഗ

The Royal Hospital Chelsea, London
Spring 1749

Striding up the sweeping driveway from the Thames to the grand facade of the Royal Hospital in its pastoral setting at Chelsea, John Hunter could have been forgiven for lapsing into a relaxed frame of mind. The spring sunshine, which was bringing out the blooms in the hospital's glorious gardens, had brought an end to the first long winter in his brother's anatomy school. Londoners had just celebrated the end of the War of the Austrian Succession—actually concluded six months earlier—with a spectacular fireworks display and a concert by Handel. And the soldiers who had limped back from the battlefields across Europe were nursing their wounds in the tranquillity of their riverside home at Chelsea.

Founded by Charles II and opened in 1692, the Royal Hospital and its gardens had been designed by Christopher Wren as a retirement home for elderly or infirm soldiers.[1] For the nearly five hundred veterans who now lived there, attired in their distinctive uniforms of scarlet jackets, blue breeches, and gold-edged black hats, the hospital provided a veritable haven. Approaching its impressive front portal, Hunter turned left, past the benches where the war heroes sunned themselves, and then proceeded through an archway into a second courtyard. With trepidation, he walked toward the low brick building, the hospital's little infirmary, on the south side of the quadrangle. After a winter confined with the cold, silent corpses of the dissecting room, he was to spend the summer in his first serious job: tending the ailing veterans at Chelsea under the watchful eye of the country's greatest surgeon.

Most likely, he discovered William Cheselden dressing a soldier's old wounds in the infirmary, where the fresh spring air was instantly overpowered by the noxious smells emanating from the communal toilet bucket. Although his post at Chelsea amounted to semiretirement, Cheselden, the venerable surgeon who had once so infuriated the Company of Barber-Surgeons by staging rival dissections on his dining room table, was still justifiably renowned.[2] Having taken the helm of the newly formed Company of Surgeons in 1746, a year after he helped force the split from the barbers, the popular anatomy teacher was famous throughout Europe. His textbook *The Anatomy of the Humane Body* was the standard work for students, while his fabulous atlas of human bones, *Osteographia,* was a collector's item. But although his talents with a pen were widely respected, it was Cheselden's skill with a knife that had made him the most celebrated surgeon of his age. Principally a man of action rather than words, he was best known for his pioneering operation to remove bladder stones.

Known as lithotomy, the operation to extract a bladder stone had been performed since ancient times; as of the eighteenth century, little had changed in the procedure since ancient times.[3] Without anesthesia or antiseptics, operations were generally crude, frequently agonizing, and often deadly. Consequently, it was only when the pain of a stone pressing inside the bladder became unbearable that a sufferer would submit to a lithotomy. Unfortunately, however, bladder stones were extremely common, even among children, in the eighteenth century and therefore many sufferers had little alternative but to endure the grueling procedure, which was not only exceedingly risky but also highly humiliating. In order to allow the surgeon access to the necessary area, patients were trussed up like a chicken and placed flat on their backs, with their ankles tied to their hands and their buttocks exposed. Usually the surgeon would insert a finger into the rectum, probe the stone so that it bulged outward, then cut straight through the muscles of the perineum, between the genitals and the anus, and withdraw the stone from the bladder with forceps. Sometimes the surgeon would insert a metal probe into the patient's urethra—the canal carrying urine from the bladder—to help locate the stone, and sometimes would withdraw it this way, too.

The diarist Samuel Pepys underwent a lithotomy in 1658, probably of the first type, when he had a stone the size of a tennis ball cut out by a surgeon from St. Thomas' Hospital. He kept the stone ever afterward in a case.[4] Others were less lucky. Two out of five patients died from the procedure in Paris in the eighteenth century, and success rates were unlikely to have been much different in Britain. The wonder is that mortality figures were not higher. Without adequate anatomical knowledge, inept surgeons could blunder about for up to an hour, while their shrieking patient expired on the table due to loss of blood. Unhygienic conditions in the home, where wealthy patients had their operations—sometimes on the kitchen table—but especially in hospitals, where the poor underwent surgery, frequently led to fatal infections. In the days before bacteriology was understood, surgeons were blissfully unaware that they themselves were often the cause of their patients' demise as a result of passing on deadly germs from their grimy hands and encrusted tools. But even survival was not without its risks, for many victims were rendered incontinent or otherwise permanently mutilated by a careless or ignorant slip of the knife.

With such poor prospects, patients understandably took the utmost care in finding a skilled surgeon for any operation. For lithotomies in mid-eighteenth-century England, there was really only one name: Cheselden. Having initially practiced the standard lithotomy method, Cheselden had altered his technique after learning of the successful work of a French traveling practitioner, Frère Jacques de Beaulieu—the Frère Jacques of the nursery rhyme. Rather than cutting through the middle of the perineum, as conventional surgeons did, Frère Jacques approached an inch or so to one side. His "lateral" operation thereby provided safer access to the bladder and reduced injury to other organs. While orthodox surgeons pointedly ignored the example of someone they regarded as a blatant quack, Cheselden had no such scruples. He adopted Frère Jacques's operation in 1725, and with the anatomical expertise he had developed on his dining table, he perfected the technique, to international acclaim.[5]

Cheselden routinely performed the lateral operation in less than a minute, and on occasion in half a minute, dramatically reducing the risk of fatal bleeding and thereby cutting death rates to less than one in ten. Even this scale of fatalities he blamed on the fact that once news of his success had spread, "even the most aged and most miserable cases expected to be sav'd by it." As well as saving the poorest from almost certain death in London's charitable hospitals, the surgeon was heavily in demand by private clients. He reputedly charged his paying patients up to five hundred pounds—equivalent to a modern fee of £41,000—for his services. At a time when physicians were unquestionably at the top of the medical hierarchy, and when many of his surgical colleagues were blithely hacking and sawing, with little inclination to change their brutal methods, Cheselden was one of a tiny handful of surgeons who garnered any respect in early eighteenth-century society. It seemed only natural, then, that when William Hunter sought a teacher of surgery for his young brother John, it was Cheselden to whom he turned.

Having quickly mastered all his older brother could teach him by working on a dead body, John Hunter knew that if he was ever to employ his knowledge on the living, he needed some bedside training. William himself had walked the wards as a surgical pupil at St. George's Hospital in his early career; now he was anxious to obtain a similar position for John. As he declared in his lectures, no doubt with his bright young brother in mind, "Were I to place a man of proper talents, in the most direct road for becoming truely great in his profession, I would chuse a good practical Anatomist, and put him into a large hospital to attend the Sick, and dissect the dead."[6] Having made Cheselden's acquaintance through his late mentor, James Douglas, and having made a favorable impression by presenting him with two specimens, William now recalled the favor and secured John a coveted position shadowing the best surgeon of the age.[7]

Arriving at the Royal Hospital in the spring of 1749 for his first taste of practical surgery, Hunter would have followed Cheselden on his ward rounds, attended his consultations with private clients, and assisted in operations both at Chelsea and privately. It was not the most regular route to a career in surgery. Most surgeons who enjoyed any training at all in the first half of the eighteenth century began learning their craft at the age of fourteen or fifteen, apprenticed to a master surgeon for up to seven years.[8] With large fees ponied up by their parents, these adolescent boys—girls were barred from this elite world—were bound by the strict rules of their indentures. They had to live in their master's house, work all hours without pay or holiday, and were forbidden from marrying, drinking, theatergoing, or gambling. Only at the end of this arduous initiation were they likely to study any anatomy—if at all. The adventurous few might spend some months at anatomy lectures in Edinburgh or Paris; others might attend one of the cursory courses offered in London. And only after their long apprenticeship and a smattering of anatomy would the typical trainee walk the wards of a London hospital, enrolled as the pupil of an established surgeon.

Again, a hefty fee was due for this privilege—as much as fifty pounds for a year's pupilage at St. Thomas' Hospital in 1750. At the end of this lengthy training, a would-be surgeon took the examination at Surgeons' Hall to qualify for a diploma, which would allow him officially to practice. Finally, he could start as a junior surgeon on the first rung of the hospital career ladder—a necessary prerequisite to building up a thriving private practice—provided he had the right connections. As in all realms of life in Georgian Britain, nepotism was the surest route to advancement in London's voluntary—or charitable—hospitals. Once set up in a hospital job, a young surgeon would embark on a steady, if protracted, path to a life-long position as a senior surgeon with a lucrative private clientele.

Inevitably, however, in the unregulated, chaotic medical world of the eighteenth century, innumerable surgeons attained status and fortune despite bucking the system; John Hunter would be one of them. Many young men, and even a few women who stepped into a deceased husband's shoes, became successful surgeons despite their scant experience or training. They skipped the apprenticeship, ignored anatomy lessons, spent at most a few months walking the wards, and launched into practice without ever securing a diploma from the Company of Surgeons. Indeed, for most of the first half of the century, although the Company of Surgeons admitted candidates for its oral examination only after they had completed a full apprenticeship and could demonstrate a knowledge of Latin, there was no requirement to have studied anatomy or to have obtained any surgical experience. After 1745, the government was so desperate to recruit surgeons to the army and navy that anyone willing to try their hand as a surgeon's mate was ushered in to take the examination before being let loose on the nation's unfortunate soldiers and sailors—and, once demobbed, the general public.

For all the pomposity, it mattered little how long a novice surgeon was apprenticed, where he walked the wards, or whether he obtained the Company's diploma, while the masters, teachers, and examiners clung to medieval procedures based on the theories of the ancients, with barely a nod toward anatomical research. In any case, the Company exerted its feeble influence only within the metropolis; elsewhere, launching a surgical career was even more of a free-for-all.

In John Hunter's case, time was firmly against his signing up as an apprentice at the mature age of twenty-one, while his liking for carousing in the taverns and boisterous behavior at the theater were plainly at odds with an apprentice's sedate lifestyle. Yet London's charity hospitals were unlikely to have accepted him as a surgical pupil without the hard-won certificate of apprenticeship. Signing up with Cheselden at Chelsea was a happy compromise. The elderly surgeon was willing to accept the industrious young anatomist without the requisite apprenticeship merely on a recommendation—and presumably a generous fee—from William. And if walking the wards at Chelsea did not offer Hunter the usual path to a surgical career, it certainly served as a more than adequate introduction to the agony and brutality of Georgian surgery.

As Cheselden's pupil, Hunter would have learned to perform all the standard treatments of the day: letting blood, lancing boils, dressing sores, treating venereal disease, and tending to minor injuries.[9] Major operations were a rarity, given the risks and the anguish they entailed, and even those that were performed were usually constrained within realistic limits. Surgeons almost never attempted to open a patient's chest or abdominal cavity, knowing they could not combat the internal bleeding or postoperative infection that would almost certainly ensue. Consequently, surgery was usually restricted to the limbs, the head, and the more accessible internal areas, such as the bladder, breasts, and testicles. Certainly, throughout the summer of 1749, John Hunter would have assisted in performing the most common operations of the day both on the pensioners at Chelsea and on members of Georgian high society. As an assistant to Cheselden, it was Hunter's job to secure patients to the wooden operating table, dose them with laudanum or straight alcohol, and hand his master the knives, probes, and saws for lithotomies, amputations, eye surgery, hernia operations, and other brutal procedures. Standing at the elder surgeon's side, Hunter witnessed the considerable apprehension Cheselden felt before embarking on any operation. As Cheselden admitted, "No one ever endured more anxiety and sickness before an operation."[10] While Hunter himself would never quail before an operation, he firmly adopted Cheselden's reluctance to wield the knife unless absolutely necessary.

Inevitably, Hunter aided Cheselden with his trademark lithotomies, helping to truss up victims in the oven-ready position, assisting in the delicate procedure, and dressing the subsequent wound. Amputations were another stock procedure. For these, Hunter helped to hold down a struggling patient as Cheselden removed a leg in three swift actions. First, he cut through the skin with a long, curved knife in a single circling motion. Then he cut through the fat and muscle down to the bone, tying major blood vessels to prevent hemorrhage. Finally, he hacked through the bone with a fine-toothed amputation saw, pulled the skin over the bone, and bandaged the stump.[11] Hunter would have learned the primitive operation of trepanning, another Cheselden favorite, firsthand, too. The procedure had changed little since first practiced in neolithic times: Just as in the Stone Age, Georgian surgeons drilled into a patient's head to remove portions of skull. Although crude, it seemingly helped some patients by removing blood clots and relieving pressure on the brain. A surprising number of people, even in neolithic times, survived the ordeal. Equally successful, on occasion, was the operation of couching. The technique, which entailed cutting out or displacing a lens in which a cataract had developed, also had its origins in ancient times. Cheselden's expertise enabled him to restore sight in several patients, including one man who was even able to read after the surgeon supplied him with spectacles.[12]

Working at Chelsea with Cheselden, even in his advanced years, was a unique opportunity for the young John Hunter. More important, however, than any manual skills he gleaned, Hunter benefited from Cheselden's singular approach to his craft. Almost any other surgeon of the time would have taught his pupil strict reverence for the ancient methods and theories, while firmly cautioning against any challenge to established ways. Cheselden, conversely, was eager to observe and experiment, reluctant to operate without a good chance of success, willing to learn from his anatomical pursuits and even from unqualified healers, and adaptable enough to amend his methods accordingly. Hunter would pursue this same creed throughout his career, systematically applying the lessons he learned in the dissecting room and operating theater to future practice.

But time was running out for Cheselden. After Hunter spent a second summer at his side, the veteran surgeon fell ill in 1751, and he died in 1752. For the intervening season, Hunter enrolled as a pupil with Percivall Pott, the up-and-coming staff surgeon at St. Bartholomew's Hospital, which was next to the Smithfield meat market in the city.[13] As Cheselden's star faded, so Pott's was rising. At thirty-seven, Pott was not only a keen anatomist but, like Cheselden, also a cautious surgeon. He even managed to avoid having his own leg removed by appealing for conservative treatment when thrown from his horse. But even if John was allowed to spend his summers with the living, each winter his attendance was required back in the dissecting room for the start of the autumn term.

With demand from pupils increasing all the time, the Hunter brothers had moved in September 1749 from their temporary lodgings into a large house at 1 Great Piazza, Covent Garden, a residence that for the first time provided permanent premises for the anatomy school as well as a shared home. The comfortable house, which was located under the vaulted arcade in the northwest corner of the Italianate square, possessed a basement, three floors, and an attic. The brothers set up their dissecting room at the rear of the house, where stables and a garden provided a conveniently discreet entrance for nightly deliveries.[14] There was space for a lecture theater and preparations room, too. But the elegant home belied the squalor of the square, which echoed with the cries of street vendors and market-stall holders by day and the carousing of sailors and other pleasure seekers patronizing the brothels by night. With London's gin craze at its height, anyone crossing the square for a morning dish of coffee in Bedford's coffeehouse or an evening's entertainment at Covent Garden Theatre was likely to stumble over men, women, and children slumped in a drunken stupor among the rotting vegetables.

If not the most fashionable address in London, at least the spacious quarters allowed the brothers to take in resident pupils for the first time. The fresh-faced eighteen- and nineteen-year-olds who joined them not only provided a useful source of extra income but helped create a lively community of enthusiastic scholars, effectively forming the first medical school in the capital. Many stayed an entire winter to attend both the autumn and the spring courses by day and to participate in animated debates on anatomy by night, as well as forming graveyard exploration parties as required. Many would go on to become leading anatomists and medical practitioners in their own right; among them, John Hunter made some of his firmest friends.

Having dutifully learned his craft at William's shoulder in his first year, from the autumn of 1749, John took on the vital job of demonstrator and instructor. Master of the dissecting room, he presided over all the anatomical research of students just a few years his junior. From first light until dusk, he was invariably to be found stooped over the dissecting bench, teasing apart blood vessels, nerves, and fibers with numbed but nimble fingers as the acrid stench of the rotting cadavers invaded his nostrils. When the pupils crowded in for practical dissection, he helped them tackle the delicate knife work they needed to master on their individual corpses.

The timetable of pupils' dissection program was dictated strictly by necessity, according to the process of decomposition.[15] Since the guts were the first parts to putrefy, the students would begin by slitting open the abdomen, folding back the flaps of skin and fat, and examining the organs of digestion: the stomach, the more than thirty feet of intestines, and the smaller organs, such as the spleen, gallbladder, and pancreas, packed tightly into the abdominal cavity. Next, they would open the chest, sawing apart the rib cage to expose and remove the lungs, the lobes of which were blackened by London's ubiquitous winter smogs; mastery of dissection demanded not only intricate skills with a knife but brute strength with a hacksaw. The lungs discarded, the pupils could examine the heart, which even the most novice anatomist now knew, having studied Harvey, to be the center of the human physiological system. Hunter would then help the pupils with the tricky task of inspecting the other main organs—the liver, bladder, kidneys, reproductive organs, and the brain—encouraging them to probe the cavities with their fingers, follow the vessels with their knives, and weigh the organs in the palms of their hands. Finally, they could work on the muscles, which decayed slowest, and the bones, which Hunter showed them how to wire together to create their own articulated skeletons.

But they had to work fast. Exposure to air quickly dried the tissues, making them brittle and difficult to work with, while the decaying flesh proceeded to swell, discolor, and generate foul odors. The necessity of working whenever possible in winter was made indisputably clear by one account in Hunter's casebooks of a postmortem in warm weather. "He was become extremely putrid," Hunter recorded, adding, "The kidneys were almost reduced to a pulp with the Putrifaction." More typically, opening a body on an "extremely cold day" with no fire in the grate, he noted that "the cellular membrane immediately under the Skin crackled upon pressure."[16]

While the dissecting room acquired a permanent stench, so, too, did the hands and clothes of the pupils and their teachers. After handling human remains all day without gloves—although they usually donned protective aprons and sleeve covers—the only washing facilities for their bloodstained hands, their fingernails clogged with putrid flesh, was a bucket of cold water pumped directly from the foul Thames. A later manual for anatomy students would offer welcome advice on dispelling some of the discomforts: "The student should endeavour to prevent the bad effects of sitting for hours in a cold dissecting room; the most effectual way is to put on an additional flannel jacket, and carpet shoes over his boots. . . . A cap should be worn in preference to a hat, which is not only inconvenient, but also quickly acquires a bad smell. . . ."[17]

Anatomy was invariably a sensory experience. As well as being subjected to the all-pervading odor of decay, the crackle of the dried membranes, and the need for intense visual inspection, students were urged to feel the textures of the different parts and even to taste the body fluids. Without sophisticated methods of scientific analysis, at a time when microscopes were primitive and visually unreliable, anatomists were forced to depend on their innate senses. Hunter frequently employed his sense of taste in dissection, and he encouraged his pupils to do likewise, as he recorded matter-of-factly: "The gastric juice is a fluid somewhat transparent, and a little saltish or brackish to the taste." And he would even observe, "The semen would appear, both from the smell and taste, to be a mawkish kind of substance; but when held some time in the mouth, it produces a warmth similar to spices, which lasts some time."[18]

Yet no matter how competently an anatomist employed all five senses, his skills were wholly transitory without the ability to maintain his knife work, both for teaching and research purposes. Carefully executed preparations played a crucial role in eighteenth-century medical education; John Hunter excelled at the art. Specimens of bones and body tissues were vital not only for showing complicated or minute networks of blood vessels and other structures that the pupils were too inexperienced to distinguish for themselves but also for displaying diseases or conditions that the future surgeons might encounter in the living. In an age when medical practitioners still commonly blamed ailments on an imbalance of humors, morbid anatomy—the study of diseased or damaged organs—was in its infancy. Linking the appearance of disease in the dead to symptoms of illness in the living, and thereby diagnosing medical conditions with accuracy, would be critical to later improvements in treatment. The Hunter brothers, and their nephew Matthew Baillie, were fundamental in developing this approach.

William counseled his students to collect as many specimens as they could. These should include not only a complete skeleton but also several skulls, preparations of blood vessels, two specimens of children's trunks showing the abdominal organs from back and front, and "as many preparations of the organs of sense and generation, and of the particular viscera as he can easily procure." And while pupils should not rely solely on preparations, they were certainly convenient, since "the more they are used there will be less expense and trouble with fresh subjects."[19] More preparations meant fewer nocturnal outings.

Essentially, specimens could be divided into "wet" or "dry" preparations. The latter were relatively easy to make: Stripped bones and sometimes muscles or organs were dried in the air, often by hanging from the rafters of dissecting rooms, then varnished to preserve them from decay. Dry preparations were of limited value, however, since they usually became shriveled and discolored in time. The anatomist's chief goal was to preserve his handiwork in such a way that the beauty of the human anatomy was retained in its most natural and supple form.

The seventeenth-century scientist Robert Boyle is credited with discovering that alcohol could be used to preserve organic tissue. He reputedly preserved a linnet and a snake for four months in spirit of wine. Although gin, rum, brandy, and whiskey were sometimes substituted, spirit of wine would remain the preservative of preference until superseded in 1893 by formaldehyde, which is still in use today. But the individual recipes anatomists cooked up for their potent alcoholic cocktails were kept closely guarded secrets—much in the way medieval alchemists had vied in their futile quest to find the philosopher's stone.

Guided initially by his brother, John Hunter learned the basic techniques of making up wet and dry preparations, but he was soon surpassing William's prowess at both and quickly developed his own unique methods to create exquisitely lifelike specimens.[20] A critical aspect of the art, which Hunter deftly demonstrated for the pupils, was the use of injections to highlight vessels in vivid color, demonstrating their various ramifications. The knack was to inject a liquid that was sufficiently viscous to fill an entire vascular system, including the tiniest vessels, which were otherwise invisible to the naked eye, without bursting the fragile structures and leaking into other parts. Often it was a process of trial and error. One of Hunter's favorite tricks was to inflate certain parts, such as bladders or lungs, with bellows or blowpipes to enlarge the vessels. As he explained: "When I was injecting the lungs of a man, the injection did not run freely; I then inflated them, and found that the injection immediately ran with freedom."[21] Usually, he first injected the arteries with warm water to flush out the blood and drive out any air. Only then did he fit a pipe into the open end of a vessel and inject the fluid by syringe.

Like his predecessors in the art, Hunter developed a veritable cookbook of recipes for the concoctions he used. They included oil compounds made from hog's lard, tallow, or butter, and water-based solutions, mixed with glues or resins from plants or animals. Sometimes he used wax, heated to a fluid for injection, then allowed to cool to a solid state in the vessels, or mercury, because of its propensity to flow easily. Mercury was also useful in investigations, revealing vessels and connections that were otherwise impossible to distinguish.

Brilliant hues were added to the injection fluids to display the vessels clearly and show the handiwork to best advantage. Hunter used artists' pigments such as vermilion, king's yellow, blue verditer, and flake white. Sometimes bristles were employed to demonstrate where particular vessels began and ended. The hairs from a rhinoceros tail were ideal, Hunter advised, but if these were in short supply—as they must have been most of the time—goose quills would suffice. With artists' hues and rhinoceros bristles to buy, as well as hefty government taxes to pay on both alcohol and flint-glass jars, the art of making preparations was a hugely expensive business. At least the pig's bladder used to seal the jars, along with a layer of lead, could be cheaply obtained from a butcher.

Many injected preparations were simply bottled in alcohol and displayed in the school's preparations room to demonstrate particular structures or rare conditions. Others were submitted to a process of corrosion to destroy the organic flesh and leave a cast of the wax-filled vessels. Usually, this was achieved by lowering the injected part into a bucket of acid, which ate away the tissues, leaving an intricate and often stunningly beautiful network of colored vessels. A skillfully corroded preparation of the injected airways of the lungs, for example, emerged as a delicate, vividly stained "bronchial tree."

Though the specimens were primarily intended to display anatomical features, aesthetics were always an essential aspect, of which Hunter was acutely aware. He recommended bloodred vermilion for run-of-the-mill injections in dried preparations, but in an extremely vascular part, such as the inner surface of the stomach, the anatomist should simply "chuse the

Colour that pleases the Eye most." By contrast, bones, which emerged as "dark or dirty brown," were not deemed pleasing, for their color was "a bad one." A particularly fine or rare preparation, expertly injected in vivid color, could fetch a handsome price at auctions of anatomical parts, where amateur collectors competed for items with professional anatomists. Both Hunter brothers would become regular customers at such auctions. But on one occasion, when John found that no amount of money could secure a preparation he coveted from a fellow practitioner, he finally snapped, "Well then, take care I don't meet you with it in some dark lane at night, for if I do, I'll murder you to get it."[22] He was probably joking.

By far the majority of the exquisite preparations John Hunter painstakingly created during his long hours in the Covent Garden dissecting room would remain firmly in William's hands. Ever the astute businessman, William scrupulously asserted propriety over handiwork generated under his roof. Later assistants would find their specimens similarly claimed as his property. As far as William was concerned, he financed the work and therefore he owned it. Ultimately, most would end up in his collection, which he bequeathed to his alma mater, Glasgow University.[23] While initially John accepted this one-sided arrangement as part of his filial dependence on William, it was not long before he began to question his unsung contribution to his elder brother's reputation. One instance in particular would rankle for decades to come.

The trouble began with an unprecedented opportunity in the winter of 1750, after John had returned from his second season with Cheselden, when an exceptional delivery arrived at the back door of the Covent Garden school.[24] The atmosphere in the dissecting room was tense as the sack was opened. The corpse that fell out belonged to a woman toward the end of her ninth month of pregnancy; on the verge of giving birth, she had died suddenly, from an unknown cause, with her unborn child still intact within the womb. Since pregnant women were never hanged at Tyburn, and women of childbearing age were unlikely to die just prior to giving birth, anatomists hardly ever managed to dissect women in the final stage of pregnancy. Despite his midwifery training, William had never before seen a fully developed baby still in the womb. Until now, he had had to study full-term pregnancy in animals. Almost unable to contain his excitement, he later recalled, "A woman died suddenly, when very near the end of her pregnancy; the body was procured before any sensible putrefaction had begun; the season of the year was favourable to dissection; the injection of the blood-vessels proved successful; a very able painter, in this way, was found; every part was examined in the most public manner, and the truth was thereby well authenticated." In other words, circumstances were perfect—without William having to contribute anything to the event.

The body had been "procured" by John from his usual underworld associates, and it was John, as usual, who conducted most of the expert knife work required. William would later grudgingly admit that he was "assisted by his brother Mr John Hunter," adding, "whose accuracy in anatomical researches is so well known, that to omit this opportunity of thanking him for that assistance, would be in some measure to disregard the future reputation of the work itself." He was equally dismissive of his "painter," a young Dutch artist named Jan van Rymsdyk, whose expert artwork in Covent Garden would never be surpassed.

Working together in the odorous dissecting room, Jan and John labored over the critical task of opening the pregnant body and accurately depicting its contents. The job required extremely delicate knife work to unveil the uterus step by step without damaging the small body nestled inside. First, John made a cruciform cut in the woman's abdomen and peeled back the four corners of skin to reveal the bulging womb. Carefully, he injected the arteries and veins crisscrossing the surface with different-colored waxes, then he opened the uterus to expose the thick lining inside. William promptly stepped in and named this "the decidua," since it was shed after childbirth, while claiming for himself the discovery that it derived from the womb rather than from the fetus. Finally, John slit open the membrane sac to reveal the fully developed child, its plump inverted body wedged tightly, its fingers curled, its dark hair glistening wetly on its head, awaiting the moment of a birth that would never happen.

At each stage, van Rymsdyk captured the revelations in ten stunning red-chalk drawings. Simultaneously, they evoke awe at the simple beauty of the baby that would never draw breath and shock at the butchered body of its mother, who had breathed her last. Whereas previously anatomical pictures of babies in the womb had shown curiously adultlike figures floating in a shapeless void, for the first time van Rymsdyk portrayed the intimate relationship between mother and child in a completely naturalistic style. Immediately, William commissioned engravers to convert the sketches into ten copper plates.

But before the plates had even been completed, there was more excitement in the dissecting room. The corpses of two more women near the end
of their pregnancies—most probably patients from the lying-in hospital
where William worked—arrived at the back door. Again, John and Jan were
set to work dissecting and sketching under William's exacting supervision.
The pair had to work especially quickly on the second "subject," which arrived when "the weather happened to be very unfavourable"—presumably
in the summer of 1751. But the third corpse "occurred very opportunely,
which cleared up some difficulties," William recorded. In all, a total of five
pregnant bodies, and one of a woman who had died two hours after giving
birth, were delivered to the school between 1750 and 1754.[25] All were investigated and sketched. In one extraordinary drawing, van Rymsdyk even
included the image of a nine-paned window reflected in the membrane
covering a five-month-old fetus. It was the very window providing light for
Jan and John to conduct their delicate tasks in the Covent Garden dissecting room.

At this point, however, it struck William that rather than publish ten
pictures of a full-term pregnancy, spectacular as that would be, it would be
far more dramatic to depict the pregnant womb at every stage, in order to
produce the first complete study of pregnancy. Everything was in place: His
willing brother was adept at procuring and dissecting the corpses, and his
talented artist was skilled at depicting the intricate anatomy. The project
would take nearly a quarter of a century. It would be 1774 before William
published his lifetime's goal, a huge atlas entitled *The Anatomy of the
Human Gravid Uterus Exhibited in Figures,* which traces the development
of the child in the womb backward from the chubby full-term baby of 1750
to a three-month fetus, with its tiny fingers, ears, and other features still
softened in development. The thirty-four plates of almost photographic detail, printed in massive "elephant"-folio dimensions at huge expense, were
a triumph. The book would be acclaimed one of the greatest anatomical
works ever. The vision was William's, but the crucial value of the work was
in the meticulous dissection carried out by John—and later assistants of
William's—and in the splendid artwork of van Rymsdyk. In all, sixteen of
the thirty-four plates depict dissections conducted between 1750 and
1754, performed by John.

By now, the yoke of William's punishing work program was beginning
to chafe. As an impressionable, enthusiastic youth, John had been happy to

labor under his brother's exacting regime, learning his craft in William's shadow and performing William's every bidding, with only the occasional snippet of praise as reward. Now in his mid-twenties, with his dissecting talents and preparation skills undeniably superior to William's, he was fast coming to resent his elder brother's unthinking assumption of both his time and his achievements. By early 1754, the cracks were beginning to show.

In May, rather late in the dissecting season, Colin Mackenzie, an anatomy assistant working for William Smellie, burst into the Covent Garden school, looking for John. Like the Hunters, Smellie and Mackenzie had obtained a full-term pregnant corpse to investigate the anatomy of the uterus. Mackenzie, who had become friendly with John, had already injected the blood vessels of the placenta, but the intricate pattern of circulation was still too complicated for him to be able to interpret the results. It was John he sought to dispel the confusion. Together, they rushed back to Smellie's dissecting room, where John carefully cut open the placenta to examine the injected arteries and veins. What he saw was a revelation. Having tried for years to trace the minute vessels, at last the pattern was plain to him. "After having considered these appearances," John later enthused, "it was not difficult for me to determine the real structure of the placenta and course of blood in these parts."[26]

Suddenly, it was clear that the maternal and fetal blood supplies, which scientists had previously assumed were connected, remained separate. Although they came close enough together in the placenta to exchange nutrients and oxygen, they never actually connected. Preserving the dissected parts, John returned home, eager to tell William of his discovery. At first, William laughingly dismissed his young brother's excited announcement. Finally, persuaded to accompany him back to Smellie's dissecting room, he was forced to admit the truth of John's discovery. Indeed, William even managed to obtain certain of the preparations John and Colin had made, which he promptly showed at his lectures, and, as John pointedly added later, "probably they still remain in his collection."

Not only did William now assume ownership of John's dissection work and his lovingly prepared specimens, but, as usual, he even appropriated this first major discovery. In his *Gravid Uterus,* as well as in his lectures, William calmly claimed precedence for discovering the circulation in the placenta. Not one reference did he make to the joint investigation by his brother and Mackenzie. For the moment, under William's roof and in his employ, John had no option but to keep his own counsel. Accordingly, not only the specimens he had painstakingly created with Mackenzie but also many more he now began to prepare as part of his own investigations into teeth, bone growth, animal life, and, not surprisingly in the noisome dissecting room, the nerves of smell would remain William's property.[27] Only a handful of John's earliest endeavors stayed in his possession. Somehow escaping William's clutches, eventually they would take their place in John's own museum, their delicate knife work perfectly preserved, their glorious colors as brilliant as ever they were when glowing on the Covent Garden dissecting bench.

At least the living could begin to benefit from Hunter's growing expertise. With the training he had received under Cheselden and Pott, and a further summer, in 1754, as a surgical pupil at St. George's Hospital, Hunter now began to treat his first patients.[28] Though it was many years before he would receive his surgical diploma, entitling him officially to practice, the inestimable benefits of long winters poring over human anatomy set him easily ahead of his contemporaries. Determined that each of these encounters should be a learning experience, he meticulously recorded his observations and the outcomes of his work. Many of the patients he saw during these earliest years went unnamed and their interventions undated in his records. But the five volumes of case notes that survive from his forty-year career, later collected in the *Case Books,* reflect a vibrant cross section of Georgian life. The fortunes and fates of those he treated, from soldiers to sailors, artists to housepainters, from lords, lawyers, and politicians to servants, waifs, and paupers, were recorded in John Hunter's slanting script.

Among his first patients—and the earliest intervention he recorded— was an anonymous chimney sweep who had been admitted to a ward at St. Bartholomew's.[29] Having contracted gonorrhea, the youth had developed a urethral stricture—a blockage in the urethra—that made voiding painfully difficult. Establishing the patient's medical history, Hunter brought to bear all his natural scientific curiosity—embarking on the experimental approach to surgery that would characterize his whole life—on the sweep lying in pain and frustration in his bed at St. Bartholomew's.

Initially, Hunter attempted the classic approach to unblocking a stricture, presumably learned from Pott, which entailed attempting to push a "bougie"—a cylindrical bung made of wax or sometimes lead—into the ure-thra to force a way through. When this failed, he characteristically decided to experiment and, importantly, to record his results. Hunter conjectured that he might shift the blockage by burning a way through, using a caustic salve on the end of a bougie, and that when the resulting scab fell off, this would create a free passage. The method, called "escharotics," had been used before for other ailments but had fallen into disuse.

First, he inserted a bougie loaded with "red precipitate"—mercuric ox-ide—but this only caused severe inflammation. So then he had a silver can-nula—a hollow rod through which a probe could be pushed—specially made and used this to insert a second bougie impregnated with mercuric oxide. This served no better. Finally, with remarkable forbearance on the part of the sweep, Hunter fastened a piece of "lunar caustic"—silver ni-trate—onto the end of the cannula and probed the urethra a third time. "After doing this three times at two days interval, he came to me and told me that he had made water much better; and in applying the caustic a fourth time, my canula went through the stricture; a bougie was afterwards passed for some little time till he was perfectly well," Hunter jubilantly recorded. It was a victory for experimental medicine. His approach—trying a traditional method, analyzing the outcome, forming a hypothesis aimed at improvement, and implementing his results—would become a standard practice throughout his career. Ultimately, it would form the foundation for his scientific revolution of surgery.

CHAPTER 5

The Professor's Testicle

Covent Garden, London
Autumn 1755

The start of the autumn term was invariably a fractious time as William and John prepared for the arrival of their new pupils at the Covent Garden school. But tensions were more strained than usual in October 1755 as the brothers watched a fellow young Scot slide into his seat. At twenty-two, Alexander Monro, Jr., was older than many of the students packed into the lecture room, although younger than John, now twenty-seven, and William, thirty-seven. Yet Monro *secundus,* as he was known, was already a prominent figure on the Scottish medical scene. Having earlier qualified as a physician, he had just been appointed joint professor of anatomy and surgery at Edinburgh University with his illustrious father, Alexander Monro, Sr.[1]

The youngest son of Monro *primus,* Alexander had been something of a child prodigy. While his older brother Donald was making his way as a physician in London, Alexander had shown an early aptitude for anatomy. By the age of eleven, he was helping his father, who had been William's own anatomy tutor, to carve up corpses in the university dissecting room. Plainly, Monro *primus* was grooming his youngest son to succeed him. If the Hunter brothers suspected that the young professor was somewhat overqualified as an anatomy novice, they were right.

In truth, William's feelings were mixed. Having himself learned the rudiments of his craft at the feet of Monro *primus,* he was understandably flattered that his old master now judged him sufficiently proficient to teach his own son. Appealing as this did to William's vanity, he reasoned that imparting his knowledge to Monro *secundus* could be judged "a particular honour conferred on me."[2] Certainly, William had recently come up in the world. Having been granted an honorary medical degree from Glasgow University in 1750, he had finally gained entry to the elite Royal College of Physicians and could officially use the title "Dr."[3] And so William extended a guarded welcome to Monro *secundus,* assuring him that he would see many interesting experiments and preparations at the school because, he boasted darkly, "in London we have commonly a greater plenty of subjects than at Edinburgh." That this flurry of research activity was abundantly John's dominion he made plain when he added, "In the dissecting room you will find a great deal of that sort of work going on through the whole winter, under my brother's direction." But William had serious doubts, too, about the precocious new pupil who was so eager to hear his lectures. In the fiercely competitive world of eighteenth-century anatomy, he had every reason for concern.

On the treacherous high seas, British adventurers were risking their lives to claim uncharted territories for king and country, beating off European rivals in the struggle for global domination. Success brought not only immediate fortune but lasting fame: The victors' names would be forever commemorated in some remote mountain or coastal feature. The exploration of the human body was no different. Across Europe, anatomists vied to discover previously unmapped parts of the body, staking their claim to a piece of the human interior. Intrepid anatomists could be assured of immortality through the parts they described; if they did not themselves bestow their names on their discoveries, they could be certain their disciples would arrange that honor. So in the sixteenth century, the followers of Italian professor Gabriello Fallopio ensured his name would live forever after he described the tubes to the uterus. His compatriot and contemporary Bartolomeo Eustachio likewise had his name commemorated in the tube running between the nose and the ear. And in the following century, striking back for England, the anatomist Thomas Willis left his name to the Circle of Willis, the loop of arteries at the base of the brain.[4]

Often rival anatomists' claims and counterclaims could be just as difficult to determine as those of pioneering colonialists. The advent of printing and the development of high-quality plate engraving helped in enabling competing researchers to publish their findings. But in a period long before authoritative scientific journals, and with anatomy books beyond the means of many ordinary practitioners, anatomists also chose alternative ways of broadcasting their achievements. A skillfully prepared specimen was one way of demonstrating an anatomical discovery, while announcing a new idea in lectures might also be regarded as equivalent to publication. Whichever route they took, the potential for controversy was considerable.

William was determined to claim a piece of anatomical territory, and a place in history, for himself. Intelligent and well read, he knew precisely which areas of the human body were ripe for further exploration. Already he had set John to work pursuing these goals. In truth, establishing the school as a center for pioneering anatomical research was as much a matter of financial security as professional pride. In the precarious free-enterprise society that prevailed, the brothers were entirely dependent on their own talents and resources to support themselves—and, since she had joined them in 1752 after their mother's death, their sister Dorothy, too.[5] William's rising status as a fashionable accoucheur had begun to bring in a respectable income. But while he enjoyed the fine clothing, refined company, and other small luxuries that elevation to the nouveaux riches brought him, William was scrupulously careful with his money. Though he loaned cash to friends like Smollett, he kept a tight rein on the family budget. Guests invited to the Hunters' home were rarely offered more than two dishes at a time when Georgian dinner tables usually groaned with delicacies.[6] Dining out, William was equally frugal. One friend remembered his sparing repast at weekly coffeehouse gatherings of Scottish physicians: "He had no dinner, but supped on a couple of eggs, and drank his glass of claret."[7] *Prudence* was William's byword, as John would wryly affirm, remarking that "whatever he was really attached to he was in the strictest sense a miser."[8]

Bound by William's harsh financial regime, shackled to his demanding research program, John was straining at the leash. By now, he had developed his own anatomical interests to pursue—as well as his own ideas about his future. He had fallen in obligingly with his brother's schemes to date, doing his dirty work in the graveyards and fulfilling his orders in the dissecting room, but in the summer of 1755, he finally rebelled when William packed him off to Oxford University, presumably with the notion of preparing him for a career as a physician.[9] Enrolled as a "gentleman-commoner"—an undergraduate without a scholarship—he survived two months before storming back home. Later he would hotly declaim, "They wanted to make an old woman of me; or that I should stuff Latin and Greek at the University," before jabbing his thumbnail emphatically on the table and adding, "but these schemes I cracked like so many vermin as they came before me."[10]

In a further step toward independence, John had taken over some of William's lectures, initially when his brother fell ill in the spring of 1753 and routinely beginning the following year as his brother's wealthy patients called him increasingly away.[11] John would never match William's poise at the podium. When he stood before the pupils with William's lecture notes in front of him, he became tongue-tied and stumbling. But he recovered his usual liveliness in answering the students' questions after lessons and he was popular with the pupils, who liked his approachable manner, good-humored company, and passion for staying up late to debate anatomy topics of the moment.

There was much to discuss. In the two thousand years since anatomists had begun exploring the human body, many secrets of its structure and functions had been unlocked, but much remained unexplored or unexplained.[12] The earliest-known anatomical research took place in ancient Greece—the term *anatomy* derives from the Greek verb meaning "to cut" or "to dissect"—but moral objections to invading the human body had restricted investigations to animals. Some puzzling beliefs about the human body resulted. The first recorded human dissections were conducted in Alexandria by two enterprising physicians, Herophilus and Erasistratus, in the third century B.C. Not only did they dissect human bodies in public; they were also reputed to have experimented on live victims, probably condemned criminals. Whatever their ethical approach, the Alexandrian pair at least overturned some of their forerunners' animal-based errors, accurately describing several structures of the human body.

After Alexandrian preeminence, human dissection fell into decline as moral taboos prevailed again. Under Roman rule, the study of anatomy was further set back by Claudius Galen, a flamboyant surgeon to the Roman gladiators, who served several emperors in the second century A.D. Though Galen's prolific theories would match those of the Roman engineers for endurance, his accuracy proved sadly inferior. Bombastic and arrogant, Galen popularized the doctrine of Hippocrates, the Greek father of medicine, that all illness stemmed from an imbalance of bodily humors, but unlike his predecessor, he advocated copious bloodletting as a panacea for every ill, launching the dangerous fashion that would remain in vogue well into the nineteenth century.

His views on anatomy were equally influential and equally wrong-headed. Basing his research entirely on animals, he promulgated errors that would be accepted unquestioningly for more than a millennium. Having never seen the inside of a human body, he mistakenly described the human liver as possessing five lobes, just as in the pigs he opened, rather than two. An irrepressible showman, like so many anatomists down the centuries, Galen's favorite public stunt was to cut the throat of a pig, then dramatically silence its squeals by severing its vocal cords. Yet the sheer weight of his works—he wrote sixteen books on the pulse alone—was sufficient to silence any critics as effectively as he had the pigs.

After the Dark Ages, Galen's flawed writings formed the syllabus of the first medical schools, founded in Italy from the twelfth century onward. Human dissection did not revive until the fourteenth century, when limited demonstrations were staged in purpose-built theaters; beginning in Italy, these spread across the Continent. For the first time, onlookers could perceive the truth about the human form for themselves—if only they were willing to look. True revelations were rarely voiced, however, since the professors read aloud from Galen's erroneous works while an assistant pointed out the relevant, and irrelevant, parts of the corpse on the table.

The revival of all things classical during the Italian Renaissance gave renewed impetus to Galen's seemingly indestructible works, so that while human dissections continued, so did the blind repetition of ancient misconceptions. But at last, the Flemish anatomist Andreas Vesalius dared to speak the unspeakable. After taking the chair in anatomy at Padua University with a brief to preach the customary Galenic creed, his investigations led him to realize that his erstwhile hero had never opened a human body. Determined to establish the facts for himself, Vesalius embarked on a relentless program of human dissection. The results, published in 1543 in the seven books of his sumptuously illustrated work *On the Fabric of the Human Body,* set out the first systematic map of human anatomy. And even if he introduced misconceptions of his own, at least Vesalius established the principle of direct observation of the human body as a way of furthering scientific understanding. Finally, anatomists began to take autopsy—literally "to see with one's own eyes"—at face value.

With Galen's errors exposed, new discoveries came fast and furious as enthusiasts carved up the human body with gusto. Fallopio, Eustachio, and their contemporaries staked their claims to parts they discovered, while Realdo Colombo showed how the blood moved from the right to the left side of the heart via the lungs, and Hieronymus Fabricius discovered the one-way valves inside veins. Arguably, it was but a small step to the landmark discovery of the circulation of the blood published by William Harvey in 1628.

After Harvey's momentous revelation, it might have seemed to the young anatomists gathered in William's lecture theater that little of great significance was left to uncover. Indeed, most anatomists since Harvey had focused their energies on the minutest aspects of the body, using early microscopes to discern the capillaries linking arteries and veins, the red corpuscles of the blood, and even spermatozoa in semen—all invisible to the naked eye. Not William; he still anticipated discovery on a grand scale, seeking nothing less than the description of an entire system of the human body equivalent to, if not more significant than, Harvey's on the circulation of the blood. This trophy, just waiting to be plucked, was the exploration of the lymphatic vessels. It was this prize William hoped to claim for his own, and it was precisely for this reason that he eyed young Monro with mistrust.

Some of the earliest dissectors had spotted the vessels in various parts of the body, notably the intestines, which seemed similar to veins but contained a transparent or white fluid.[13] Yet since the minutely fine vessels were exceedingly difficult to trace, anatomists were still unable to chart their geography. Many assumed they were simply extensions of the veins, while their purpose remained an enigma. William, however, was convinced he had all but unlocked the secrets of the lymphatic vessels and that acclaim for their proper discovery lay just within his grasp. He was certain that the lymphatic vessels formed a complete and independent system throughout the body—the lacteal vessels that could be seen in the stomach were simply one part of this network—and he had taught this view in lectures as early as 1746. Their function, he argued, was to absorb fluids all over the body. Yet although he had broadcast his theory for nearly ten years, and had set John to work exhaustively investigating the lymphatic vessels, he had never published his views nor obtained any proof for his beliefs. With rival anatomists working equally frantically to explain the lymphatic vessels, he knew the race would be close.

Dutifully laboring in the dissecting room, John had scant idea of the discoveries of his predecessors, beyond what he had gleaned from his brother's lectures, and little interest in emulating their small triumphs by naming a new organ or muscle. Rather than seeking to build on the ancient knowledge of others, as William did, John approached the human body afresh, with the same spirit of discovery he had applied while roaming the Lanarkshire countryside. It was, he would often stress, bodies, not books, that he preferred to read. And so, with knife and forceps in his hands, he simply set out to establish the facts for himself. As in childhood, there were always so many questions to answer. How do bones grow—from the middle or the outside edges? Is the substance of teeth alive or dead? How can blood regenerate itself?

Bent over the rotting corpses in his blood-spattered apron throughout each long winter, John traced every vessel, probed every cavity, and followed every fiber in a systematic exploration of the human body. Driven by tireless curiosity and a compulsion to improve the surgery he had witnessed in hospitals, he embarked on a solo voyage of discovery to create his own map of the human interior. Comparing organs, bones, muscles, and tissues in diseased and healthy states in hundreds of bodies, he examined how the different parts worked when healthy and what happened when they went wrong. Given the sheer number of hours he spent poring over an endless stream of corpses, he had more opportunities to study the human interior than any other anatomist in Britain at the time.

Whenever aspects of the human anatomy proved too intricate or complex to determine, John turned to animals, whose simpler structures could often provide clearer explanations, conducting experiments on living creatures bought from markets, animal dealers, and private menageries. Initially, this consisted mainly of dogs, horses, donkeys, fowl, and other domestic beasts easily obtainable in the markets and shops of teeming Georgian London. But soon his tastes became more exotic and he began to search farther afield to satisfy his growing obsession, applying to circuses and traveling showmen for the corpses of performing animals when they died. He even managed to arrange a deal with the keeper of the royal menagerie at the Tower of London to receive the corpses of all the rare beasts that died there.[14] Before long, the bodies of a seal, a monkey, a leopard, an opossum, a mongoose, two crocodiles, and assorted other creatures rarely seen within British shores had fallen into his hands. Painstakingly, he dissected their bodies and preserved parts of their anatomy during moments stolen from William's busy schedule.

While William rather despised the investigation of what he regarded as lesser species, for John, animal anatomy became a consuming passion. His hours dissecting and comparing human organs, bones, muscles, and structures with the same parts in simpler life-forms would prompt almost unthinkable questions about the development of life on earth. There were no such awkward questions for William. Despite rejecting a theological career, he had no doubts that the spectacular design of the human form was unequivocally God's work. For John, it was always nature, supreme and omnipotent, that deserved worship for the wondrous design of the human anatomy. Precisely how nature had performed this fantastic trick was the question that absorbed him more and more as he studied both human and animal forms.

So John had begun a remarkable series of investigations and experiments leading to a trail of discoveries for which, often, William simply adopted the credit. Sometimes John repeated experiments previously done by others. Earlier in 1755, he had kept a dog breathing artificially by cutting its windpipe and inserting the muzzle of a pair of double bellows into the hole. The experiment had first been performed in 1664 by the indefatigable Robert Hooke for the benefit of the Royal Society.[15] Sometimes, he would reconstruct experiments when he doubted their conclusions. Hearing William describe investigations on the development of chicken embryos by the Swiss anatomist Albrecht von Haller, John determined to try these for himself.[16] Embryo research had become a highly controversial topic by the eighteenth century. Most anatomists believed, like Haller, that every living being was perfectly formed in miniature in its earliest embryonic stage and thereafter simply grew in size. To John, this notion was simply ridiculous, and so he launched his own trials in autumn 1755.[17]

First, he attempted to watch a chicken embryo develop from conception to hatching by incubating an opened egg in warm water, but the embryo always died after a few hours. Then, like Haller, he gathered eggs from hens' broods and opened them at intervals to observe the development of the chick in stages. For sufficient numbers of fertilized eggs to be readily at hand, he must have kept hens in the backyard at Covent Garden; while

William ate his meager coffeehouse supper of two eggs, John squandered dozens at home. Since he also needed to view the embryos at frequent intervals in their earliest stages, he worked day and night, creeping down to the yard in the dark to steal the warm eggs from the clucking hens.

Studying the chicks growing inside their shells required meticulous care. First, Hunter gently cracked open each egg near its top, then detached a portion of shell about "the breadth of a shilling" before removing the shell completely to expose the insides. Then, with a pair of forceps, he gingerly peeled away the membrane covering the embryo and examined the tiny creature while still alive in a bowl of warm water. Finally, he preserved each embryo in spirits, placing it on a piece of ebony to distinguish the minuscule parts. "In this way I have been able to bring parts distinctly to view that before appeared to be involved in a cloud," he wrote, adding that a microscope helped reveal the smaller details. First invented in circa 1600, microscopes by the eighteenth century could magnify by up to ninety times, but the images they produced were usually distorted and blurred.[18] Hunter generally preferred to trust to his own eyes. Now he repeated the same process on eggs at different stages until, at the point before the chick was ready to hatch, he could "hear it pip and chirp in the egg." As well as making careful notes of his observations, he enlisted the trusty van Rymsdyk to sketch the specimens, recording every stage over nearly three weeks until the moment of hatching.

Straining to see the minute changes through his primitive microscope, Hunter noted the development first of the chick's brain and spinal cord, then its major organs and limbs, and finally its features, including the eyelids and the eyetooth it would use to break out of the shell. He even observed the tiny blood islands before they developed into connected arteries and veins. It was clear, to Hunter's trained eye, that the chick's various parts were transformed from simple structures into more complex ones in the egg; plainly, it did not simply grow from a perfectly formed miniature creature. Although he could not discern the very first beginnings of life—the minuscule bundle of simple cells—he correctly surmised that these parts were simply too small to detect. Though he would not publish his findings until much later in the century, by which time a German anatomist, Caspar Wolff, had published the same, correct, conclusions detailing embryonic development, he had understood the process in 1755.

These remarkable experiments, which he would repeat on various animals, were the start of a remorseless campaign to pinpoint the first moments of life and explain the awe-inspiring process of generation.

At the same time as he wrestled with controversies of the moment, Hunter was following his own lines of inquiry. In the summer of 1754, he had become engrossed in tracing the paths of the twelve pairs of nerves leading from the brain.[19] Working during the height of summer, Hunter sawed the head from a corpse in an effort to determine the end point of the delicate nerves emanating from the skull. "I steeped the head in a weakened acid of sea-salt till the bones were rendered soft," he noted, "and that the parts might be as firm as possible, and at the same time free from any tendency to putrefaction (it being summer), the acid was not diluted with water, but with spirit." Gently, he teased apart the fibers to follow the first pair of nerves to their destination in the nose. Keen to record this first discovery of the nerves of smell, he employed van Rymsdyk's skills once more to depict his handiwork, while pickling the parts in spirit of wine. William promptly announced the discovery and presented the specimen in his lectures, but for once, John managed to hang on to his preparation. It would become the earliest-known specimen in his own fledgling collection.[20]

Other investigations John pursued were directed by William, their results invariably deemed William's property. One typical case concerned the injection of the internal structure of the testes—an experiment young Monro claimed to have performed in the summer of 1753. In November 1752, almost a full year earlier, William had successfully injected mercury through the vas deferens, the duct that carries sperm from the epididymis, where newly formed sperm mature, toward the urethra. He showed his preparation to the pupils the next day. But he did not trust himself to undertake the next crucial step—to cut open the injected testicle and reveal the intricate labyrinth of mercury-filled tubes, the seminiferous tubules that manufacture sperm, lying inside. This, he knew, was a challenge of which only his brother was capable. After two weeks' intensive labor, John rushed to present William with his handiwork.[21] "He shewed me the Testis opened, and the tubular internal substance very generally filled with mercury," recorded William, who was so delighted at the success that he showed the preparation to students that very evening, and in every ensuing course.

Undoubtedly, the Hunter brothers had beaten their competitors to this goal—and William had six witnesses, all pupils, prepared to sign an oath confirming that they, too, had seen John's work in 1752. But the wily young Monro was one step ahead. Although he only injected the testes the following summer—and even then did not achieve John's all-important next stage of displaying the internal structure—Monro had the perspicacity to publish his accomplishment in an Edinburgh medical journal in 1754, and duly claimed precedence for the work. Still this was not enough for the ambitious Monro. He also claimed, during that same summer of 1753, to have discovered the tear ducts—another achievement William had assumed for himself and taught in lectures. And most crucially of all, Monro declared that while making his experiment on the testes, in that frantically busy summer of 1753, while he was yet a medical student of twenty, he had suddenly discerned that the lymphatic vessels formed a complete bodily system. Eager to lay claim to this, the most dramatic anatomical prize awaiting discovery, Monro rushed into print again, publishing his climactic conclusion in his graduation thesis in the summer of 1755. "Valvular lymphatic vessels, in all parts of the body, are absorbent veins; they do not emanate from arterial twigs, as is generally believed," he explained, exactly repeating the doctrine William had been teaching in lectures for seven years, though without giving any justification for announcing his theory.[22]

Writing to William that same summer with the request to attend his lectures, Monro had coyly enclosed a copy of his thesis. William was mortified. As he read through Monro's theory, he immediately suspected the young professor had somehow stolen his landmark discovery, most probably from hearing or reading reports of his lectures. Pupils' lecture notes, he knew, were often circulated among their friends, and there was a regular traffic of students between London and Edinburgh. With leaden sarcasm, William would later remark, "Shall we call the year 1753, *fortunate* or *unfortunate* for Alexander Monro, jun. Professor? Surely it was a *remarkable* year. He was then a *student* of anatomy, and in *that one year* made *three discoveries*. . . . If he goes on at the same rate, he will become a prodigy."[23]

So as Monro now listened so attentively in class, William was convinced the precocious new pupil had plagiarized his work and quite probably planned to commit further academic theft. Even so, he was prepared to overlook this apparent misdemeanor, because in the autumn of 1755, William was confident that he—with the help of his compliant young brother—was still ahead of the game. Grand theories were all very well, but it was proof of the nature of the lymphatic system that would really take the anatomical world by storm.

Already, under William's impatient urging, John had taken another vital step in the exploration of the lymphatic system. By the time Monro's thesis had hit the streets, John had succeeded in the delicate task of injecting some lymph glands, or nodes, and the lymphatic vessels emanating from them, with mercury.[24] The carefully executed preparation was displayed to pupils as further evidence of William's theory. John's trick, to the delight of the fascinated pupils, was first to inflate the vessels with a blowpipe before blowing mercury into them. He could highlight the lymphatic vessels in the testicles by the same method—and commonly performed the stunt on the male organs of dead dogs and horses. Now growing increasingly confident in his skills, and less and less inclined to remain in William's shadow, he even harbored ambitions to stake his own claim to solving the biggest anatomical enigma of the age. Having found "so easy a method" of displaying the elusive lymphatic vessels, he set his sights on tracing the system throughout the entire body and even aspired to publish the first "complete description and figure of the whole absorbing system." By the time Monro arrived for the start of the autumn term in 1755, John was itching to pick up his investigations where he had left off.

Working tirelessly throughout the winter, John soon succeeded in producing a handsome preparation clearly displaying the lymphatic vessels running from the thigh up to the thoracic duct—the most prominent lymph vessel in the body—"all finely filled with mercury." The faithful van Rymsdyk recorded his handiwork. Dutifully taking notes, the young Monro attended William's lectures and observed John's preparations with scrupulous interest. When he packed his bags at the end of the winter course and left the school for a study tour of Europe in the spring of 1756, the Hunters had all but forgotten his potential threat to their reputations.

Spring brought fresh challenges. In May, John secured his first proper hospital job, as a house surgeon—a junior doctor on the first rung of the medical-career ladder—at St. George's, where he had already served time as a pupil.[25] Like the handful of other London voluntary hospitals founded at the time, St. George's had been established to treat the "deserving" poor, partly in a genuine spirit of charity and partly on the more mercenary principle that returning honest laborers to a working life benefited the nation as a whole.[26] Since anyone who could afford medical fees opted for treatment in their own home and those who were considered too destitute for the hospitals' charitable aid were thrown on the mercy of family and friends, this left a precise, though substantial, pool of potential patients. Funded by subscriptions from donors, who thereby earned a place on the governing board, the voluntary hospitals offered free medical care to those found eligible, while their medical staff worked for nothing. The opportunity for a physician or surgeon to enhance his reputation through charitable work, while practicing his rudimentary skills on the uncomplaining poor, was considered sufficient recompense for his occasional attendance.

Yet for a patient to gain admission to a voluntary hospital could be nearly as hard as obtaining an audience with royalty. As well as having to prove merit for charitable care to the satisfaction of the governors, a prospective patient had to convince the board that his or her condition was neither infectious nor incurable. Since infectious diseases were rampant in poverty-stricken London, and since medical practitioners were incapable of curing most illnesses, it was virtually impossible to comply with such stringent conditions. Patients admitted to St. George's therefore suffered a variety of contagious diseases, including scarlet fever, whooping cough, and chicken pox, as well as conditions from which they were distinctly unlikely to recover. Once a patient had wheedled his or her way through the doorway, regulations were equally strict, with bans on smoking, gambling, drinking gin, bringing in food, and swearing, though the observance of these rules was as lax as those on admission: In 1750, the governors lamented the fact that "not only victuals, but also spiritous liquors, are too often introduced into the wards."

Working as St. George's most junior surgeon, John Hunter began to apply his considerable knowledge in anatomy to the patients who turned up at its doors with limbs mangled in traffic accidents, skulls fractured in falls, bites from mad dogs, and assorted tumors, abscesses, and sexual diseases. Among his duties, it was the junior surgeon's job to treat any casualties admitted when the staff surgeons were absent, prepare patients for operations in the first-floor theater, and bandage postoperative wounds in the drafty wards.[27] For this privilege, he had to pay for his board and lodging, living in a little cottage within the grounds.

But as well as the valuable experience of hands-on surgery, the job brought something else of particular use to John Hunter—the keys to the "dead house," or mortuary, where the least fortunate patients of St. George's awaited interment in the nearby burial ground. Often their wait would be cut short. As well as recording operations at St. George's, Hunter's casebooks detail numerous autopsies he conducted on patients in the late 1750s. While some of these may have been approved by families—postmortems were certainly not a legal requirement—many more were plainly performed on bodies purloined from the dead house or hospital graveyard.

By rights, Hunter's post should have gone to another aspiring surgeon, John Gunning, who had already completed a full three years of training at the hospital. No doubt William, who numbered several governors of St. George's among his influential friends, had pulled a few strings. It was a slight that Gunning would remember all his life. In the event, Hunter spent only five months in the job—instead of the typical spell of between one and two years—whereupon Gunning nimbly stepped into his shoes.

With winter steadily approaching, Hunter had to abandon the living for the dead once more and return to his duties in the dissecting room, his energies devoted to solving finally the riddle of the lymphatic system. An empty house awaited him in Covent Garden, as well as further controversy. During the summer, William had joined the fashionable migration to the West End, establishing a home with his sister Dorothy in a smart town house on Jermyn Street.[28] Their new abode was not only more in keeping with William's status as a physician but safely distant from the stench and distasteful activities of the dissecting room. Living alone in the Great Piazza, awaiting the next flurry of pupils, John had become something of a minor celebrity among London's growing community of medical students. As a record class of around a hundred pupils crowded into the lecture room for the autumn course, his latest research was the buzz of the town.[29]

As part of his study of the male reproductive system during the previous winter, John had investigated how the testes descend in babies in the womb. The task, initially set by William, had been prompted by another research paper from the busy Haller. In this paper, published in 1755, Haller had argued that the testicles suddenly dropped at the moment of birth as a result of the baby taking its first breath. John had immediately realized the folly of this notion. From his many dissections of fetuses at different stages, he knew that the testes underwent a gradual descent from high in the abdomen to the scrotum as the fetus developed. By the time of the baby's birth, they had generally already descended. Since he was almost certainly working on stolen bodies, or fetuses that had miscarried, charting this progress precisely was not easy. "It is the more difficult to ascertain the exact time of this motion, as we hardly ever know the exact age of our subject," he frankly admitted. Nevertheless, he correctly surmised that the descent was normally complete by about the eighth month. Furthermore, he noticed that the process of descent sometimes resulted in a congenital hernia, in which a portion of gut protrudes through the abdominal wall and into the scrotum because a section of the baby's intestines has been pulled down with the descending testicle. While William had initially suggested this hypothesis, too, it was John, as always, who obtained the proof.

As John proudly handed around his immaculate preparations displaying the progressive stages of descent and some examples of congenital hernia to the pupils crammed into the lecture theater, the students were suitably impressed. No sooner had they seen for themselves this exciting anatomical discovery than they raced off to pass on the news to their fellow students and teachers on the London wards. Before long, the school was besieged by fellow anatomists eager for a peek.

Among the first visitors was Percivall Pott, John's former tutor at St. Bartholomew's, who arrived one autumn morning unannounced at the Covent Garden house. Having first inquired after William, though Pott almost certainly knew the elder brother was present now only for evening lectures, Pott was invited in by John, who was more than willing to describe his findings. Pott's interest in the topic was no surprise: He had himself published a book on different types of hernias, *A Treatise on Ruptures*, just a few months earlier. Uncowed by his former tutor, John immediately countered Pott's view, which accorded with Haller's, that the testes descended only as the baby took its first breath. He promptly brought out his preparations to prove his point, and afterward the pair sat together in the parlor, discussing issues of mutual anatomical interest. Pott's interest could not have appeared more convivial.

So as the new year of 1757 dawned, prospects for the Hunter brothers seemed rosy. The reputation of their school for pioneering research had been firmly established and the discoveries the brothers had magnanimously shared with fellow anatomists seemed to place them at the center of a vibrant and mutually beneficial intellectual community. But it was not to last. Before the year was out, the heady atmosphere of cooperative inter-action would be shattered by venomous rows, which split the anatomical world apart. In bitter exchanges of abuse, even by the standards of typically acrimonious eighteenth-century polemics, the brothers quarreled first with Pott and then with Monro *secundus.*

William first gained a hint of the machinations under way when Pott announced his intention to publish a treatise on congenital hernias shortly after his friendly parlor chat with John. When the new publication, *An Account of a Particular Kind of Rupture Frequently Attendant Upon New-born Children,* hit the streets in March 1757, William was apoplectic.[30] Pott shamefacedly stated that congenital hernias were an accidental by-product of the descent of the testes in the fetus, just as John had explained to him—although he erroneously repeated Haller's notion that they only fully descended after birth—without crediting either of the brothers. "It hardly contained one new idea," William fumed. "It was what any of my pupils might have written . . . and yet neither my brother's name, nor mine was mentioned."

Fortunately, as always, William had friends in high places, or at least a loyal chum with the power of the press in his grasp. The trusty Smollett, indebted to William for his elastic loans, leapt to his benefactor's defense in the March issue of the *Critical Review,* which the novelist had founded a year earlier. Reviewing Pott's book, Smollett roundly accused its author of plagiarizing the findings of both Haller and the Hunters. It was the begin-ning of what William, who automatically defended his brother's discoveries as if they were his own, would call a "paper war." Pott returned fire robustly in the *Review,* insisting he had arrived at his conclusions solely through his own inquiries, having neither read Haller's work nor heard of John's discov-ery. But his initial denial that he had ever visited the Hunters' school was a mistake. Two of the pupils wrote letters to confirm that John had indeed shown the surgeon several preparations, described his view on the descent of the testes, and outlined his proof on congenital hernias. Pott was cor-nered, and he knew it. He recovered his memory sufficiently to claim grudgingly that John had shown him "a single preparation," while still deny-ing ever discussing hernias. Eating humble pie, he asserted that he had never intended to cheat the Hunters of their rightful claim and vowed never to speak on the subject again.[31] William was willing to accept this olive branch, although he could not resist rehearsing the argument in a pompous pamphlet, devoted largely to his quarrels with Pott and Monro, later published in 1762. It was in that same pamphlet that John's discoveries on the descent of the testes would be detailed. It was his first published paper, consisting of a ten-page description with three plate engravings, in which John even named a new anatomical part—the "gubernaculum," from the Latin for "steering oar" or "rudder," after the fact that it "steers" the route of the descending testicle.

Yet if William was prepared to excuse Pott, he would not let young Monro off the hook so lightly. The brothers may have forgiven Monro's earlier trespass on their territory, but they would not overlook his next transgression. Having spent his intervening years in anatomy studies abroad, the precocious Monro took it upon himself to publish a major work on the lymphatic vessels in Berlin in the summer of 1757. Expanding his earlier thesis, Monro repeated the view that the lymph vessels form a complete circulatory system designed to absorb fluids from the tissues and return them to the blood. Not once did he acknowledge the work of the Hunter brothers. Once again, Smollett leapt to the brothers' defense, denouncing Monro's pamphlet in the September issue of the *Critical Review* as a blatant theft of William's views, which had been "publicly made by Dr Hunter to his pupils, for the space of eleven years."

But the powerful Monro family was not as easy as Pott to subdue. First, Monro *primus* responded to the attack with a letter to Smollett's magazine, insisting that his son had written his student thesis "in absolute ignorance of Dr Hunter's having any particular opinion concerning lymphatics." He had been "surprised," insisted pater Monro, on attending the Covent Garden lectures, to hear William teach the same view. William immediately countered with the support of six witnesses, all now esteemed lecturers or surgeons in their own right, who confirmed he had taught his doctrine in lectures since 1746. Furthermore, they pointed out that notes of these lectures had circulated in Edinburgh for several years. Then, in December, big brother Donald, about to take up a post as physician at St. George's Hospital, weighed in with his support. Finally, young Monro, now safely back home in Edinburgh, joined the fray with a highly abusive sixty-nine-page pamphlet attacking the Hunters.

It was a provocation that went too far. When in 1762 William finally responded with his own pamphlet, *Medical Commentaries, Containing a Plain and Direct Answer to Professor Monro Jun,* he lambasted each of

Monro's claims in impeccable courtroom style. Accusing Monro of "the open violation of truth and candour," William proved his precedence with incontrovertible evidence, publishing two damning letters from the highly respected chemist Joseph Black. Black confirmed that Monro had shown him his unpublished student thesis back in 1755, whereupon the chemist had immediately warned him that the theory on the lymphatic system had already been expounded by William Hunter. Black had even accused Monro of filching his views from reports of William's lectures circulating in Edinburgh. There was no question but that Monro had been aware of William's theory before he published his thesis and had simply appropriated it from students' notes, most probably taken by his brother Donald when he had studied with William.

Though the rift between the two Scottish families would never be healed—and Donald would avenge the fraternal slur by joining later attacks against John at St. George's—it did neither side much harm, either. Monro *secundus* went on to teach anatomy to increasing acclaim at Edinburgh, eventually retiring at the age of seventy-five in favor of his son, known, of course, as Alexander Monro *tertius*. In all, the Monro dynasty would hold the Edinburgh anatomy chair for a total of 126 years. The Hunters continued their dynamic relationship, too, with William delivering his pithy lectures every afternoon and John overseeing every other aspect of the school, with mounting success—although John declined to formalize the partnership when William finally made him the offer in 1758, "on account of his aversion to public speaking, and his extreme diffidence of his own abilities and skill."[32]

William would later dismiss the whole nasty episode involving Pott and the Monros with the casual comment: "It has likewise been observed of anatomists, that they are all liable to the error of being *severe* on each other in their disputes . . . and for anything that we know, the passive submission of dead bodies, their common objects, may render them less able to bear contradiction."[33] He would certainly do his utmost to make sure that state of affairs continued, going on to squabble with numerous colleagues, including his own brother. In fact, as so often in such anatomical wrangles, really neither camp deserved full credit for the discovery of the lymphatic system. An earlier British anatomist, Francis Glisson, had actually suggested a century earlier that the lymphatic vessels formed a system to recycle fluids, although his theory had languished in obscurity.[34] And it would

be many more years before the other vital roles of the lymphatic system, fil-
tering bacteria and fighting infection, were understood. If none of the
squabbling anatomists could really claim credit for explaining the structure
and functions of the lymphatic system, there still remained the glittering
prize of proving this doctrine through scientific experiment. That proof was
finally provided by John in a dramatic set of experiments, controversial
even then, on living animals in the winter of 1758.

By now, John had not only dissected large numbers of dead animals but
was performing regular experiments on living creatures, as William freely
reported: "At this time my brother was deeply engaged in physiological en-
quiries, in making experiments on living animals, and in prosecuting comparative anatomy, with great accuracy and application. It is well known that
I speak of him with moderation, when I say so."[35] In what Benjamin
Franklin would call the "age of experiments," it was common practice to invite parties of fellow professionals, students, and interested amateurs to observe and verify significant scientific events. The artist Joseph Wright
would powerfully record such a scene in his *An Experiment on a Bird in the
Air Pump*, which depicts a group of adults and children watching in mixed
fascination and horror as a cockatoo is suffocated in a vacuum.[36] Without
the technology of sophisticated visual aids, such experiments were often
the only way to shed light on areas of human ignorance. And so, when John
Hunter tied a yelping dog to his dissecting table on November 3, 1758, a
crowd of onlookers had been invited to Covent Garden for the show. The
party included three physicians and four surgeons, as well as several of the
brothers' pupils.

John's aim in this showpiece experiment was to prove once and for all
the theory that only the lymphatic vessels—in this case, the lacteals in the
intestines—and not the veins were capable of absorbing fats. The question
was still open to debate, since Monro *secundus,* for one, had declared himself unconvinced that the veins were not involved in absorbing fluids. As
his observers craned to watch, John slit open the dog's belly, whereupon
"[t]he intestines rushed out immediately." While the bleeding animal
writhed and squealed, John made a hole in its upper intestines, then
poured warm milk through a funnel into the gap. As the guests peered forward, he proudly demonstrated the lacteals turning white as they conveyed
fat from the gut, while the veins remained filled with blood. But this was
not conclusive enough for his purposes, so John made a second cut lower

down in the tortured animal's intestines and repeated the milk test. Again, he made sure all his guests could see for themselves the fruits of his labors, recording:

> In both these experiments we could not observe that the least white fluid had got into the veins. After attending to these appearances a little while, I put all the bowels into the abdomen for some time, that the natural absorption might be assisted by the natural warmth; then took out and examined attentively the two parts of the gut and mesentery upon which the experiments had been made: but the lacteals were still filled with milk and there was not the least appearance of a white fluid in the veins.

In all, Hunter performed a total of four experiments on the unfortunate mongrel, until "the gut at last burst" and the animal died. He went on to conduct similar experiments before similar parties of guests, using milk, dyed blue starch, and warm water mixed with musk, on four more living animals—three sheep and an ass—from November until August of the following year. The notion, proposed a century earlier, that lymphatic vessels alone absorb fats and fluids had been proved beyond all doubt. Inevitably, the achievement was quickly adopted by William as yet another notch in his burgeoning reputation. John would always be careful to award William the credit for devising the theory of absorption. He was aggrieved that William did not likewise acknowledge his role in providing the crucial proof, remarking bitterly that William had not "done justice to my Experiments."[37] But if the evidence of Hunter's experiments was now broadly accepted, the morality of such apparently barbaric acts on innocent living creatures was not.

Hunter was by no means alone in performing numerous acts of vivisection at the time. Throughout Europe, enthusiasts conducted experiments on creatures of every description. In the Age of Enlightenment, often the only way in which anatomists could enlighten themselves on how the living human body functioned was by dissecting living animals. At a time when brutal deeds were daily inflicted on all manner of creatures in London's cockpits and bull-baiting rings, animal experiments were broadly condoned, though not by all. In the very year that Hunter tortured his dog, three sheep, and an ass, such practices provoked a passionate assault by no less a figure than Samuel Johnson. Writing in his own magazine, the *Idler*, Johnson launched a ferocious tirade against acts of vivisection by unnamed anatomists: "Among the inferior professors of medical knowledge is a race of wretches, whose lives are only varied by varieties of cruelty; whose favourite amusement is to nail dogs to tables and open them alive; to try how long life may be continued in various degrees of humiliation, or with the excision or laceration of the vital parts," stormed the formidable animal lover.[38] Continuing in similar vein, Johnson argued that such "acts of torture" were designed more for their perpetrators' amusement than to extend scientific knowledge. "I know not," he declared, "that by living dissections any discovery has been made by which a single malady is more easily cured."

Although Johnson nowhere identified the subject of his attack, it is easy to imagine John Hunter, by now renowned for his anatomical experiments, as his target—especially since Johnson added the jibe that "he surely buys knowledge dear, who learns the use of the lacteals at the expense of his humanity." Yet Johnson wrote his article three months before John began his famous round of experiments on the lacteals. In fact, it was probably Haller, the indefatigable Swiss, who was conducting more experiments on living animals than probably any anatomist in Europe, who was the real subject of Johnson's ire.[39] Indeed, Hunter himself would condemn the endless repetition of experiments to no purpose; it was always advancement of medical knowledge, not showcase experimentation, that drove him. As he remarked, "I think we may set it down as an axiom, that experiments should not be often repeated which tend merely to establish a principle already known and admitted; but that the next step should be, the application of that principle to useful purposes."[40]

Whether or not Hunter was in Johnson's sights at the end of the 1750s, his reputation had certainly captured interest far and wide. When William Shippen, a prominent physician in Philadelphia, decided on furthering his young son's medical education in 1758, it was to Britain, and to John Hunter, he naturally turned. Having already served a four-year apprenticeship with his father, young "Billey" crossed the Atlantic to spend the winter in London, walking the wards and studying the human body "with the finest Anatomist for Dissections, Injections, etc in England."[41] No sooner had Shippen arrived than he set out to enjoy all the pleasures "dirty London" could afford, activities that he carefully recorded in his animated diary. But his social life came to an abrupt end on October 2, 1759, when he "moved [his] trunk to Mr. Hunter's," becoming a resident pupil in John's Covent Garden house for the brothers' autumn course. From then on, Shippen's packed schedule left little time for social activities. On one typical day, October 9, Shippen recorded, "Rose at 6, operated till 8, breakfasted till 9, dissected till 2, dined till 3, dissected till 5. Lecture till 7, operated till 9, sup'd till 10 then bed."[42] On other evenings, he would sit with John, debating the latest turns in anatomy, so that daily entries commonly ended, "chatted till 10 with Mr. Hunter upon anatomical points." As their friendship grew, Shippen spent more and more time at Hunter's shoulder in the dissecting room, in awe of this charismatic teacher, who seemingly had a knack of inspiring young pupils with a fascination for anatomy.

When he returned to Philadelphia in 1762, after also studying medicine at Edinburgh University, Shippen imported wholesale the Hunterian approach. In November of that year, he established the first anatomical lectures delivered in the American colonies. Not only did he promote the Hunter brothers' hands-on model of dissection; he assiduously copied their reliance on body snatchers for research material. Shippen's school would face attack by outraged crowds on several occasions; at one point, its proprietor narrowly escaped with his life.[43] Three years later, he cofounded America's first medical school at what was initially called the College of Philadelphia and later the University of Pennsylvania, along with a fellow former pupil from the Hunters' school, John Morgan, who stayed in Covent Garden in 1760. Together, Shippen and Morgan established Pennsylvania as the cradle of modern American surgery, but it was the Hunter brothers who were its inspirational parents.

That autumn course of 1760 was the last in which the two brothers worked side by side. The strain of twelve winters of tireless labor in the rank atmosphere of the dissecting room had begun to take its toll on John's health. He had taken to his bed in early 1760.[44] It is unclear what ailed him, and in any case, eighteenth-century diagnostic skills were always vague, but early biographers suggested he had contracted pneumonia, while William recorded that his brother suffered a "very indifferent state of health, the effect of too much application to anatomy, which obliged him to be much in the country." Hunter himself later recorded, "I left anatomical pursuits in the beginning of the Summer 1760," although he did not explain why. Yet he was not too ill to undertake postmortems in both May and June, and to treat a patient at St. George's in July. He was back living in Covent Garden by the autumn, when John Morgan was there, and he would later hint that Morgan had pinched a research idea from him while lodging in his house.

That same autumn, John reached a momentous decision: to part company with William and chart his own course in the medical world. The wild, unkempt country lad of twelve years earlier had grown into a self-assured, spirited young man, ambitious to make his own way in society. Having dissected, by his own reckoning, some thousand human corpses in his years at the school, he was without doubt the most experienced anatomist in Britain, if not Europe. Yet his limited training in surgery and lack of formal qualifications severely restricted his chances of securing a permanent hospital job or setting up private practice in highly competitive London. That left him with one obvious place to pursue his talents: the battlefield. In October 1760, like so many aspiring surgeons before him, John Hunter signed up as a surgeon with the army and awaited orders throughout the winter.

Without the help of his brother, William taught his last complete course that autumn, then gave up the lease on the Covent Garden house for a solitary existence in Jermyn Street, Dorothy having since married and returned to Scotland. At last, the house in the Great Piazza, the scene for more than a decade of countless human dissections, experiments on live animals, heated anatomical debates, cozy parlor chats, and secretive night-time deliveries, was empty and quiet, its neighbors' peace finally restored. And in March 1761, thirty-three-year-old John Hunter set sail from Portsmouth to face the muskets and cannons of the French troops.

CHAPTER 6

The Lizard's Tails

ᏬᏭᏬᎯᎧ

Spithead, Off Portsmouth
March 1761

Commodore Augustus Keppel surveyed with dismay the troops crammed on board the squadron of almost 130 ships anchored at Spithead. Disheveled, ill-disciplined, and decidedly unfit, many were plainly sick or lame, while one whole regiment was so poorly clothed that the soldiers stood in rags.[1] Already he had proposed leaving the two worst regiments behind. But this, he knew, was impractical, for Keppel had been handed his "secret instructions" from the newly crowned George III only days earlier. The orders were clear: His mission, impossible as it might seem, was to capture the French island of Belle-Ile.

Having been at war with France and its allies for five years, Britain had already won important campaigns under its popular war secretary, William Pitt.[2] The struggle for overseas supremacy, which would later become known as the Seven Years War, had begun badly. But after early setbacks, Pitt's audacious strategies had helped Britain seize control in key territories, so that by the end of 1760, there was little doubt that France had been pushed to the brink of defeat. All Pitt needed was one more push to secure the victory he knew was within grasp. The conquest of Belle-Ile, in its strategic position eight miles south of Brittany, was the objective he believed would win the war. With eight thousand troops commanded by Maj. Gen. Studholme Hodgson, under the escort of a powerful fleet supervised by Keppel, Pitt was determined his daring joint-forces expedition would prove a success. And on March 29, 1761, just four days after the commodore had received his secret orders, the British force set sail for the coast of France.

Crossing into French waters on board the hospital ship *Betty*, innocently unaware of its intended destination, John Hunter was already hard at work tending the assorted ailments that were an inevitable fact of life in the cramped and unhygienic conditions at sea. Bribed or press-ganged into action, many soldiers and sailors were in poor shape before they ever engaged battle. The Greek father of medicine, Hippocrates, had counseled, "He who wishes to be a surgeon should go to war," and centuries later the battlefield still provided the perfect training ground for would-be surgeons.[3] Not only did wartime present surgeons with the opportunity to brandish their knives fully and freely on parts of the body they never normally encountered; the inevitable failures this engendered provided numerous chances for instructive autopsies. By the time Hunter signed up, a spell as a surgeon in the services bestowed the automatic right to practice in civilian life. With this in mind, Hunter had secured a commission as an army staff surgeon on October 30, 1760, most likely through William's influence with Robert Adair, the army's deputy surgeon general.[4]

No doubt John's friend Smollett, who had worked as a surgeon's second mate in the navy during the previous war with France, had warned him of the conditions he could expect.[5] Describing his seagoing experiences in his first novel, *The Adventures of Roderick Random,* Smollett wrote that his eponymous hero had found that "[t]he sick and wounded were squeezed into certain vessels, which thence obtained the name of hospital ships . . . and the space between decks was so confined, that the miserable patients had not room to sit upright in their beds. Their wounds and stumps being neglected, contracted filth and putrefaction, and millions of maggots were hatched amid the corruption of their sores."

Flagging with seasickness, Hunter found that his discomfort was little helped when his destination finally became clear. The floating army arrived on April 6, within sight of its target, and the anxious troops gazed at the soaring cliffs and fortified battlements defending Belle-Ile. Surveying the jagged coastline merely confirmed what Keppel and Hodgson had already surmised: The island appeared to be impregnable. Nevertheless, the commanders attempted a landing two days later on the island's southern coast. After the fleet knocked out the large guns defending the bay, more than three thousand soldiers rowed through the smoke in flat-bottomed boats to storm the French positions. Although some of the redcoats succeeded in landing, and a hardy band of about sixty grenadiers—the army's crack troops—even mounted the cliffs, they were speedily mown down by the French muskets above, while their fellows flailed helplessly in the surf below. The assault was quickly abandoned, but not before hundreds of troops had been shot or taken prisoner by the French or drowned in the storms that then blew up.

Fearfully reporting the disastrous defeat back to England, Hodgson estimated that five hundred soldiers had been killed, wounded, or taken prisoner during the assault. "Purcell is killed, and Maclean had his arm broke with a musket shot, and is taken prisoner," he wrote. "My secretary, who was at the poop, on board the Dragon, looking through a glass to satisfy his curiosity, has got a shot in his forehead."[6] Those soldiers who could be rescued from the lashing waves were hauled on board the *Betty*, waiting their turn with the sailors injured by cannon fire and splintered timbers, to be patched up by Hunter and his fellow surgeons. While the storm tossed and dispersed the wooden ships, Hunter battled to save the wounded, bleeding, and dying men in his primitive surgery.

Nothing in his experience of performing operations in London's hospitals had prepared him for this. As men lay groaning in the long straggle waiting for treatment, he worked at full speed, sawing off smashed limbs, digging out musket balls, prizing free shards of wood and splintered bone, and bandaging ragged wounds. His traditional officer's red jacket conveniently concealed the bloodstains as his urgent appeals for more supplies were lost in the tumult of the raging storm and the frenzied shrieks of his patients. With no anesthetics to numb the pain, many patients died of shock as the amputation knife cut through their flesh, or bled to death while the surgeons probed around in their wounds to extract debris. Their bodies were given hurried burials at sea. Others would join them days later as sepsis took hold in their wounds while they lay in the hammocks of the foul-smelling sick bay.

As two weeks of gales prevented any further attempt on the island, Pitt sent reinforcements for a renewed assault, which was launched on April 22. This time, a battalion of men succeeded in scrambling up the rocky slopes on the southern coast, where they overpowered the French forces at the summit. By nightfall, British troops had control of the surrounding area, the French had beaten a retreat to the island's fortified capital of Le

Palais, and the conquest of Belle-Ile was declared a triumph. British casualties had been few this time—a total of fifteen soldiers killed and twenty-
nine wounded. There were also several wounded French prisoners needing
treatment.[7]

As soon as the remaining troops could be landed and supplies unloaded onto the beach, the surgeons and surgeons' mates set up basic first-
aid stations in cottages and chapels abandoned by the retreating French. In
these makeshift field hospitals, Hunter set to again amputating limbs and
probing gunshot wounds. Often there was little he could do to relieve the
desperate men. Where the poor grenadiers or their French opponents had
been shot at point-blank range or bayoneted in the chest or abdomen, their
chances of survival were slim. Even if the surgeons could stitch their
slashed organs together and stanch the bleeding, which was unlikely, the
internal infection that inevitably followed would lead to almost certain
death. That did not stop them making valiant efforts, however, for they
even performed crude emergency surgery right on the field of battle. But
given the conditions Hunter was forced to work under, the results were
hardly surprising, as he later noted: "Such cases will seldom or ever do
well."[8] In addition to the casualties of the successful assault, there were
still the victims of the earlier attack, who were now carried gingerly ashore
and laid on straw mattresses in their rough shelters.

While Hunter and his colleagues struggled to save those they could,
the army advanced on Le Palais, forcing the three thousand inhabitants
and the troops defending them to take refuge in the city's fortress, the
Citadel, where they were bombarded mercilessly for six weeks. Now on
standby for further British casualties, Hunter rushed to the aid of Brig.
Thomas Desaguliers, who was concussed by a bursting shell; he recovered
sufficiently for Hunter to attend him nineteen years later on his deathbed.[9]
Finally, the weakened islanders surrendered. Their resilience having won
them the respect of the British victors, they were escorted safely to the
French mainland.

The long siege had brought the final toll of British killed or wounded
to more than seven hundred—official reports home rarely bothered to dif-
ferentiate between the dead and the nearly dead—but just as the rudimentary hospitals were overflowing with surgical cases, an outbreak of "severe
fever and flux" attacked those troops that had not already been injured by
the French artillery.[10] Having full possession of the island now, the army

medical department set up a general hospital in Le Palais itself, where both the wounded and the infected were laid side by side on the straw palliasses. Contagious diseases swept through the troops, so that the soldiers who survived the primitive operations performed in dirty surroundings frequently succumbed to the sickness raging all around them. In the overcrowded, dirty conditions of typical army camps, where sewage overflowed from pits, linen went unwashed for months, and water was rarely clean, infectious diseases such as dysentery, typhus, malaria, and smallpox spread out of control. The army physician Richard Brocklesby, who had served in Germany during earlier campaigns, reckoned that typhus—or "camp fever"—commonly claimed eight times as many lives as enemy action.[11]

Despite the fact that fevers and infections dominated the military sick list, army surgeons always vastly outnumbered army physicians. There was logic in this, since any competent surgeon could liberally dose a soldier with a concoction of the usual ineffective medications, while few physicians knew how to wield an amputation saw, even if they deigned to stoop to such base levels. Only one physician, Edward Blythe, had accompanied the huge expedition to Belle-Ile, and even then he scurried away immediately after the siege, despite the incidence of sickness being at its height. Meanwhile, half a dozen surgeons labored together in the general hospital at Le Palais, assisted by twelve nurses and twenty surgeons' mates.[12] The usual strict division of labor was not an option. "I am obliged to be Physician and everything here," John Hunter complained to William in one of six letters he sent to his brother from Belle-Ile, adding, "and I think I do as well as the best of them."[13] Naturally enough, army physicians commanded a significantly higher salary—double the daily rate of ten shillings awarded to army surgeons, who were only on a par with the apothecaries and who even had to fork out from their own pockets to purchase medicines and other supplies.[14] Since practice in treating the fevered patients varied little from the standard care back home—letting blood, administering noxious purges or enemas, and prescribing unpalatable medications—differences in outcome were rarely noted. In the absence of any effective antibiotics, and with only the most basic ideas about hygiene, there was little the beleaguered surgeons could do to combat internal diseases, apart from doling out the usual remedies and hoping for the best.

Hurrying between the sweating and vomiting soldiers laid out on their foul mattresses as flies buzzed around the wards in the sweltering midsummer heat, John Hunter could offer his patients no more than the typical contents of the army medicine chest, although he was at least liberal with the opium to relieve the worst of their discomfort, while also advocating restraint in bloodletting. When it came to surgery, however, he developed his own maverick style of treating war wounds, a method that quickly attracted the scornful attention of his colleagues. Writing home to William in September, not long after the successful siege, he told him, "My practice in Gunshot wounds has been in a great Mesure different from all others, so that I have had the eyes of all the surgeons upon me, both on account of my suppos'd knowledge, and method of treatment." Plainly, this departure from orthodoxy had not won him friends, for he added, "My fellow creatures of the Hospital are a damn'd disagreeable set. The two Heads are as unfit for the employment, as the devil was to reigne in heaven."[15]

If Hunter's unconventional methods aroused the wrath of his fellow surgeons, he was simply practicing the scientific approach to surgery he would advocate all his life. When first the casualties poured in from the initial failed assault on the island, Hunter had closely observed his colleagues' treatment of gunshot wounds. Conventional practice dictated that army surgeons open up a gunshot wound—a technique known as "dilatation," or enlargement—prize out the musket ball or shot with forceps, or more often with their fingers, then clean away any debris before dressing the wound.[16] The principle of dilatation stemmed from a belief that gunpowder was poisonous, dating back to its first use in European warfare in the thirteenth century. In the sixteenth century, the French army surgeon Ambroise Paré had at least put a stop to the practice of branding gunshot wounds with a red-hot iron or scalding oil as a supposed antidote to the poison, but methods in both British and French armies had continued relatively unchanged since then.

In theory, and in ideal circumstances, Hunter's colleagues were right in most cases to excise a missile, and especially to remove any debris such as timber shards or fragments of a soldier's filthy clothing embedded in a wound, in order to prevent sepsis; often it was the debris entering the body with the bullets that led to fatal deep-wound infection. In practice, however, since the circumstances they worked in were far from ideal, their doctrine probably increased the death and suffering. Not only was the act of incising flesh within a wound exceedingly painful and traumatic before the advent of anesthetics, causing huge shock and loss of blood; it frequently introduced fatal infection, since the military surgeons often treated their casualties on muddy, manure-ridden battlefields, digging around in their patients' wounds with dirty, blood-smeared knives, forceps, and fingers. Surviving surgery was no guarantee of success. Often deadly bacteria were passed from bed to bed when pus-soaked dressings were changed. Cross-infection was so common that surgeons simply regarded it as a normal stage in recovery, or decline.

Ever a passionate believer in the healing powers of nature, and always adamant that surgeons should intervene as little as possible, Hunter realized that conservative treatment—what he called "being very quiet"—was the most effective remedy for gunshot injuries. Observing that meddling with the wound generally led to a worse outcome than leaving the injury alone, he believed the wounded soldiers and sailors had a better chance of survival by letting nature take its course, even to the extent of leaving bullets embedded permanently in their wounds. In his later *Treatise on the Blood, Inflammation and Gun-shot Wounds,* published thirty years later from notes made on the island, he explained, "It is contrary to all the rules of surgery founded on our knowledge of the animal oeconomy to enlarge wounds simply as wounds. No wound, let it be ever so small, should be made larger, excepting when preparatory to something else."[17] While his contemporaries viewed infection as an unavoidable stage in healing, Hunter regarded it as a failure of treatment. "Suppuration may be considered a resolution but it is a mode of resolution which we mainly wish to avoid," he would later write.[18] Not until 1867, when Joseph Lister published his successful trials using antiseptics to fight infection, would Hunter's radical view finally be accepted.

In a chance incident that stands as a prototype controlled experiment, Hunter soon found the evidence he needed for his controversial beliefs. On the day the British landed, five French soldiers had been shot in the exchange of fire but had managed to hide out in an empty farmhouse, where they lay low until discovered four days later. One had been hit in the thigh by two musket balls, one of which was still lodged in his thighbone; a second had been shot in the chest and was spitting blood; the third had been hit in the knee; the fourth had been hit in an arm; and the fifth was only slightly wounded.[19] Neglected through accident rather than design, their injuries had healed significantly better than those of their British counterparts who had been subjected to the surgeon's knife. "The first four men had nothing done to their Wounds; indeed very little was done to the men themselves; for they lay in an uninhabited house for more than four days, with hardly any subsistance," Hunter noted. "The wounds were never di-lated, nor were they dressed all this time, excepting once by one of our Surgeons. All of them healed as well, and as soon (if not sooner) as the like accidents do in others who have all the care that possibly can be given of them: indeed they did surprisingly well, for the man that was shot through and through the breast, recovered perfectly: as also did the others." Further proof of his beliefs arrived in the shape of a British grenadier who had been shot in the arm and taken prisoner by the French during the first attack; he, too, had received only superficial care. Having escaped a fortnight later, he surprised his surgeons by revealing his wounds healed and his elbow only a little stiff.[20]

While his colleagues dismissed these examples as mere curiosities, Hunter adapted his methods to suit his observations in his first systematic application of scientific evidence to practice. He was not dogmatic. He still advocated opening a wound in certain circumstances—to tie severed arter-ies or to extract pieces of smashed bone, for example—and he adopted this doctrine when treating an officer wounded by a musket ball that shattered his cheek, recording that "with a pair of small forceps I extracted all I could of the loose pieces of bone."[21] Despite the infection that inevitably fol-lowed, the officer recovered. But when a second officer arrived at the hos-pital with a musket ball lodged in the same place and Hunter perceived no splinters of bone, he recommended leaving the wound to mend of its own accord, noting, "My advice was complied with, and the wound did well, and rather better than the former, by healing sooner." Even if his methods were raising eyebrows, he was obviously not being opposed.

Hunter advised similar restraint in the army surgeon's other main stock-in-trade—amputations. Customarily, military surgeons would hack off mangled arms and legs right beside the field of battle, often laboring un-der fire as they cowered behind a hedge, rather than risk causing further damage by moving injured soldiers to a first-aid tent. But Hunter observed of such patients that "few did well."[22] As a consequence, he recommended delaying surgery until the patient could be transported to more comfortable surroundings and his condition was stable, in order not to "run the risk of producing death by an operation." Again, his doctrine suited the unsavory circumstances he worked in; only under more sanitary conditions would early amputations generally increase survival chances.

Despite Hunter's unorthodox methods, he won the respect of his superiors. When William Young, the director of the hospitals on Belle-Ile, returned to England in early 1762, Hunter was appointed to take his place, becoming chief surgeon and director of all medical operations on the island. Writing with obvious pride to William in March, he announced, "My titles and business are at present many, I am called the Surgeon-general, Deputy Purveyor, and Inspector and Director of the Regimental Hospitals." And he added, with blatant designs on gaining further advancement, "I could wish that you would give a hint of this to Mr Adair who I hope is my friend."[23]

In truth, by this point there was not a great deal of medical care to superintend on Belle-Ile, since Hodgson had sailed with many of the troops, while most of the soldiers left behind to defend the island had little to do but drink, gamble, and speculate about where they were destined for next. Hunter hoped it would not be the West Indies, as widely rumored, recalling from his choppy voyage across the Bay of Biscay that "the Sea plays the Devil with me."[24] Much preoccupied, as in all his letters, with the state of his earnings—which were being deposited with William in London for safekeeping—he preferred to stay on the island and remain in charge of the hospital. This would guarantee that his daily pay stayed at the double rate of twenty shillings and, equally important, would double his half pay once he returned to civilian life. But since five shillings of his daily rate went to his surgeon's mate, and he had to keep a horse in order to inspect the first-aid camps stationed around the island, he was not much better off, at least in the short term.

With his prospects hanging in the balance and talk of a new campaign to Portugal in the air, John petitioned William relentlessly to secure him promotion in his next assignment, wherever that should be. Even if relations between the brothers had become strained during their intense working partnership, John knew William could always be relied on to advance family fortunes. And even though Hunter was angling to remain in Belle-Ile, he was evidently homesick. Successive letters implored William to tell him "how anatomy is to go on this winter" and to send news of Dorothy and her new husband, the Reverend James Baillie, who were now living in

Scotland. At one point, he remarked forlornly, "We seem to be a people almost distinct from the rest of [the] world."[25] If William was somewhat tardy in his replies, he could perhaps be forgiven, for when Queen Charlotte, the new king's bride, had fallen pregnant at the end of 1761, it was to William that her care had been entrusted. Renowned for his conservative management of childbirth, William presided over her delivery the following August, waiting decorously in an antechamber while a female midwife described the progress of the birth of the new prince, the future George IV. William was made physician in extraordinary to Queen Charlotte the following month; he would attend the deliveries of all fifteen of the royal couple's children.[26]

By the spring of 1762, with many of the troops having left Belle-Ile to pursue British war interests elsewhere, island life slowly began to return to normal. Seabirds settled back on their nests in the sheer cliffs and lizards basked on the rocks edging the white beaches. Effectively a captive on this solitary rocky island, where Alexandre Dumas would later maroon two of his musketeers in *The Man in the Iron Mask,* Hunter spent his leisure hours riding across the moors and cliff tops or walking along the shores and creeks, searching for local wildlife.[27] He had already snatched moments between battles to haul fish from the sea and cut open their bellies on deck to test how their blood coagulated. He had also netted conger eels, having seen a "vast number" in the seas around the island, and dissected them, in the forlorn hope of discovering their spawn.[28] During the winter, he had captured hibernating lizards and force-fed them with worms and pieces of meat, then sliced them open, hoping to understand how their digestive system shut down during their long sleep. At a time when fellow naturalists were still struggling in vain to understand basic principles of life, this suspended animation proved a consuming subject. Hunter would always be keen to explore the boundaries where life seemed to merge with death.

The lizards he kept captive in his quarters offered an even more remarkable peculiarity, which Hunter discovered to his amazement one day when picking one up by its tail: The tail came off in his fingers and the stumpy creature promptly ran away. He tried the same trick on others and realized their tails were expressly designed to detach at a certain point when caught by predators. "The separation is so easily effected," he noted, "that if a lizard be caught by the tail, it will leave it in your hand by the strength of the animal only."[29] Watching the severed tail on his temporary dissecting bench, he saw that "the tail continues to move and writhe for some time, and when these motions have ceased, as it were from fatigue, they recommence when the tail is pricked or otherwise irritated."[30]

With his customary zeal, he began experimenting in earnest on the island's lizard population, catching more reptiles in order to break or cut off their tails, dissecting the animals to discover how their tails became so easily detached, and then pickling some of the mutilated creatures in alcohol to take back home. What astounded him even more was that the tails of his amputated lizards regrew; some even generated double tails where the first had broken off. Both these peculiarities would afford him fertile new territories for future research: The ability of lizard tails to regrow would lead him to conduct controversial experiments attempting to stimulate regeneration in humans; the double-tailed lizards provided him with a classic example of nature deviating from the norm, which started him pondering on the ways in which organic life might have developed on earth.

John Hunter's peaceful springtime pursuits on Belle-Ile were short-lived, for May brought the new orders he had been waiting for. In July, he was bound for Portugal with a contingent of troops sent to defend Britain's oldest ally against its neighbor, Spain, which had just joined the war. Stepping off the *Betty* in Lisbon with his medicine chest and his stock of bottled lizards, Hunter was dispatched immediately to the front line.[31] For the next six months, he struggled in blistering heat and squalid conditions to treat the sick and wounded sent back from the fighting by the regimental surgeons. The battle casualties were few—the campaign was little more than a sideshow at the tail end of a costly war, of which all sides had wearied—and a truce was declared in November. But infectious diseases wrought far more damage among the troops than the Spanish artillery could manage. At one point, Hunter pronounced himself "allmost knock'd up" with exhaustion at the challenge he faced tending the sick in makeshift emergency bases set up near the front line.[32] Even after hostilities ended, he was kept busy until early in 1763 evacuating the soldiers from the "flying hospitals" back to Lisbon and ultimately home.

At least, as the military activities wound down, he could return to his more abiding love—investigating the natural history of his adopted home. Already, on his arrival in Lisbon, he had conducted an ingenious experiment to test whether fish could hear. Hunter had previously discovered the inner organs of hearing in fish during his animal investigations in Covent

Garden, but their existence would still be disputed for many more years by fellow anatomists. Spotting an ornamental fish pond on the Lisbon estate where he had been barracked, he persuaded his host to allow him to perform an experiment to demonstrate his earlier discovery. "Whilst I lay on the bank observing the fish swimming about, I desired a gentleman who was with me to take a loaded gun and fire it from behind the bushes," Hunter recorded. "The moment the report was made the fish seemed to be all of one mind, for they vanished instantaneously, raising a cloud of mud from the bottom. In about five minutes afterwards they began to reappear, and were seen swimming about as before."[33] With his usual tardiness, it would be twenty years before he reported his findings to the Royal Society, by which time a rival, the Dutch anatomist Peter Camper, had published the discovery.

Now at the end of the campaign, packing up supplies near the Spanish border, he had time to stand back and survey the austerely beautiful landscape of the region he had traversed so many times. He captured more lizards for his growing collection, and he probably watched the eagles, vultures, and kites soaring above the São Memede Mountains along the frontier, too. But when he stood on the edge of the vast Alentejo plain, looking south as far as the eye could see across the expanse of open countryside studded with cork trees, olive groves, and hilltop villages, he began to entertain thoughts that would one day lead him into territory almost as dangerous as his wartime escapades.

His strict religious education at the village school run by the Scottish kirk had taught him unequivocally not only that the world had been created by God in six days but that its major geological features had been shaped by the great flood of Noah's time. But refusal to accept authorized beliefs was a way of life to Hunter. Quite simply, he could not accept that such an enormous plain had been created by a single flood lasting forty days. Only an immense period of erosion by the sea could have produced such an expanse, he concluded.[34] The sight of peculiar natural stone structures, like "inverted pyramids," convinced him the sea had not only covered the area for a long period of time, sufficient to wash away the lower strata of the stone columns, but had afterward "left it gradually." Similarly heretical views had been expounded by Robert Hooke, the remarkable seventeenth-century scientist, but they had since been conveniently forgotten. For the time being, Hunter kept his views to himself. It would be nearly thirty years before he committed his thoughts to paper.

It was February 1763, just prior to the signing of the Treaty of Paris, which brought the Seven Years War to an end, before Hunter managed to return to Lisbon himself. After briskly discharging most of the remaining soldiers left in the capital's temporary hospitals, he sailed for home with the remainder of the forces in April. When he unloaded his army trunk onto the quay at Portsmouth, it contained more than two hundred specimens he had lovingly preserved from the ravages of war and Continental heat during his two years' absence. Army lifestyle had at least supplied the plentiful quantities of alcohol he needed for pickling his treasures. As well as fifty specimens of lizards and their tails—including his precious double-tailed lizards—he also repatriated several body parts belonging to unfortunate soldiers who had died during the campaigns. Among them were a shoulder blade, a piece of skull, and a thighbone, all showing the impressions of French musket balls, as well as a section of intestine taken from a soldier who had died in Portugal of dysentery.[35]

Back in his civilian clothes once more, John Hunter set out for London. Like his fellow demobilized army veterans, he had no job, no prospects, and only his army half pay on which to survive.

CHAPTER 7

The Chimney Sweep's Teeth

ꙮ

London

Spring 1763

When the Reverend James Woodforde, a Norfolk country parson, passed a miserable night suffering from a toothache, he knew the consequences would be unpleasant. Developing a toothache was no laughing matter in eighteenth-century Britain. As the slave trade brought cheaper sugar imports from the West Indies and sugar became an inseparable addition to the copious amounts of tea consumed in salons and drawing rooms, wealthy Georgians developed a collective sweet tooth. The results were drastic. Many an aristocratic debutante's smile was spoiled by black and rotting stumps, while the pain of tooth decay, abscesses, and gum disease spelled misery for young and old alike. Although the young king, George III, surprised royal watchers by displaying a set of teeth that were "extreamly fine," a more typical example of Georgian dental care could be seen in his financial adviser, the duke of Newcastle. According to Horace Walpole, the tireless commentator on society, his "teeth are jumbled out and his mouth tumbled in."[1] Missing, blackened, and diseased teeth were not confined to the upper classes; as a love of all things sweet spread throughout society, so did dental decay.

There was little effective remedy. While some optimists resorted to charms, amulets, and far-fetched concoctions, there was often no alternative but to have a decaying tooth removed—usually painfully, ineptly, and with much loss of blood. Both physicians and surgeons looked down on tooth extraction, considering it far beneath their status. The physicians were happy to administer enemas into a patient's rectum, but not to delve into their foul-breathed mouths; for their part, the surgeons had decreed that, alone among surgical acts, tooth extraction could be carried out by barbers. So most dental work was performed by a motley selection of un-qualified and untrained opportunists, not only barbers but also itinerant travelers, wig makers, and blacksmiths.[2] The outcome was fairly pre-dictable, as the Reverend Woodforde attested when he called his local blacksmith to dispatch the cause of his agony:

> My tooth pained me all night, got up a little after 5 this morning, & sent for one Reeves a man who draws teeth in this parish, and about 7 he came and drew my tooth, but shockingly bad indeed, he broke away a great piece of gum and broke one of the fangs of the tooth, it gave me exquisite pain all the day after, and my Face was swelled prodigiously in the evening and much pain. Very bad and in much pain the whole day long. Gave the old man that drew it however 0. 2. 6. He is too old, I think, to draw teeth, can't see very well.[3]

The field of dental care was plainly ripe for improvement. Yet hardly any reputable medical practitioners were prepared to take on the challenge.

Returning to London in April 1763 at the age of thirty-five, John Hunter was homeless, jobless, and without any obvious prospects. There was no place for him at his brother's anatomy school, as William had found a new assistant to help run the lessons, which he had restarted in a ware-house in Piccadilly. And even if John had wanted to return to his fraternal partnership, he soon wrecked that chance, quarreling with William within two months of his return, most probably over the ownership of preparations John had left in his brother's care.[4] Although William, who had now ac-crued a sizable fortune, would help his brother financially several times af-ter his return from the army, this was clearly not going to be one of them.[5] At the same time, John had scant hope of securing a hospital job without the patronage of several worthy governors, and little chance of finding a footing in private practice, since this was dominated by a handful of estab-lished surgeons. Struggling to make ends meet on his army half pay of ten shillings a day and taking lodgings in seedy Covent Garden, he turned to one of the few occupations open to him and teamed up with one of London's best-known dentists, James Spence.[6]

It was a logical move. There was no shortage of wealthy clients requiring the attentions of a surgeon with the anatomical knowledge and skills to extract rotten teeth and treat gum disorders. Hunter could be assured of a steady flow of income. But it was not without its risks, since dentistry was held in singularly low esteem within the medical hierarchy. Hunter's bitter biographer, Jessé Foot, would describe his dalliance in dentistry in ornate detail, precisely because he hoped such an association would bring scorn on Hunter's head. In fact, the alliance Hunter forged with Spence, a fellow Scot, was simply ahead of its time. Attending Spence's premises to offer advice to his clients, Hunter had established nothing less than a modern system of professional consultation. For Foot, such a coalition was unthinkable. Certainly, Hunter's surgical rivals "were above submitting to consultation with dentists," he sneered.[7]

Yet James Spence was no charlatan, as Foot patently knew. Only the year before Hunter returned from the army, Foot had himself enjoyed Spence's talents. Then an impressionable young surgeon's apprentice with a painful twinge in a troublesome tooth, Foot had set off for the Gray's Inn Road shop, since "[t]here was no one so high in fame for extracting teeth as the elder Spence."[8] The practice was easy to find, for in the window hung a painted hand, adorned at the wrist with lace ruffles and daintily holding a large replica tooth. Venturing inside the shop, which doubled as a barber's, Foot was pleasantly surprised. "The barber's blocks were as white as soap-suds could make them," he recalled, "and the blood basons were as shining as if they had been directly brought home from the scowerers. The teeth exhibited as specimens in the shop, were as white and polished as ivory." Meeting the shop's owner, Foot was equally impressed. After the "dreaded process" of having his tooth yanked out, he was delighted when Spence refused, as one medical professional to another, to charge him a penny, then invited him into a back room to demonstrate a curious electrical machine that could make figures dance, bells ring, and gunpowder ignite. Spence attracted other admirers, too. In 1766, he would be appointed "operator for the teeth" to George III. Both his sons followed in their father's footsteps, becoming eminent dentists in their own right.

Working profitably alongside the Spence family for at least five years, John Hunter advised on every aspect of dental treatment. The partnership enabled him to put into practice the comprehensive knowledge of the human skull and jaws he had acquired from studying dead bodies in Covent Garden. In addition to offering his expert opinion on extractions, he advised on inserting fillings, made of gold or lead, as well as scaling, filing, cleaning, and treating gum disease. Children were frequent visitors to the joint practice. Unusually conforming to prevailing doctrine, Hunter recommended cutting through teething babies' gums with a blunted lancet to allow first teeth to emerge. Should the gums regrow, as they frequently did, he simply repeated the procedure on the howling infants, recording, "I have performed the operation above ten times upon the same teeth."[9]

While the chief motive for Hunter's union with the Spences was financial, he treated his excursion into dentistry as he approached all experiences in medicine—with rigorous investigation in order to improve understanding and practice. His surgical experience in the army was fresh in his mind, yet it was another memory from his service years that surfaced as he helped the Spences wrench out rotten teeth from fetid gums. Examining the extracted teeth with their long roots, commonly known as "fangs," his thoughts returned to his experiments on lizards in Belle-Ile and their strange ability to regenerate their tails. He had observed the same tendency in other animal tissue and was fascinated by the ability of skin and bone to mend and grow. Keen to investigate further, Hunter launched a series of bizarre grafting experiments on living animals. First, he cut off the spur from the foot of a cockerel and fixed it into the fowl's comb, where it appeared to continue growing.[10] Next, he removed one of the testes from a rooster and implanted it in the bird's belly; it, too, seemed to flourish. Warming to his theme, he transplanted the testis of another cockerel into the belly of a hen, with seemingly similar results.

Encouraged by his success so far, Hunter now found a human volunteer—probably one of Georgian London's ubiquitous down-and-outs—who agreed to have a healthy tooth extracted, presumably for a generous fee. This, Hunter transplanted into a cockerel's comb, as he excitedly recorded:

> I took a sound tooth from a person's head; then made a pretty deep wound with a lancet into the thick part of a cock's comb, and pressed the fang of the tooth into this wound, and fastened it with threads passed through other parts of the comb. The cock was killed some months after, and I injected the head with a very minute injection; the comb was then taken off and put into a weak acid, and the tooth being softened by this means, I slit the comb and tooth into two halves, in the long direction of the tooth. I found the vessels of the tooth well in-

jected, and also observed that the external surface of the tooth adhered everywhere to the comb by vessels, similar to the union of a tooth with the gum and sockets.

The carefully bisected specimen would become one of his most prized preparations. It remains to this day in his museum—half a perfect human tooth sticking incongruously out of a scarlet cockerel's comb.[11]

Although the tooth was firmly fixed, it had not in reality bonded with the cock's comb, contrary to what Hunter believed. In fact, the tooth was dead—there was no connection to a blood supply—and he had to admit that subsequent attempts failed. By contrast, the testicle he transferred to the hen's belly appeared to have been accepted, in a remarkable early organ transplant, and had even acquired blood vessels. Possibly the donated testicle was not rejected—although rejection would be expected without modern medicines to encourage acceptance—because the hen and cockerel were closely matched genetically through inbreeding.[12]

Hunter repeated his strange transplantation experiments, and others in a similar vein, on numerous occasions. Only a few years after he began his alliance with the Spences, a former pupil from his Covent Garden years wrote to a fellow ex-classmate to report "many surprising things" being performed by their former tutor.[13] "The cutting of spurs from the heel of a cock and making them grow in his head, J. Hunter has often performed and has such cocks at present," wrote the incredulous ex-pupil, who was staying with Hunter. The very next day, the surgeon even planned to graft a cow's horn onto the forehead of an ass before "several gentlemen," his visitor reported.

Whether or not the horn grew on the donkey's head, the success—or apparent success—of his grafting experiments soon prompted Hunter to attempt further transplanting trials, this time from human to human. By now, he was convinced it was possible to transplant living human teeth. It was already commonplace to buy false teeth in order to fill unsightly gaps; Samuel Pepys's wife, Elizabeth, had some teeth "new done" as early as 1664.[14] But it was in the eighteenth century, as decay spread and vanity soared, that the market in replacement teeth really boomed. Artificial teeth were often made of elephant or hippopotamus ivory, but they were usually expensive and ill-fitting. An alternative—both cheaper and more lifelike—were real human teeth pulled from dead bodies, either from cadavers stolen from graves or from soldiers' corpses on battlefields. These were not to everyone's taste, however, for, as one dentist pertinently noted, "most people have a dread of teeth which have been obtained from a corpse."[15]

There had been previous attempts to transplant living human teeth from one person to another, but these invariably failed in the long term. Most practitioners had simply abandoned the practice. Ambroise Paré, the French army surgeon, recorded one of the earliest-known efforts in 1562, when a wealthy woman had a tooth extracted and replaced by another, it having been supplied by one of her ladies-in-waiting. By the middle of the eighteenth century, tooth transplantation was being regularly performed in Paris by a dentist who used young boys as his donors. In Britain, too, transplantation was available before Hunter's time. It was advertised in Norwich in 1762, a year before Hunter's return to England. Yet concerns about safety, efficacy, and, for some at least, morality remained. It was Hunter, from 1763 onward, who rejuvenated, popularized, and legitimized tooth transplantation in Britain, initiating a fashion that would last well into the next century.

For Hunter, the principle of tooth transplantation presented above all an irresistible scientific challenge. It was the propensity of living tissue to grow, mend, and adapt that fascinated him. This was a key part of the puzzle of organic life, and, regardless of ethical considerations, he was determined to investigate. Teeth mystified him. Though hard like bone, they stopped growing once they had erupted through the gum, and his strenuous efforts to inject them failed to discern any blood supply. In this case, his instruments were simply too crude to trace the blood vessels and nerves in the living pulp. Still, he was certain that teeth possessed what he termed a "living principle"—an ability to form and grow. The animal freaks he had created convinced him that successful tissue grafting was perfectly practical—he may even have been the first to coin the term *transplant*—and eager, as always, to put research into practice, he set out to demonstrate the phenomenon on humans.

Advertising for donors, Hunter attracted queues of impoverished girls and boys clamoring for a few pennies in return for their healthy teeth. With his well-to-do customers seated in a comfortable chair, Hunter yanked out a healthy front tooth from one of his ragged donors and promptly implanted it in the gaping mouth of his patient. Deftly, he tied the new implant to its neighbors with silk thread, or sometimes seaweed, then sent the gappy youngsters away with mouths bleeding and fingers clasping their meager earnings. The pitiful scene was graphically captured by the artist Thomas Rowlandson, who showed a fashionable practitioner—actually a disciple of Hunter's—pulling a tooth from the jaws of a grubby young chimney sweep in his satirical cartoon *Transplanting of Teeth*.[16] As in the picture, Hunter recommended lining up several potential donors at once, "for if the first will not answer, the second may."[17] The donors should be young and preferably female, since their teeth were generally smaller, he advised, although the young women should first be inspected for signs of ill health. And the operation should be performed as quickly as possible, he stressed, "as delay will perpetually lessen the power upon which the union of the two parts depends."

Hunter performed live-tooth transplants on a number of occasions, though not as frequently as some writers would later assume.[18] More significantly, he endowed tooth transplanting with a scientific seal of approval, which launched a veritable craze. The procedure was quickly adopted by other dentists in London and, despite unease over the ethics of the operation, spread across Europe and even to America, reaching a peak in the 1780s. The transaction was always a one-way exchange within the social hierarchy, however, as a female dentist in York made abundantly clear when she offered to "transplant teeth from the front jaws of poor lads into the heads of any Lady or Gentleman without putting both patients to any anguish."[19] One New York dentist was prepared to pay as much as five guineas in 1782 for healthy front teeth—"slaves teeth excepted."[20] A young and impoverished Emma Hart, who would become better known as Lady Hamilton, almost forfeited her renowned beauty—and quite possibly her future relationship with Lord Nelson—when tempted to sell her front teeth during the same period in London. On her way to the dentist, a friend persuaded her to preserve her smile.[21]

Although transplantation became widely accepted, in common with the general exploitation of the have-nots for the benefit of the haves in the eighteenth-century world, there were nevertheless isolated voices raised in protest within the profession. Thomas Berdmore, who was appointed George III's dentist in 1766, along with Hunter's partner, James Spence, denounced the procedure in a treatise on teeth two years later. Not only was the operation "precarious, ineffectual, and dangerous in general," it was also "immoderately expensive," he argued, "for it is not to be supposed that any young person will sell a handsome sound Tooth, to be torn out of his head, without being extremely well paid for his loss and pain."[22] William Rae, another Scot, who moved to London to practice dentistry in the 1770s, was an even more ardent critic, despite having been a pupil and remaining a close friend of John Hunter. Giving the first-ever series of lectures on dental surgery in 1780, actually at Hunter's suggestion, he insisted that tooth transplanting was both unsuccessful and immoral. "In the first place," he warned, "it is cruel to take the teeth of a poor creature, whose necessities may induce him to part with it as a means of procuring subsistence; and, in the second place, we are obliged to take them from poor people, who are very often diseased, and generally with the *lues venerea*."[23] Although Hunter insisted that the lues venerea—syphilis—was not passed on through tooth transplants, Rae was invited to give his next series of lectures at Hunter's own lecture rooms. Hunter not only allowed but welcomed dissent from those whose opinions he respected.

Ultimately, however, the transplantation of live teeth would fall into disrepute. Although transplanted teeth could adhere for several years—one of Hunter's patients attested he had received three, which stayed firm for six years—they never bonded permanently and they could certainly pass on disease. Even Hunter's confidence was shaken when a wealthy female patient developed signs of venereal disease after he had given her teeth from a young donor. Explaining the case in 1771 to William Cullen, by then a professor at Edinburgh University, Hunter maintained that "the girl from whom the teeth were taken had all the appearance of a sound person," but he acknowledged that the recipient's own physician had confirmed his patient had contracted syphilis.[24] Despite the concerns, the operation would finally be abandoned only when aesthetically acceptable false teeth, fashioned from porcelain, appeared on the market in the early nineteenth century.

Although clinically a failure as far as teeth were concerned, the transplantation methods Hunter recommended were based on sound principles, which would remain relevant in future human organ transplants. He clearly understood the need for donor tissue to be as fresh as possible and was careful to match for size—both key factors in later transplants.[25] Not surprisingly, he had no notion of the need to match blood types or of the body's natural rejection—today countered through drugs—and he was unconvinced that donor tissue could transfer disease. Although finally his tooth transplants were discredited, he remained fixated by the idea of regeneration, convinced that living tissue possessed an innate tendency to grow and unite. This notion—his "living principle"—would continue to preoccupy him as a fundamental area of research.

Working with the Spences in the ignominious field of dentistry in the 1760s prompted Hunter to produce his first major work of research—a treatise on the teeth and jaws—which he would publish in two parts in 1771 and 1778, *The Natural History of the Human Teeth* and *A Practical Treatise on the Diseases of the Teeth*. It was the first work to consider the teeth in a truly scientific manner, describing in detail the anatomy and physiology of the teeth and jaws, and would be considered one of the most important books in the history of dentistry.

At the same time, Hunter's steadily rising reputation for accuracy and sound judgment gave dentistry a major boost in status. His comprehensive descriptions of the anatomy of the teeth, facial muscles, and bones of the jaws, with sixteen exquisite drawings by van Rymsdyk, are considered remarkable for their accuracy even today. Some later practitioners would regard him as the father of scientific dentistry. Inevitably, he made errors. His reliance on stolen bodies misled him over the ages at which adult teeth emerged, and he was plainly misguided over the feasibility—and ethics—of transplanting live teeth. Yet he lucidly described several dental disorders, recognized the destructive nature of plaque and recommended its removal by cleaning or brushing, professed the benefits of eating fruits and salads, and gave teeth the names by which they are generally known today. Previously, teeth were known as incisors, canines, and molars. Hunter established four classes of teeth—incisors, cuspids (canines), bicuspids (premolars), and molars—which remain the standard terms used by dentists in the United States.[26] His work rapidly became a best-seller, going through fifteen editions, including one in the United States, and being translated into French, German, Dutch, and Italian.

Despite the disdain some heaped on dentistry, Hunter's work with the Spences and his treatise on teeth helped to establish the charismatic surgeon as a rising figure in the harshly competitive London medical world, as well as an eligible bachelor.

CHAPTER 8

The Debutante's Spots

London

December 1763

By the flickering light of candles, the small party peered at the strange painted symbols on the wooden casket placed on the table. The Egyptian mummy, which had been plundered from the royal pyramids almost a century earlier, had been gathering dust in the Royal Society's museum ever since. But on December 16, at the London house of John Hadley, professor of chemistry at Cambridge and a physician at St. Thomas' Hospital, the moment had come for the casket to reveal its secrets.[1] As the nails restraining the lid were levered off, the half dozen men gathered in the room held their breath in apprehension. Among them were John and William Hunter. Having mended their summer tiff in the interests of family unity, they now stood shoulder-to-shoulder, as they had in their Covent Garden partnership, ready to unwrap an Egyptian princess.

Gingerly picking at the rotten brown linen, the brothers peeled away the layers from the emaciated body. Neglected through its decades of storage in the Royal Society, the mummy's head was already detached from its torso and the feet from its legs. Eager to uncover the mystery of the Egyptians' preservation techniques, John leaned forward to taste the white powder that smeared the bandages. While most of the bones had long since crumbled to dust, the gathered guests were delighted to see that the ribs and spine had survived and the left foot was completely preserved. "The toes being carefully laid bare, the nails were found perfect upon them all, some of them retaining a reddish hue, as if they had been painted," recorded Hadley for his report to the Royal Society. Even the prints on the big toe could plainly be seen. They were carefully reproduced in a sketch presented with notes of the findings at the next society meeting.

With his passion for curious experiments attracting interest, Hunter was fast becoming a familiar personality in medical and scientific circles. As well as extracting and transplanting teeth, he had built up a small clientele of patients referred for other surgical problems and also taught students at private classes in anatomy and surgery to eke out his income.[2] Although his blunt manners and refusal to conform to social niceties offended some of his more stuffy medical contemporaries, he won firm friends and loyal supporters among other men with a scientific bent, who were impressed by his energy, breadth of interest, and ebullient manner.

Fellow enthusiasts in natural philosophy—as scientific studies were generally known—flocked to collaborate with him. Whenever it became known through the grapevine that Hunter was planning to conduct a postmortem or attempt a new experiment, he could be sure of an audience. Gradually, he was becoming a key figure in a small but growing band of professionals and amateur enthusiasts committed to pushing forward scientific boundaries.

Many of these people were members of the Royal Society, Britain's most prestigious body devoted to scientific investigation. Founded in 1660, the society had been established under the auspices of Charles II, with the help of such luminaries as Robert Boyle and Christopher Wren. By the mid–eighteenth century, it had flourished into an eclectic grouping of professionals in a variety of fields, along with well-to-do and often aristocratic gentlemen professing a passing interest in experimental science. They met weekly to exchange ideas and published their findings in the *Philosophical Transactions*. Although Hunter had not yet produced sufficient research to warrant membership in this exclusive gathering, he was being fêted by some of its most illustrious fellows, as well as being invited to play a prominent role at some of their more unusual occasions, such as the mummy unwrapping.

T *here were more,* significantly fresher bodies, to open that winter. As Londoners shivered—that year, the Thames froze over—Hunter was in his element: It was perfect dissection weather. After his army break from regular anatomy, he had returned to his dissecting bench with gusto.[3] With bodies no easier to obtain, the majority were still provided by his old friends, the body snatchers.

So the anonymous corpses of London's poor continued to find their way to Hunter's dissecting room. But increasingly now, they rubbed shoulders with the rather better-nourished bodies of some of Britain's richest and best-known individuals. Fellow surgeons, and especially physicians, who themselves did not possess the skills or training to dissect human bodies, began to seek out Hunter's expertise to open the bodies of patients who had died in unexplained circumstances. This was a momentous development. Although public dissections of criminals hung at Tyburn still aroused widespread repulsion, there began a gradual movement toward general acceptance of autopsies conducted in private and with decorum in order to understand the causes of death.

Partly, this trend was prompted by the determination of a few enlightened surgeons and physicians who came to realize the significance of autopsies in determining how a patient had died. Rather than simply dismissing all illness as a mysterious imbalance of humors, they grew to appreciate that signs in a body after death could be linked to symptoms in the living. Consequently, they persuaded more liberal-minded families to allow postmortems on relatives who had died. The Italian anatomist Giovanni Battista Morgagni had professed the importance of autopsy examinations when in 1761 he published the findings of some seven hundred postmortems, which demonstrated how certain diseases could be traced in organs after death.[4] Plainly, at least some conditions had specific causes, which might one day even be prevented. This was a crucial first step toward a search for effective remedies for different diseases, although the real shift toward clinical medicine would not emerge until the following century. Morgagni's work was translated into English in 1769, reinforcing a trend that had been developing independently in Britain. Partly, too, the rise in postmortems was driven by changing attitudes within society. The Georgians' avid fascination with their health encouraged people to seek answers when their loved ones died; since advancing knowledge in anatomy could sometimes now provide those answers, autopsies gradually became more acceptable. This was self-interest as much as familial concern: Knowing the cause of a relative's death might help avert a similar fate for the survivors. Results of postmortems were often sent to family members and occasionally reported in newspapers.

John Hunter frequently performed the required autopsies. From the time he set up in London as an independent practitioner in 1763, his expertise in conducting postmortems was in regular demand. Nobody was more aware of the importance of postmortems in improving medical care. Indeed, when he was refused permission to perform an autopsy by the relative of one patient he had treated, despite his protestations that it would advance medical understanding, he lashed out: "Then I heartily hope that you yourself and all your family, nay all your friends, may die of the same disease and that no-one may be able to offer any assistance."[5]

Among Hunter's most notorious cases was the autopsy of William Chaworth, who was killed by his cousin, Lord Byron, in an argument in January 1765.[6] Byron, who was the fifth baron and the great-uncle of the poet George Byron, had met Chaworth at the Star and Garter tavern in Pall Mall on January 26. Here, they pursued an ongoing petty quarrel over how best to deal with poachers, and as the argument became heated, they requested an empty room in which to settle their differences. In the scuffle that ensued, the baron stabbed his cousin in the stomach. While Chaworth clung to life, two of the most prominent surgeons of the day were called to the scene in an effort to save him. Neither Robert Adair, the army surgeon, nor Caesar Hawkins, the king's surgeon, could patch up the victim's deep wound. Eventually, he was carried to his apartments near Piccadilly, but there was little hope for his recovery. Hunter was called in the next morning to see the dying man, but it was too late even for his skills to make any difference. Twelve hours later, after "violent vomitings and retchings," Chaworth expired.

Now Hunter's autopsy expertise was pressed into action. His fellow surgeons asked him to open the body to inspect the site of the wound; perhaps the examination might reveal how death from such an injury could be averted in the future. Recording both Chaworth's dying moments and his postmortem in his casebooks, Hunter noted that the wound had led to widespread inflammation within the abdomen—much as he would have expected. Given the clear evidence of his cousin's violent death, Byron was convicted of manslaughter before the House of Lords, but, as a peer of the realm, he escaped punishment, other than being fined. The remorse he suffered over the fatal argument prompted him to live the rest of his life in seclusion.

Many other autopsies Hunter recorded were performed at the express request of families anxious to understand the reasons for the loss of a loved one. One such case was an autopsy on a ten-month-old baby, "a Son of Mr Abbott's," who had died suddenly from an unknown cause. "This was the fourth Child that died in that way, and all about that Age," Hunter wrote in the casebooks; "their parents were desirous of knowing if there could be any possible cause for it."[7] Sadly, he could find no obvious reason why this child and his three siblings had all died before their first birthdays; possibly a congenital defect was the cause. In a similar tragic mission, he performed an autopsy on "Sir William Lee's Child," an eighteen-month-old girl who had died in the throes of a fever and convulsions. The family's physician had blamed her death on teething. Here, too, Hunter was unable to confirm the cause of death, which could have been due to any one of the infectious diseases circulating in Georgian London, or, indeed, to the remedies prescribed by the family's physician. The infant had, after all, been dosed with magnesium and crabs' eyes.[8]

Before long, Hunter's popularity at houses of bereavement was bringing in much-needed extra income to supplement his fees from dental consultation and his army half-pay. By 1764, he had advanced his finances and his reputation sufficiently to be considered something of a catch. Since returning to civilian life, he had maintained several army friendships. One colleague who remembered him fondly was Lt. Robert Home, a fellow army surgeon whom Hunter had treated for a swollen eyelid on Belle-Ile. Back from the battles, Home lost no time in inviting the intelligent and energetic young surgeon to his home in Suffolk Street off Pall Mall to meet his wife and accomplished daughters.[9] Hunter was immediately captivated by twenty-two-year-old Anne, the eldest of the seven children.

It was not an obvious match. Poised and elegant, highly educated and self-assured, Anne inhabited a high-society world of fine manners and witty conversation, an environment that was completely alien to John. While he had always balked at books, even despised the written word, she had already shown literary flair and was considered an accomplished poet in the Romantic style. Some of her first poems would be published in an Edinburgh journal in 1765. While John had no interest in fashionable society, Anne delighted in the company of female intellectuals such as Elizabeth Montagu, Elizabeth Carter, and Mary Delaney. She was soon numbered among the Bluestockings who met at literary gatherings, in emulation of the Parisian-style salons or Italian *conversaziones* dedicated to the art of sparkling conversation. With her classic English rose complexion, blond-haired, blue-eyed Anne was regarded as something of a beauty. One contemporary described her as "very handsome, tall, singularly dignified and lady-like in appearance," while another pronounced her "a very pretty woman and a very aggreeable [sic] one."[10]

While Anne's attributes were fairly evident, what attracted her to the thirty-six-year-old surgeon who paid repeated visits to the family home that summer was less obvious. Not only was Hunter well known for his coarse language, disdain for etiquette, and casual dress, he was far more inclined to discuss curious experiments on live animals and dissections of dead bodies than lyrical poems on love and nature. Neither was John conventionally handsome, in the polished manner of his brother William, for example, whose fine and delicate features were captured in a painting circa 1765, showing him in full gray wig and blue velvet frock coat decorated with gold brocade.[11] At five two, short even for a Georgian, and with his stocky shoulders and short neck, John presented a somewhat incongruous contrast to the tall, graceful army officer's daughter. Yet when Anne's young brother Robert painted John at some point during their engagement, he revealed a debonair young man with an open and intelligent face, sensual mouth, and dreaming eyes.[12] Eschewing a wig, as he always would, his unruly hair was held in place by a black three-cornered hat, and he sported a white cravat, a handsomely embroidered damson red waistcoat, dark velvet breeches, white silk stockings, and a dark blue jacket. A world away from the self-confident man of business shown in William's portrait, John was set in a wild woodland scene, his hand resting gently on the head of a large mastiff, entirely at one with the natural elements he worshiped and looking every inch the ideal prince for a romantically inclined young woman.

Admittedly, Anne might have been slightly alarmed had she known that the great brute resting its heavy jowls on John's knee was actually half dog, half wolf. Hunter had been given the hybrid as a puppy in about 1766 by a menagerie owner known as "Wild Beast Brookes," who had bred a litter of nine puppies from a male wolf and female dog. While Hunter clearly adored his unbiddable half-breed as much as it adored him, it scared the life out of passersby. Eventually it was stoned to death by a mob who mistook it for a rabid dog.[13]

A second portrait by Robert, painted at about the same time, shows a far more practical man, however.[14] Seated at a desk with a quill in one hand, the long, tapering fingers of his other hand resting on his knee, John is writing notes, perhaps about the skull of the small mammal placed in front of him. His coppery hair is curled simply at the sides, his eyes look confidently ahead, and this time his embroidered waistcoat is covered with a brown overall to protect his day clothes from the blood and grime of the dissecting table. In the background can be seen the painted casket of an Egyptian mummy.

In fairness, Anne was not at her best when first John began paying her his attentions. He had been called to see the young woman as a patient in the summer of 1764. Her father had been bewildered by the strange symptoms his daughter was displaying—Anne complained of numbness down her left side and had developed bluish spots on her leg—and so he called in his young army friend for a second opinion. These were not the usual circumstances in which romance might be expected to flourish. With typical eighteenth-century attention to bodily functions, John noted in his casebooks that Anne had passed a long roundworm in July, while in August "[s]he was much troubled with wind in her stomach especially at night."[15] Anne was put on a course of worming medicines and purgatives, supplemented by cold baths. After nine cold dousings, she was pronounced recovered, or at least she had had her fill of the surgeon's medicine—although not of the surgeon. The intimacy of the sickroom had worked its magic and that year the pair became engaged.

Despite their divergent worlds, their contrasting interests and their different educational backgrounds, Anne and John forged a seemingly perfect partnership. In an era not noted for sexual equality, there was between them strong mutual respect. While neither ever truly engaged in the other's world, each allowed the other sufficient freedom to pursue individual interests. Even the ever-critical Foot praised the union, remarking, "To her he was directed, not only by personal attractions, but also mental endowments, which she possesses in a very eminent degree."[16] Hunter quickly became a regular visitor to the Home household and a favorite with all the family. Another of Anne's young brothers, Everard, who was an eight-year-old pupil at Westminster School when first John began to woo his sister, particularly admired the brilliant and exuberant surgeon and was soon keen to follow in his footsteps. But it would be a long engagement. Hunter's financial circumstances were still too shaky a foundation for starting a family. Equally, his passion for experiment—particularly one highly risky self-experiment he was about to begin—may have provided another reason for the seven-year betrothal.

Fresh financial disappointment came in January 1765, when Hunter failed to secure a staff job at St. George's Hospital, a position that would have set him up with a lifelong surgical career. Having subscribed to the hospital for several years, he had become a governor in 1760, not long before he sailed for Belle-Ile, and had dutifully attended board meetings since his return from the army. When one of the surgeons resigned in December 1764, John had high hopes for recouping his outlay. In the event, his former rival John Gunning, who had been assistant surgeon at the hospital now for three years, secured the post, but not before Hunter had garnered a large share of the votes.[17]

Nevertheless, his finances were growing. In June 1765, Hunter was able to buy a substantial piece of farmland at Earls Court, then a quiet village about two miles beyond the London sprawl, where he began building a sizable house. At the point of making the purchase, he was still lodging in Covent Garden, according to the deed of covenant, although by March of the following year he had moved to a smart rented town house at 31 Golden Square, in the heart of London's fashionable West End.[18] The villa and grounds Hunter began to develop at Earls Court would provide not only a welcome retreat from demanding city life but a dedicated biological research center. Here he would keep his exotic animals, breed strange hybrids, and perform his relentless experiments, free of interruption and interference from nosy neighbors.

Now that he possessed suitable premises and funds with which to finance his passion, he could indulge his obsession with experimentation to the full. In the grounds and laboratories he established at Earls Court, Hunter would pursue his myriad interests, not only investigating the entire human body but also every species of animal life. Still fascinated by the notion of his "living principle," he began in the winter of 1766 to investigate durability of life and suspended states of animation. Deciding to test the commonly held notion that fish and reptiles could regain life after being frozen, Hunter devised a series of experiments that entailed freezing an assortment of animals.[19] Wrapped up against the cold, he persuaded the physician George Fordyce, a former pupil from the Covent Garden school who had remained a close friend, to help conduct the experiments. When

another friend, a Dr. Erwin, who was a teacher in chemistry at Glasgow, arrived unexpectedly, he, too, was roped in to help.

The three experimenters began by placing two carp in a glass vessel,
which they lowered into a tub that was wrapped in woolen cloths for insu-
lation and filled with ice and snow. To their consternation, the ice around
the fish kept melting, leading Hunter to realize that animals generated
heat. The discovery would prompt him to measure standard body heat in
different life-forms in later experiments. As the carp continued to swim
freely, the trio shoveled in more snow until, by now almost frozen themselves, they were forced to give up. While the three friends repaired inside
to warm themselves before an open fire, their victims finally succumbed.
"They were frozen at last, after having exhausted the whole powers of life
in the production of heat," Hunter recorded. Now the real purpose of the
experiment could begin. Full of anticipation, the enthusiasts thawed out
the frozen fish in the hope they would regain their former vigor. Naturally
enough, the carp remained as stiff and lifeless as ever.

Undeterred, Hunter repeated the experiment on a dormouse. The
three friends settled down to watch as the little creature began to feel the
drop in temperature. "The atmosphere round the animal soon cooled,"
noted Hunter, "its breath froze as it came from the mouth; a hoar-frost
gathered on its whiskers, and on all the inside of the vessel, and the external points of the hair became covered with the same." Once again the
threesome were defeated by natural body temperature. Although the
mouse's feet froze, it survived the extreme cold, living to preen its whiskers
once more. A second dormouse was less lucky. This time working alone,
Hunter soaked its fur in water, better to conduct the cold, and it was
quickly overcome. "The animal dying, soon became stiff," he noted, "and
upon being thawed, was found quite dead." With a miraculous resurrection
now seeming increasingly unlikely, Hunter pressed on, taking a toad and
then a snail as his deep-freeze victims.

Finally, he had to admit defeat. The results were not just a blow for
folklore—never one of Hunter's big considerations—but also to his naïve
ideas for making a fortune, as he later confided to his pupils:

Till this time, I had imagined that it might be possible to prolong life to
any period by freezing a person in the frigid zone, as I thought all action

and waste would cease until the body was thawed. I thought that if a man would give up the last ten years of his life to this kind of alternative oblivion and action, it might be prolonged to a thousand years; and by getting himself thawed every hundred years, he might learn what had happened during his frozen condition. Like other schemers, I thought I should make my fortune by it; but this experiment undeceived me.[20]

Successive freezing and thawing were plainly not the key to immortality, but Hunter was loath to surrender his fascination with reviving the dead.

Regeneration remained an interest in February that year, when the surgeon became his own patient. One morning at about four o'clock, Hunter snapped his Achilles tendon while jumping up and down on his toes. Never one to forgo the chance to investigate a new condition, he recorded his self-treatment under the heading "Mr Hunter's Case."[21] He noted he had been "jumping and lighting upon my toes without allowing my heels to come to the ground," when his tendon suddenly snapped. "I stood still, without being able to make another spring; and the sensation it gave me was as if something had struck the calf of my leg; and that the noise was the body which had struck me, falling on the floor, and I looked down to see what it was, but saw nothing."

The accident provided a perfect case on which Hunter could practice his natural approach to treatment. Confident that nature would mend the torn tendon without human intervention, he declined any surgical aid or physicians' remedies, and after a brief rest with his leg bound, he endeavored to walk on the injured foot. For this purpose, he adapted an old shoe by inserting pieces of leather to raise the heel, and as his movement gradually returned, he reduced the height of the heel accordingly. Just as he expected, the tendon soon mended and he could walk perfectly normally. The lesson strengthened his conviction in nature's healing powers, and, in a period when fascination with health commonly bred hypochondria, he would frequently exhort ailing patients to abandon their medicines and their beds in favor of exercise and fresh air.

In one instance, while staying with friends at a country house, he discovered the mistress of the family had been confined to a wheelchair since fracturing her kneecap four years earlier. Hunter prescribed a series of intensive daily exercises—essentially early physiotherapy—in which the patient sat on the end of her dining table and attempted to move her toes.

After one month, she could wriggle her toes slightly, and after several more months of regular exercises, she was finally able to walk again.[22] Regardless of the fact that his noninterventionist approach frequently meant he went without a fee, many more sickly Georgians would be brought up sharp by Hunter's commonsense doctrine.

The tendon injury also afforded Hunter new inspiration for research. He experimented on several dogs by cutting their tendons, allowing them to heal naturally, then killing the hounds at various stages to examine the regrowth.[23] Already highly respected for his knowledge of the human body, Hunter was becoming widely esteemed for his understanding of animal anatomy, too. In the early years after he had resettled in London, he began to compile the first catalog detailing the specimens in his growing natural history collection.[24] Most of the pickled or dried preparations he recorded were parts belonging to animals, either dissected on his army campaigns— like the two-tailed lizards from Belle-Ile—or bought from circuses, menageries, and auctions in England. They ranged from domestic creatures such as dogs, cats, and donkeys, through native wild animals such as badgers, hedgehogs, and deer, to exotic beasts from around the world, including leopards and monkeys. Hunter had already obtained the liver, kidney, and tongue from a lion in the Tower of London menagerie—possibly an elderly lion named Pompey, which died there in 1758. He had also acquired the skeleton of an orca, or killer whale, which had beached itself at the mouth of the Thames in 1759. Hunter had the twenty-four-foot-long carcass towed to Westminster by barge and then dissected its huge hulk on the spot with the help of some of his pupils.[25] In addition, he obtained the body of a crocodile, the third he had dissected, which died in a London show during the winter of 1764–1765. "It was at the time of its death perhaps the largest ever seen in this country," he wrote, "having grown, to my own knowledge, above three feet in length, and was about five feet long when it died."[26] With adventurers relentlessly exploring uncharted lands overseas, new species were regularly turning up on British shores. Increasingly, whenever a hitherto-unknown animal came to the notice of amateur naturalists, it was taken to Hunter for examination.

One such creature became the subject of Hunter's first paper presented to the prestigious Royal Society.[27] News of the slimy olive-colored aquatic animal that looked like a cross between a newt and a large eel caused a stir when it first came to the notice of the scientific community in the 1760s. Possessing both a pair of legs and feathery gills at the front, and no hind legs at all, the creature of roughly two feet in length had been discovered in a swamp in the southern part of North America; it was known by locals as the "mud iguana." A Scottish physician and amateur naturalist, Alexander Garden, who lived in South Carolina, had captured several of the beasts and shipped their preserved bodies to a friend, John Ellis, who was one of Britain's most eminent naturalists. Aware the strange animal could be a newly discovered species, Ellis sent a specimen to another friend, the renowned Swedish naturalist Carl Linnaeus. Having established the modern categorization system of plants and animals, in which he described more than four thousand known animal species, Linnaeus was regarded as the chief authority on naming species. The professor wrote back excitedly to Ellis in 1765 from Uppsala, where he held the chair of botany, confirming that the animal represented not only a new species but also a new genus. He promptly named it *Siren* (Latin for "mermaid") *lacertina* ("little lizard"). Assuming that the specimen Ellis had sent was the larva of an animal that would eventually develop four legs—in fact, the *Siren lacertina,* or greater siren, continues in a permanent larval state, retaining just two legs—Linnaeus noted, "There is no creature that ever I saw that I long so much to be convinced of the truth as to what this will certainly turn out to be." Ellis needed no further encouragement. A stalwart of the Royal Society, he immediately set about describing the external appearance of the new animal for fellow members. But Ellis had no anatomy skills; he needed an expert capable of investigating and describing the internal structure of the animal. For this, he turned to John Hunter.

The greater siren was no stranger to Hunter. He had seen a specimen before, as well as its two sister species, the dwarf siren and the lesser siren, when in 1758 he purchased a job lot of curiosities originating in South Carolina. At Ellis's behest, he now dissected the new specimens to produce a full description of the creature. Bent over the peculiar animal with his knife, Hunter discovered that along with the siren's single pair of legs, it possessed both external gills like a fish and internal lungs like an amphibian. He sent his report to Ellis, who appended it to his own. Together, they were read to the Royal Society on June 5, 1766.

In his first ever communication to the esteemed body of scientists, Hunter boldly suggested that the new creature represented a missing link in the chain of animals between fish and amphibians. "This tribe of animals are widely different from all hitherto known," he proclaimed. "They are compounded of two grand divisions of the Animal Kingdom." He added, "They hold with respect to respiration a middle rank between Fish, which breath water and those immediately above them who breath air, viz, those call'd amphibia, placed in this respect between the two, filling up the scale." He also noted that the siren's tiny eyes were well adapted to its environment, remarking, "This smallness of the eye best suits an animal that lives so much in the mud."

The "Great Chain of Being" Hunter invoked was not a novel idea in mid-eighteenth-century scientific circles. Many naturalists subscribed to the idea of a ladder or scale ascending from the most primitive creatures to the most complex—humans. Rather than an early indication of evolutionary theory, the chain represented a fixed system, with every species having been designed by God in an immutable state. But for Hunter, the missing link he had described would come to mean something more than a mere rung in a ladder of creation designed by God. His deliberations set him on a trail that would eventually force him to clash even with his newfound friends in the society.

For now, though, he was warmly welcomed into the fold. Eight months later, on February 5, 1767, Hunter was elected a fellow of the Royal Society, recognized "as a person well skilled in Natural History and Anatomy."[28] Heading the list of the seven who proposed his acceptance was John Ellis, while others included Daniel Solander, the Swedish botanist and disciple of Linnaeus who worked in the new British Museum, and Richard Warren, who was George III's physician. Hunter had become a fellow almost three months before his older brother, despite William's having contributed a paper as far back as 1743.[29] Within a few years, John Hunter would become one of the central figures within the Royal Society, not only the society's favorite anatomist but also a linchpin of scientific debate. In the meantime, he preserved the siren's heart in a jar.[30]

CHAPTER 9

The Surgeon's Penis

ᘖᘎᘓ

London

May 1767

John Hunter had all before him. He was engaged to marry an accomplished society hostess, he was established as a rising surgeon with a growing private practice, he had just been accepted into the elite company of the Royal Society, and, at thirty-nine, he was an energetic, intelligent, and healthy man in the prime of life. It seemed nothing could go wrong. And then one Friday in May 1767, he embarked on an astonishing and foolhardy experiment.

Hunter had determined to solve one of the foremost questions taxing medical practitioners of the day. Yet this riddle could not be examined in his usual way, using animals as subjects; only an experiment on a human being could provide the answers he sought. Accordingly, he took up his knife, dipped the blade into a festering venereal sore, and deliberately jabbed the tip first into the end and then into the foreskin of a man's penis. He recorded the entire process: "Two punctures were made on the penis with a lancet dipped in venereal matter from a gonorrhoea; one puncture was on the glans, the other on the prepuce. This was on a Friday; on the Sunday following there was a teasing itching in those parts, which lasted till the Tuesday following."[1]

Hunter never revealed in any of his written works the name of the human guinea pig whom he subjected to his reckless experiment. The man's identity would perplex Hunter's disciples, his enemies, and numerous scholars for the next two centuries and beyond. Conceivably, he could have used a volunteer plucked from his local neighborhood, someone he persuaded to submit to his knife in return for a generous fee, just as the local paupers had given up their teeth. He could perhaps have hoodwinked a trusting but unwitting patient, for Hunter would certainly experiment on patients without their express permission in other cases. But all evidence points to the conclusion that the victim of the bizarre experiment was Hunter himself.[2]

Certainly, Hunter had good reason to focus his investigative skills on sexually transmitted diseases in the middle of the eighteenth century. Like any busy surgeon of the time, he quickly found that venereal complaints comprised a large proportion of the ailments he treated on his daily rounds. Syphilis, gonorrhea, and a host of other venereal infections were rampant in Georgian London, largely as a consequence of relatively liberal attitudes toward sexual freedom, along with the widespread and broadly condoned use of prostitutes. Sex was big business in eighteenth-century London, little different from the thriving trades in tea, coffee, and spices. As a commodity, it was widely available, whether in dark alleys leading off the Strand, in the seedy brothels, "bawdy houses," and bagnios around Covent Garden, or by arrangement with any number of women prepared to offer sexual favors in return for a comfortable lifestyle. Customers eager to pay for illicit sexual relations spanned all walks of life, from the soldiers and sailors swarming the streets on leave to the dukes and dandies who gratified their physical desires after an evening's gambling in West End clubs. Prices varied widely. According to James Boswell, a connoisseur of London prostitutes, the cost ranged from the "splendid Madam at fifty guineas a night, down to the civil nymph with white-thread stockings who tramps along the Strand and will resign her engaging person to your honour for a pint of wine and a shilling."[3] And just as potential buyers of spices or tea would peruse catalogs describing the latest merchandise from overseas, so prospective clients for sex could flip through a handy pocket guide to select their preferred partner.

Published annually starting in 1771, *Harris's List of Covent-Garden Ladies, or Man of Pleasure's Kalendar,* provided vivid written particulars on London's many prostitutes.[4] Potential customers were advised that Miss Brown of 7 Berwick Street, Soho, was "constantly well dressed, and lives in very genteel lodgings at present." Just twenty-three, she could already boast eight years' experience. Alternatively, older clients were recommended to Mrs. Hamblin, who lived, appropriately enough, at 1 Naked-boy-court, near the Strand. Now nearly fifty-six and forced to enhance her charms with lavish makeup, she could offer her customers the benefit of thirty years' experience. At the same time, the directory provided essential details on prostitutes' well-being. Formerly satisfied customers were cautioned to steer clear of Miss Young, who had "very lately had the folly and wickedness to leave a certain hospital, before the cure for a certain distemper which she had was completed, and has thrown her contaminated carcase on the town again." Military gentlemen were likewise warned to avoid Matilda Johnson, since "it is thought by some experienced officers, that her citadel is in danger, on account of a quantity of *fiery combustible matter* which is lodged in the *covered way.*"

One well-traveled tourist who quickly acquainted himself with the brothels of Covent Garden when visiting London in 1763 was the renowned Italian lover Giacomo Casanova.[5] Infatuated by the charms of one unusually elusive young prostitute, who lured him into parting with large sums of money without once allowing him carnal pleasures, the libertine almost leapt into the Thames in despair. Although the object of his desires never relented, there were many more women eager to satisfy Casanova's boundless lust. By the time he returned to the Continent the following year, he took with him the sure symptoms of gonorrhea as a souvenir of his London sojourn. It was one of many occasions on which he required treatment for venereal disease.

Boswell was another regular frequenter of London's brothels.[6] His first sexual experience, at the age of nineteen, took place shortly after his arrival in London in 1760, with a young prostitute who plied her trade at the Blue Perriwig, near the Strand. His first encounter with "Signor Gonorrhoea," as he termed his recurring visitor, followed shortly after. Boswell duly visited a surgeon in Pall Mall to undergo the necessary treatment—a course of unspecified pills, a strict diet, and the obligatory bloodletting.

His experience did not prove a salutary lesson. When he returned to London in 1762, following a brief spell back home in Scotland, where he endured a second episode of venereal disease, he began to pay court to a young married actress, code-named "Louisa" in his diary. After a cursory few weeks' wining and dining, Boswell persuaded her to spend a night with him in January 1763 in a room above a London tavern. Describing their night of passion—no doubt with an eye on his future reputation—he boasted in his journal, "A more voluptuous night I never enjoyed. Five times was I lost in supreme rapture."[7] Louisa, for her part, declared his performance "extraordinary"—or so Boswell claimed.

But the by-now-familiar symptoms that soon ensued were by no means extraordinary. Just a week after his night of bliss, Boswell dolefully admitted in his journal, "These damned twinges, that scalding heat, and that deep-tinged loathsome matter are the strongest proofs of an infection."[8] After confirming the diagnosis with his surgeon, he confined himself to his lodgings in Downing Street to fulfill his course of treatment and escape the prying questions of his associates. Five weeks later, he was sufficiently recovered to recommence his social life, as well as his sexual adventures with prostitutes.

Chastened for some time at least, Boswell donned a rudimentary condom when next he engaged in sexual activities, this time picking up a teenage prostitute in St. James's Park—a notorious haunt for casual sex. "For the first time did I engage in armour," he noted in his journal on March 25, adding that he found this a "dull satisfaction."[9] This was scarcely surprising, since condoms of the time, though rarely used, were made from sheep's or pig's gut, secured with silk ribbon. Popularly held to have been invented by a certain Dr. Condom, a physician to Charles II, they were reusable but had to be kept in water to maintain their suppleness. Boswell dutifully washed and reused his condom—thinking of his own health, of course, rather than his partners' well-being—on several more occasions. At other times, he threw caution, and condom, to the wind, and further attacks of gonorrhea inevitably ensued. Over a sexual career that would last more than thirty years, Boswell visited brothels throughout Europe, enjoyed numerous affairs, and fathered at least two illegitimate children before his marriage in 1769. As a result, he suffered at least nineteen episodes of venereal disease, seemingly always gonorrhea, which he recorded woefully in his diaries. His death in 1795 may even have been caused or accelerated by the various complications of his venereal complaints, exacerbated by excessive drinking.

As the frequency and severity of Boswell's attacks increased, he consulted many of London's top surgeons and physicians. He may easily, since his wife and son were patients, have been seen by John Hunter, for the treatment of venereal diseases formed a major component of Hunter's work from his earliest days as a surgeon. His first recorded patient, the chimney sweep at St. Bartholomew's, had been a victim of a stricture caused by gon orrhea, and Hunter had since had plenty more opportunities to study sexual infections—on his ward rounds, in his military campaigns, and while examining the dead bodies that turned up on his dissecting bench. Inevitably, he preserved some of the relevant parts. He pickled in spirits part of the penis of "a man who died clap'd" and dried several bones and a skull bearing the long-term effects of syphilis.[10]

But now that Hunter was seriously beginning to build up his reputation and his clientele, the flow of victims of venereal complaints crossing the threshold of his new home in Golden Square grew steadily greater. While Hunter would have offered house visits for his wealthier patients, many may well have preferred to visit his surgery incognito, along with his clients of lesser means, rather than reveal their condition to family and servants.

Venereal disease was no respecter of class. Among the many patients Hunter treated for sexual infections were prostitutes and their married clients, servants and their masters, high-ranking army officers and rank-and-file soldiers, as well as a vicar, a surgeon, and a Prussian count. It was not only Mr. Anderson but Mr. Anderson's servant who came under his care for venereal infections; both Lord and Lady "L——" sought out his advice. Often Hunter discreetly concealed or coded his patients' names in his notes, but whatever their identity or background, their intimate circumstances and unfortunate symptoms were dutifully recorded, along with some of the more elaborate stories they invented to cover their indiscretions.[11]

While society generally turned a blind eye to prostitution and its customers, some of Hunter's more embarrassed patients concocted far-fetched tales in an effort to disguise the true origin of their complaint. One "Gentleman" claimed he had been sauntering innocently along the Haymarket—another popular prostitutes' haunt—when he was accosted by a woman, who invited him to join her for a drink. "He told her to be gone for a poxed Bitch," Hunter noted. "Soon after, he went aside to make water; she came behind him and took him by the Penis, saying 'If I am poxed, you shall be so too.' After quitting him, he wiped the Glans and walked on, without further notice. Next morning, he found a disagreeable heat round the Glans, especially on the orifice of the Urethra, which alarmed him (he being a married man)." Hunter recorded the sorry tale without comment. He rarely took a moralistic standpoint. But he was somewhat incredulous at the claim made by one patient, who said he had abstained from sexual relations for almost two months before an attack of gonorrhea. "How came this Clap?" Hunter queried. "He gave his Word and Honour he had not lain with a Woman since; which is more than seven weeks."

Even men of the cloth were prone to stray. Reporting the case of the Reverend Mr. Stevens, Hunter noted, "He came to London, and one evening had connection with a Woman of the Town. Two or three days after, in the morning he observed a spot on his shirt, which alarmed him much, and he immediately applied to me for a Preventative." To the errant cleric's dismay, Hunter informed him that no such medicine existed, but he advised him to steep his penis in warm water and to hold it over steam made by hot water and brandy, which, if nothing else, would provide some comforting fumes.

Hunter was well aware that venereal diseases were transmitted by sexual contact. Sometimes, therefore, he advised patients to refrain from sexual relations with a spouse while symptoms persisted. At other times, he would give the all clear for marital sex to be resumed. He realized, too, that venereal infections could be passed from a mother to her unborn child. Many infants born with syphilis were treated at the Lock Hospital, which specialized in venereal complaints after it opened in 1746. In some instances, infections were even transmitted to a child from its mother or a nursemaid while breast-feeding. In one family treated by Hunter, the wife contracted syphilis from her husband and passed it on to her newborn twins, who died soon after birth, as well as to the two-year-old daughter she was breast-feeding.[12]

Yet while methods of transmission were well accepted, understanding of venereal diseases largely ended there. Along with most practitioners of the day, Hunter believed that gonorrhea and its more serious cousin, syphilis, were simply different forms of the same disease and stemmed from the same cause. The classic symptoms of gonorrhea, commonly known as "the clap," were widely regarded as the local manifestation of this one venereal infection, affecting only the genital parts. That these symptoms frequently disappeared of their own accord, without recourse to medical aid, helped confirm this view. When the less common but far more disastrous symptoms of syphilis, known as "the pox," were observed in patients, these were attributed to the venereal "poison" having circulated around the body, or having become *constitutional*, to use Hunter's term.

The fact that such symptoms usually spread from the classic hard chancre on the genitals to a general rash, sore throat, swollen lymph glands (or buboes), and, in some cases, more serious complications again lent credence to the one-disease theory. So while Boswell somewhat miraculously managed to confine his many bouts of venereal disease to gonorrhea, he expressed due anxiety that this infection might "spread" to the rest of his body. "I thought of applying to a quack who would cure me quickly and cheaply," he remarked when another gonorrhea attack occurred. "But then the horrors of being imperfectly cured and having the distemper thrown into my blood terrified me exceedingly."[13]

There was similar confusion over the origins of venereal disease, although this was rather more understandable, since controversy continues today.[14] Symptoms of gonorrhea had been recorded since ancient times, but it was only in 1495 that the first definite signs of syphilis were reported. It was then regarded as a new disease. Since the first cases were noted after French soldiers laid siege to Naples, the venereal epidemic that quickly took hold across Europe was initially termed "the Neapolitan sickness." It soon went through a long list of alternative names, according to the nationality of its sufferers and that of their closest enemies. Accordingly, it was called "the French sickness" by the Italians, "the Spanish disease" by the Dutch, "the Polish disease" by the Russians, and "the Russian disease" by the Turks, while in Tahiti, where natives were introduced to syphilis by visiting eighteenth-century explorers, it was quickly dubbed "the British disease." For their part, the first British victims initially called the horrifying infection that reached English shores in about 1497 the "Bordeaux sickness," since that was where it seemed to emanate from, but soon they settled on the more general "French pox," as it became widely known.

As the first European cases seemingly arose soon after Christopher Columbus had returned from the New World, it was generally reckoned that the new scourge had been imported from there. And this may well be true: In all probability, the *Treponema pallidum* bacterium that causes syphilis was brought back by the first European explorers, in a decidedly uneven exchange that resulted in the devastation of Native American tribes by smallpox, measles, influenza, and typhus. However, other theories include the possibilities that the infection had long been present in Europe but had not been distinguished from leprosy, or that syphilis developed in both continents from related diseases.

Although the term *syphilis* was coined to describe the horrific symptoms of the highly virulent disease in the sixteenth century, the name would not fully catch on until the end of the eighteenth century—unlike the infection itself, which spread unchecked through European civilization. The intensity of the pox, or the "Great Pox," as it was also known—to distinguish it from smallpox—had somewhat abated by the middle of the eighteenth century, but the infection had become endemic and its ramifications could be severe. Though early symptoms of syphilis usually disappear of their own accord within a year or two, the infection can remain latent for decades. In some victims, it reemerges, and in its final or tertiary stages, can destroy bones, soft tissue, the heart, the brain, the eyes—indeed, almost any part of the body. Certainly by the mid-eighteenth century venereal disease in all its guises had become sufficiently widespread to provide unending employment for surgeons, physicians, and quacks claiming to offer miracle cures. One publication giving career advice for young Georgian men in 1747 remarked that three out of four surgeons in London were reliant for their income on treating venereal infections.[15]

If the origins of sexual infections remained obscure, even more confusion arose over the most appropriate treatment. Both orthodox practitioners and quack healers proffered an array of pills, potions, ointments, and other remedies claimed to cure symptoms of the clap, although in reality most bouts of gonorrhea resolved themselves after a few weeks. This was plainly good news for serial sufferers such as Boswell and Casanova, but it did little to dispel the muddle. Every charlatan and overambitious surgeon could claim miraculous success for his singular approach. And it was little comfort for those who went on to develop the nastier complications untreated gonorrhea could lead to, including inflammation of the prostate or testicles.

There was greater unanimity over treatment of the pox. While the earliest known sufferers of syphilis were urged to press a cockerel "plucked and flayed alive" to the affected parts, or, failing that, a "live frog cut in two," treatment had become a little more straightforward by the eighteenth century, if no less unpleasant. Mercury was the universal remedy for classic signs of syphilis, probably on account of its earlier use in treating the similar symptoms of leprosy, and it would remain the standard treatment into the early twentieth century. As with most Georgian remedies, it had little, if any, effect, although it sometimes gave the appearance of success.

The tendency of early symptoms of syphilis—the primary and secondary stages—to disappear without treatment after one or two years tended to grant mercury a seal of approval. After these early stages, however, the untreated syphilis bacterium could lie dormant for anywhere up to thirty years, and in roughly two-thirds of the victims, it would never reappear. For those whose symptoms did recur, the ensuing damage could wreak severe and sometimes fatal destruction on internal organs.

According to the various surgeons' whims, mercury was administered as an ointment applied directly to the infected parts, by an injection or on a bougie—a cylindrical plug loaded with the mixture—into the penis or vagina, or orally as pills and potions. While lay healers pretended to offer patent remedies of their own, these secret recipes, when analyzed, often consisted mainly of mercury, too. Unfortunately, the painful, unpleasant, and noxious side effects of mercury treatment were often as nasty as the condition itself. Mercury poisoning caused mouth ulcers, sore gums, loose teeth, and copious production of black saliva, sometimes several pints a day, as well as leaving a telltale metallic taint on the breath. It was hardly surprising that many victims took themselves to the "salivating" wards of the voluntary hospitals, where they could at least suffer their treatment in relative privacy, safe from the prying eyes and inquisitive noses of friends and neighbors.

With confusion obscuring the origins, nature, and treatment of venereal disease, the field was ripe for investigation. And so Hunter made venereal diseases the next subject of his relentless research. It would be 1786 before he felt sufficiently confident to publish his investigations in his major work, A Treatise on the Venereal Disease, but the experiments and observations that formed its basis were well under way by 1767. He already had firm views on the progress of sexual infections and the best mode of treatment. Drawing on his growing experience with the male patients who sought out his advice from the 1760s onward, he recorded the unwelcome signs of itching, discharge, swelling, pain, and unwanted erections his gentlemen clients described. He was aware, too, that women could often contract gonorrhea but show few, if any, symptoms, which caused no little problem for those male patients seeking assurance that a prostitute or potential mistress was clear of the disease.[16] In such instances, the only acceptable evidence that a woman was infected or "clean" was the word of a gentleman—a woman's word counted for little in such circumstances—as

Hunter made clear: "In such a case, the only thing we can depend upon is the testimony of those whom we look upon as men of veracity."[17]

As to treatment of sexual diseases, Hunter was outspoken about the ineffectuality of the majority of medications claiming to cure gonorrhea. His growing experience taught him that mercury was useless as a treatment, but he likewise poured scorn on all other remedies for the clap, whether from orthodox surgeons or untrained quacks. While syphilis required certain skills to treat, he asserted, gonorrhea could be cured by "the most ignorant."[18] The reason, he explained, was obvious: "Gonorrhoea cures itself, whilst the other forms of the disease require the assistance of art." He even questioned the benefit of medication at all, arguing, "I am inclined to believe it is very seldom of any kind of use, perhaps not once in ten cases: but even this would be of some consequence, if we could distinguish the cases where it is of service from those where it is not."[19] With this motive in mind, Hunter performed an ingenious test, giving some of his unsuspecting patients pills made of nothing more effectual than bread. He recorded, "The patients always got well; but some of them, I believe, not so soon as they would have done, had the artificial methods of cure been employed." The experiment was one of the earliest examples of a controlled trial testing a placebo against an established medication. (He would later use the same deception on his wife, tricking her into believing she was enjoying the therapeutic benefits of the best spa water by sticking labels from the Bath pump room onto bottles containing ordinary water pumped from the Thames.[20] Plainly, the veracity of the surgeon was sometimes no more to be trusted than that of his gentlemen patients.) Hunter's bread-pills test provided solid proof that leaving gonorrhea untreated was just as effective, or almost as effective, as offering the best that London's most eminent practitioners could prescribe. Not surprisingly, the conclusion did little to boost Hunter's popularity among his more mercenary colleagues when he finally published his treatise.

Syphilis was a trickier puzzle. Hunter described the symptoms of *lues venerea* in its various phases in colorful detail. Mercury offered patients the best chance of success, Hunter averred, although he was under no illusions that it provided a complete cure, being aware that the disease could lurk in the body unnoticed for years before reemerging to wreak more damage. But it was the enduring question of whether gonorrhea and syphilis were really the same disease, one of the biggest medical enigmas of the eighteenth century, that truly concentrated his investigative energies. This was to form the central theme of his treatise and would prove the most controversial aspect of his research.

Hunter already subscribed to the thesis that gonorrhea and syphilis were caused by the same agent or "poison." He based this on his view that no two diseases could occupy one body at the same time. With diagnosis still very much in its infancy—most infectious diseases were denoted simply as "fever," for example—this was an easy mistake to make. Hunter was well aware that the clap and the pox had quite distinct symptoms, and that one responded to mercury but the other did not, yet he was still certain that the two were simply different manifestations of the same infection. In common with most fellow practitioners, he believed gonorrhea represented the local stage of the disease, while symptoms of *lues venerea* emerged when the disease spread throughout the body.[21] All his experience in consultations pointed to this sadly mistaken conclusion. It was quite apparent that both conditions were contracted in the same way. More significantly, many of Hunter's patients were actually suffering from both infections at the same time; it was easy to confuse the two. Captain Duncan, for example, who complained of a "pain and running," followed by the appearance of a chancre, evidently had both gonorrhea and syphilis.[22] But hypothesizing was one thing. In Hunter's regime, there was only one way to solve a mystery—by performing an experiment.

The plan was simple, and it should have been straightforward. In the spring of 1767, he resolved to inoculate a person with gonorrhea and monitor the progress of the disease for signs of syphilis. If, as he expected, the signs of gonorrhea were followed by symptoms of syphilis, then he could prove that the two were one disease. If, as should have happened, no signs of syphilis emerged, plainly gonorrhea was a separate disease. All Hunter needed was a compliant, willing, and reliable human volunteer. Finding a donor with clear signs of gonorrhea to provide the venereal matter he required was easy: there was no shortage of patients knocking on his door with the most virulent evidence of the clap. But obtaining a subject for his experiment was a more knotty problem. He needed someone he could be sure had previously been clear of venereal infection, someone who would be willing to undergo regular examination and treatment, someone he

could observe on an almost daily basis over the long course of his trial.
The obvious subject for the experiment on that fateful Friday in May was
himself.

Within days of beginning the experiment, Hunter felt the first effects,
noting a "teasing itching" in the penis. After four days, the signs were all
too clear. "Upon the Tuesday morning the parts of the prepuce where the
puncture had been made were redder, thickened, and had formed a speck;
by the Tuesday following the speck had increased, and discharged some
matter, and there seemed to be a little pouting of the lips of the urethra,
also a sensation in it in making water, so that a discharge was expected from
it."[23] Classic symptoms of gonorrhea had appeared, exactly as Hunter had
expected. Examining the inflamed parts on a daily basis, he watched for the
telltale signs of *lues venerea*. He was not disappointed. Within ten days, to
his evident satisfaction, an ulcer or chancre appeared on the foreskin and
soon other signs of syphilis emerged. A gland in the right groin swelled,
forming the classic bubo, then two months later, "a little sharp pricking
pain was felt in one of the tonsils." On examination, this proved to be an
ulcer, another common occurrence in the primary stage of syphilis. Seven
months after the start of the experiment, "copper-coloured blotches" broke
out on the skin, clearly indicating that the syphilis had entered its secondary stage. Applying mercury to the festering sores on his genitals and tonsils, Hunter took care to use only sufficient quantities as he believed would
control, not vanquish, the disease in order to carry on with his experiment,
although in reality the symptoms probably disappeared of their own accord.
For three long years, he continued, anointing his sores with mercury and
then letting them reappear, four times on the tonsils and three times on the
skin, until at last he decided to administer enough of the poisonous ointment to eradicate the symptoms—or, more likely, the disease had simply
lapsed into its latent and noninfectious phase.

The experiment, as far as Hunter was concerned, had been a resound-
ing success. It proved, to his satisfaction at least, that gonorrhea developed
into *lues venerea*. In reality, it was a complete disaster. The experiment had
been doomed from the outset, since Hunter had plainly used infected matter containing both syphilis and gonorrhea bacteria. The person from whom
he had taken the venereal pus had evidently, like so many of Hunter's patients, been a victim of both diseases. The results of the fated trial would
set back medical progress in terms of the understanding of sexual diseases

for half a century. It would be 1838 before the American-born French-practicing physician Philippe Ricord, after conducting his own series of inoculation experiments on a remarkable 2,500 unwitting patients, categorically established that syphilis and gonorrhea were two distinct infections.[24]

Hunter's error had been easy enough to make. Judged on the limited evidence of his flawed single-subject experiment, there seemed no doubt that gonorrhea produced the characteristic chancres of syphilis, although clearly he should have based his conclusions on a much larger sample. When he published the results in his treatise, he was therefore adamant that "a gonorrhoea will produce chancres." He added, "It proves many things, and opens a field for further conjectures." In fact, it proved nothing, though the field for further conjectures was wide open.

The venereal disease experiment was not the only one that Hunter performed on himself, though it was certainly the most controversial. In one instance, he fed himself with madder, a root used to produce a crimson dye, in order to see whether it turned his urine red.[25] To his satisfaction, it did. Another time, he drank a potentially lethal poison, laurel water, to discern its effect.[26] And he was not alone in making such sacrifices in the pursuit of progress. Many scientists through the centuries have resorted to using themselves as human guinea pigs. Indeed, the Scottish surgeon Benjamin Bell, a pupil of Hunter's, eventually concluded that syphilis and gonorrhea were separate diseases—although his conclusion in 1793 was never fully broadcast—only after experiments by medical students at Edinburgh University. In these reckless trials, students almost exactly, but rather more successfully, emulated Hunter's inoculation experiment by producing classic symptoms of syphilis from pus taken from a chancre.[27]

By the time Hunter's treatise did appear, it was eagerly anticipated. According to Joseph Adams, one of Hunter's later pupils, "It was no sooner known that Mr Hunter was engaged on such a subject than the expectation of the medical world was raised to the highest pitch."[28] Still painfully aware of his inadequacies when it came to writing, Hunter met with a committee of three friends, the physicians Gilbert Blane, George Fordyce, and David Pitcairn, who checked his script "to render the language intelligible" before Hunter committed himself to print.[29] A fourth adviser, George Baker, who was one of George III's physicians and another of Hunter's medical friends, may well have helped, too, for Hunter dedicated the work to him.

The treatise would have a profound impact on the understanding and treatment of venereal disease for many decades. As far as most surgeons were concerned, the debate over the duality or unicity of syphilis and gonorrhea was at an end. Hunter had proved to their satisfaction as much as his own that both emanated from the same bacterial agent. Only in Edinburgh did the arguments rumble on, prompting Bell to prove the opposite conclusion in 1793, though this view was only widely accepted after Ricord's extensive experiments. Such was Hunter's influence that the use of mercury for both gonorrhea and syphilis diminished substantially.[30] At the same time, his detailed descriptions of the symptoms of venereal diseases increased general awareness. Certainly they had a profound effect on James Boswell, who sank into a deep depression for several weeks after reading Hunter's work in April 1786, although this did not last long enough to prevent his contracting gonorrhea twice more.[31]

Almost as controversial as Hunter's conclusions on the nature of venereal infections was his broad-minded approval of masturbation. Ever since the Swiss physician Samuel Tissot had published his essay *Onanism, or A Treatise Upon the Disorders Produced by Masturbation,* translated into English in 1766, masturbation had been condemned as the root cause of all manner of ailments, not least impotence. Hunter dismissed this nonsense preemptorily. To his mind, impotence was simply too rare a complaint to originate "from a practice so general," while those he had treated for the problem did not seem to have resorted to masturbation "more than usual."[32] He went on to say, "But this I can say with certainty, that many of those who are affected with the complaints in question are miserable from this idea; and it is some consolation for them to know that it is possible it may arise from other causes." He declared, "I am clear in my own mind that the books on this subject have done more harm than good." Nineteenth-century editors would feel compelled to add an apologetic footnote to his comments, reaffirming the view that masturbation was indeed harmful.

Hunter also proffered eminently sensible sexual advice. He advised one patient who was experiencing difficulties performing with his paramour that he should sleep beside the woman for six nights while doing all in his power to resist sexual intercourse—a remedy commonly suggested by modern sex therapists. By the end of the week, the would-be lover had fully recovered his virility, "for instead of going to bed with the fear of inability, he went with fears that he should be possessed with too much desire."[33]

Hunter's liberal and outspoken views on sex and sexual diseases would bring him vehement enemies. Foremost among them was the young surgeon Jessé Foot, who dedicated most of his life to darkening Hunter's reputation, seemingly on account of a chance remark Hunter made about a bougie Foot had invented to treat strictures caused by gonorrhea.[34] Since his impressionable youth as a surgical apprentice, Foot had climbed the slippery medical-career ladder to gain a post at the Middlesex Hospital and build a fledgling private practice. Having worked along the way in the West Indies and Saint Petersburg, Foot had returned to London determined to make his mark, especially in the lucrative field of venereal complaints. But travel had not broadened Foot's horizons: He remained firmly wedded to the dogma of traditional teachings, revered the classics, and spent his evenings poring over the works of past masters. A full sixteen years Hunter's junior, he plainly aspired to knock the controversial surgeon off his pedestal.

No sooner did Hunter's treatise on venereal disease hit the streets than Foot responded with a venomous counterattack, ridiculing Hunter's comments in a three-part diatribe published the same year. Some of his criticisms were undoubtedly shrewd. Foot accused Hunter of generalizing from too few circumstances, failing to study his predecessors and contemporaries, and writing in obscure and long-winded language, although he did not dispute Hunter's conclusion that syphilis and gonorrhea were one disease. But, preening and arrogant, he consistently peppered his remarks with snobbish, petty, and vicious cynicism. Facetiously referring to Hunter as "the Professor," he sneered, "I have before me many pages so loaded with rubbish, so many useless sets of distinctions, sections so narrative and inapplicable stand in my way, as Hercules himself would turn from." He confessed his "pleasure" at attacking his supposed rival and pronounced himself Hunter's self-appointed "voluntary watchman."[35]

The fact that Foot felt impelled to expend such immense time and energy attempting to undermine Hunter's considerable reputation says more about Hunter than about Foot. Few people were indifferent to John Hunter. As his iconoclastic views gathered notoriety, he drew around him a loosely bound coterie of devoted friends, like-minded experimenters, and fascinated up-and-coming young medical men who were spellbound by the charismatic and innovative scientist. To be a part of this dynamic circle— challenging accepted dogma, debating pressing topics of the hour, performing astonishing experiments—was to live in the full, brilliant glare of the eighteenth-century Enlightenment. For those privileged enough to be admitted to the intoxicating atmosphere of this inner sanctum, there was a constant flow of ideas to stimulate the imagination, novel theories to shock contemporaries, startling discoveries with which to astound friends and family. This was life on the edge—daring, thrilling, sometimes frightening, but always exciting. To be outside looking in could seem, in contrast, a dark, cold, and lonely place. And Jessé Foot was decidedly on the outside. He had worked alongside the famed surgeon at least once, he had considered attending his lectures at one point but then rejected the notion as too expensive, and perhaps he had even aspired to join the throng surrounding the fêted anatomist.[36] But the perceived slight about his bougie put paid to that. Foot turned away, cast out of the fold forever, destined to become a bitter, peeved, and unfulfilled young man, whose only claim to fame would be his assaults on Hunter.

Hunter's devotees charged to his defense when Foot launched his attack on Hunter's treatise on venereal disease. One pupil, Charles Brandon Trye, quickly penned an eloquent pamphlet, which chiefly took Foot to task for attacking Hunter without once offering any evidence to back up his claims. "Mr Hunter's Treatise may have its imperfections," Trye patiently countered. "Had Mr Foote attempted to have pointed them out in a manner agreeable to the true spirit and laws of criticism, his undertaking might have been praiseworthy." Instead, he noted, Foot "writes against Mr Hunter, rather than his opinions." And he pointedly contrasted Foot's pretentious aspirations to fame to Hunter's commitment to scientific progress by concluding, "That there can be any thing in Surgery or in Science novel to Mr Foote, let those determine, who are better acquainted with his erudition and knowledge. But as there are many in the profession, who have no pretensions to omniscience, to them it is possible that Mr Hunter may teach something."[37]

Hunter himself never stooped to respond to Foot's persistent jibes in public, but he wrote jocularly to Trye to thank him for "the pains you have taken to laugh at Jessé Foot" and sanctioned the publication of the reply. "Poor Jesse Foot wanted a dinner, and he thought he saw a fine animal he could live on for a while," Hunter explained, adding with typical dismissal, "every animal has its Lice."[38] And he had little time for Foot's sniping comments about his lack of a classical education, at one time commenting,

"Jessé Foot accuses me of not understanding the dead languages; but I could teach him that on the dead body which he never knew in any language dead or living."[39]

His predilection for experiments now well known, Hunter was fast becoming an important figure at meetings of the Royal Society. Among kindred spirits—the leading naturalists, inventors, engineers, and explorers of the age—Hunter flourished. At society meetings and social events, he delighted in talking botany with Daniel Solander, the talented ex-pupil of Linnaeus, who had joined the staff at the new British Museum. In between cataloging the motley collection of items languishing on the museum shelves, the mild-mannered botanist had developed a warm comradeship with Hunter. John Ellis, the naturalist who first encouraged Hunter, was a mutual friend. And Hunter had also made a lifelong friend of Joseph Banks, the well-connected aristocratic naturalist who had become a fellow of the society a year before Hunter, at the age of twenty-three. Having just returned from a botanical expedition to Newfoundland and Labrador, Banks was busy classifying his flora and fauna samples at his home in New Burlington Street, a five-minute stroll from Hunter's house in Golden Square.

Eager to continue these stimulating scientific debates after meetings ended, Hunter led a splinter group of fellow members and other amateur scientists to a nearby coffeehouse, initially Jack's in Dean Street and later Young Slaughter's in St. Martin's Lane. At these boisterous weekly gatherings, members dined on oysters and shared ideas for investigations or papers.[40] Hunter quickly became regarded as chairman of this elite club. As well as Solander and Banks, members of Hunter's circle included Dr. Nevil Maskelyne, the astronomer royal, who was bent on discovering the key to longitude; Robert Mylne, the architect of Blackfriars Bridge, who would later marry Anne Home's younger sister Mary; James Cook, the naval explorer; and James Watt, who was diligently working on his designs for improving steam engines. Other members, all household names in their fields, included the civil engineer John Smeaton, the royal instrument maker Jesse Ramsden, and the mathematician Sir George Shuckburgh.

Richard Lovell Edgeworth, the Irish inventor, was introduced to Hunter's group when visiting London in the late 1760s by James Keir, the Scottish chemist and fellow member of the Midlands-based Lunar Club.[41] Describing Hunter's group as a "society of literary and scientific men,"

Edgeworth noted in his memoirs, "Without any formal name, its meetings continued for years to be frequented by men of real science, and of distinguished merit." According to Edgeworth, Hunter kept numbers strictly limited through a rigorous initiation rite, which only the hardiest of candidates could hope to survive. It entailed subjecting each would-be member to a barrage of mockery and abuse. "We practised every means in our power, except personal insult, to try the temper and understanding of each candidate for admission," Edgeworth recalled. "Every prejudice, which his profession or station in life might have led him to cherish, was attacked, exposed to argument and ridicule. The argument was always ingenious, and the ridicule sometimes coarse." Unsurprisingly, some of the more sensitive applicants quailed before this outright assault in front of their friends and associates. But once these men were admitted, the heat of the coffeehouse wrangles was easily matched by the warmth of its friendships. As Edgeworth recollected, "I have felt, ever since I belonged to this society, the advantage of its conversation."

As well as debating current enigmas of the day, club members shared ideas for papers they were considering presenting to the Royal Society and sometimes read these aloud to the gathered guests. One initiative Hunter may well have shared with fellow members in the club's earliest days, in the summer of 1767, concerned a set of experiments in which he hoped artificially to impregnate silkworms—a prototype attempt at creating test-tube babies.[42] Keeping a female moth in confinement until she laid some unfertilized eggs, Hunter carefully dissected a male moth to extract a sample of semen with a "hair pencil," then combined the two in a covered box. The experiment proved a remarkable success: Eight of the eggs hatched at the same time as those that had been naturally impregnated.

He would later apply the results to remarkable effect, helping a couple who were experiencing difficulties conceiving a child. Since the husband suffered a congenital abnormality called hypospadias, in which the urethral opening is on the undersurface of the penis, rather than at its tip, he was unable to impregnate his wife naturally. Hunter recommended that the couple have intercourse as normal, ensuring the wife was sexually stimulated, and then that the husband's semen be collected in a warmed syringe and injected into the wife's vagina. The couple had their first baby soon after. The case is the earliest-known record of artificial insemination, although it was left to Everard Home to reveal the details to the Royal Society after Hunter's death.[43]

One Royal Society fellow who was never listed among the members of the coffeehouse club, however, was William, Hunter's own brother. He had far more refined company to impress by this stage of his career. Having attended Queen Charlotte at all three of her births to date, William was heavily in demand at the childbeds of the nobility, delivering heirs to great fortunes as well as the unwanted offspring of illicit liaisons. In December 1767, he helped the Viscountess Bolingbroke, Lady Diana Spencer, to secure a divorce by confirming that he had attended the birth of a child born of her affair with Topham Beauclerk.[44] At other times, William would take care of the illegitimate offspring of aristocrats' mistresses with gentlemanly discretion. He attended Martha Ray, the mistress of Lord Sandwich, who was later killed by a jealous rival.[45] But sometimes his vanity overcame his professional confidentiality. He sometimes entertained friends with the story of secretly delivering the daughter of a famous peer of twin babies in her family home without her parents suspecting a thing.[46]

Though the brothers inhabited overlapping but different worlds, they continued on amicable terms—for the time being. When William heard in 1767 of some curious animal bones and teeth, similar to those of an elephant but considerably larger, discovered near the Ohio River, he enlisted John's aid to explain the phenomenon. William obtained one of the tusks and one of the teeth from the Tower of London, which had acquired a hoard of the bones. The eminent French naturalists Georges Louis Leclerc, better known as the Comte de Buffon, and his colleague Louis-Jean-Marie Daubenton had already inspected similar bones and tusks and concluded that they must belong to elephants. Religious dogma suggested that all animal life created by God still existed on earth, and William was ready to accept this view, too. Just to be sure, he showed the specimens to John, whose expertise in comparative anatomy was now widely recognized. John did not hesitate to disagree with the Frenchmen: "At the first sight he told me that the grinder was certainly not an elephant's," William reported. After showing William two elephants' jaws, which contained distinctly different teeth, from his growing collection, William had to concede that his brother was right. William went on to declare in a bold and important paper to the Royal Society in February 1768 that the mysterious teeth and bones must belong to a previously unknown animal—an "incognitum"—which had probably become extinct.[47] The bones would later be identified as belonging to a mastodon, a relative of the mammoth. William delivered the eloquently argued paper, but it was John who had solved the riddle.

William's wealth was expanding as fast as his social standing. He had commissioned, at great expense, a new house with a custom-built lecture theater and museum in Great Windmill Street, which was ready for occupation that summer. When William moved, he turned the lease of his former home, at 42 Jermyn Street, over to John. It was a fit home in which a newly married couple could raise a family, but Anne would still have to bide her time.

That same summer, John finally won entry to another professional body, but one he cared far less about than his beloved Royal Society. On July 7, he passed the oral examination at Surgeons' Hall—the criteria for obtaining the diploma of the Company of Surgeons—before examiners who included his former tutor Percivall Pott.[48] It was a full fifteen years since he had first begun practicing surgery, but Hunter gave the moribund Company such little credence that he never actively participated in its business. At the height of summer in 1768, he was far more wrapped up in the preparations being made by three of his dearest friends to embark on a perilous journey, one he fervently hoped would bring him new treasures for his ever-growing collection.

CHAPTER 10

The Kangaroo's Skull

⟨ↀ⟩

Plymouth

August 1768

After five days spent waiting apprehensively while summer storms lashed the south coast, at last, on August 25, the clouds cleared and a fair wind coaxed the *Endeavour* out of Plymouth harbor.[1] Crammed with a year's supply of provisions, packed with scientific instruments, and crewed by eighty-five sailors, including a drummer, the little ship's destination was the other side of the world. The voyage was the first dedicated scientific mission of its kind. Inspired by the Royal Society, bankrolled by George III, and victualed by the Royal Navy, the expedition had two important aims: to observe the transit of Venus across the sun from the South Pacific island of Tahiti and to search for a fabled hidden continent in the Southern Hemisphere.

In command was James Cook, the good-looking and good-natured lieutenant who had made his name during the Seven Years War. Like John Hunter, Cook was the son of a humble Scottish farmer, although his family had settled in Yorkshire; like Hunter, he was forty that year. On board with him were two of Hunter's closest friends, Joseph Banks and Daniel Solander, both intent on discovering new varieties of flora and fauna with which to delight fellow naturalists. At just twenty-five, Banks displayed all the self-confidence of his privileged background and his education at Harrow, Eton, and Oxford. Having come into a fortune yielding six thousand pounds a year at the age of twenty-one, he could comfortably afford the ten thousand pounds he had committed to the expedition. Along with a retinue of seven attendants and a pair of greyhounds, Banks had packed a welter of nets, hooks, and containers for catching and preserving wildlife.

Invited by Banks to join the trip, the plump, affable Solander had jumped at the chance, promptly obtaining leave from the British Museum to join the voyage.

With the holds already crammed full, there was little space for Banks and his entourage. But this was no deterrent to the botanizing aristocrat. "No people ever went to sea better fitted out for the purpose of Natural History, nor more elegantly," the naturalist John Ellis wrote excitedly to his pen pal Linnaeus.[2] With Cook forced to share his captain's cabin with Banks and Solander, it was a cramped but happy ship that headed for the open seas.

Back in London, Hunter waited eagerly for news of his friends' discoveries. Charting the stars and seeking lost continents were of little interest to him; as far as he was concerned, the principal purpose of the expedition was to bring back new and exotic wildlife to boost his burgeoning collection. Now that he had built up his reputation as a private surgeon and an expert anatomist, his renown as a naturalist was spreading just as fast. He had become every bit as adept at procuring animal carcasses as he was at obtaining human cadavers; only the suppliers were different. Hunter was a familiar figure at traveling shows and circuses, where he promised animal keepers a bounty if they saved him the bodies of their charges when they died. He was a frequent visitor, too, at London's markets, shops, and animal dealers, where customers could buy all manner of creatures as pets, livestock, or ingredients for the cooking pot. Having obtained his wolf-dog puppy from "Wild Beast Brookes," who sold an assortment of exotic animals at his shop in the New Road, Hunter bought eels every month from a local fishmonger in an attempt to discover their method of generation. At the same time, he cultivated friendships with various aristocrats who kept rare animals in their private menageries, pestering them for carcasses.

Obtaining exotic animals was easier than might be imagined in eighteenth-century London; the capital was a cornucopia of wildlife.[3] The royal menagerie was the city's most popular tourist attraction. Officially established in 1235 with the gift of three leopards to Henry III, the menagerie had steadily expanded with successive royal acquisitions. By 1767, the year before the *Endeavour* left Plymouth, visitors could see five male lions, three lionesses, two lion cubs, a panther, a leopard, and four tigers with the genteel names of Sir Richard, Miss Groggery, Miss Jenny, and Miss Nancy.[4] How many of these exhibits ended up on Hunter's dissection bench is unrecorded, although his catalog of pickled animal parts from around the same period bears an uncanny resemblance to the menagerie's inventory. Later writings refer specifically to an ocelot, an antelope, two hyenas, and a caracal, or desert lynx, all received from the Tower.[5]

For those who could not afford the menagerie's sixpence admission fee, there was no shortage of opportunities to see rare animals. Fairs, novelty acts, and freak shows featured dancing bears, performing monkeys, and even trained bees, while traveling menageries displayed exotic creatures brought back from foreign explorations. One guidebook to London's attractions even claimed that there were 'Lions, Tygers, Elephants, &c in every Street in Town.'[6] While this boast may have been an exaggeration, it certainly contained a kernel of truth. On one occasion, Hunter burst into the bookshop belonging to his friend George Nicol and begged the loan of five guineas; it was needed, he explained, to buy "a magnificent tiger which is now dying in Castle Street."[7] Having won his friend's consent, Hunter had the tiger's corpse carted away to his dissecting table.

Though a tiger dying in a West End thoroughfare could still excite Hunter's enthusiasm, other animals were commonplace. With the livestock regularly driven through London's streets, the popular bear- and bull-baiting events, and the many more animals unloaded at the docks for consumption—turtles were shipped in alive as dinner-table delicacies—and for entertainment—monkeys were popular pets—London was fairly overrun with exotic creatures.

This abundance of wildlife did little to enhance the populace's general knowledge of the animal kingdom, however. Many Georgians still believed in the existence of unicorns, dragons, and mermen, and books offered little enlightenment. In his famous 1755 dictionary, Samuel Johnson described whales as "fish" and made no mention of giraffes. Although a giraffelike beast came under the name "camelopard"—it reputedly had extremely long legs, a long neck, and a spotted brown-and-white hide—it was still doubted that such peculiar-sounding animals could exist.[8]

Making full use of London's myriad opportunities for studying wildlife, Hunter had voraciously enlarged his natural history collection. By the time Cook, Banks, and Solander set sail, he had already accumulated assorted parts from a lion, a porpoise, a seal, a monkey, a leopard, an opossum, a mongoose, a flying squirrel, and a mole from South Carolina, as well as his lizards from Portugal and Belle-Ile, numerous examples of native British species, and the two elephant jaws he had shown his brother. Many of the rarer animals were the first of their kind to be seen in Britain. Every new animal he obtained was transported back to his home and laid out on his dissecting bench, where Hunter wielded his knife and tweezers to explore its internal structure. Interesting organs were preserved in spirits and unusual characteristics meticulously noted in a catalog.

When he moved from Golden Square to William's former house at 42 Jermyn Street, his collection of dried bones, skulls, skins, and pickled organs went along, too. As the animal carcasses arrived at his back door with increasing regularity, the fledgling collection soon took over several of the best rooms of the house. And just as John Ellis had sought Hunter's expertise in dissecting the greater siren, so other amateur naturalists now begged his help whenever they obtained previously unknown species. According to Everard Home, Hunter's future brother-in-law, "no new animal was brought to this country which was not shewn to him."[9]

But the heart of London's West End was not the ideal location for dissecting and experimenting on large and often noisy wild animals. For this, his country retreat at Earls Court, two miles beyond London's boundaries, was perfect. Having bought the farmland in 1765, Hunter had begun building a house there shortly afterward and had since added extensions to form a modest country villa with a variety of outbuildings.[10] Nobody who passed its iron gates, which fronted the east side of what would become Earls Court Road, could do so without stopping to stare. Though the two-story house was unremarkable in itself, the crocodile's jaw that yawned wide over the front porch gave an early signal of its owner's unusual preoccupations. But it was the scene surrounding the house that really drew gasps from passersby. Its typically English lawns were grazed not only by sheep, horses, and cattle but by zebra, Asiatic buffalos, and mountain goats, while neighbors could hear the roars of lions and snarls of leopards emanating from a large grassy mound within the garden. Beneath this hillock, Hunter had built three dens to house his collection of wildcats.

Visitors brave enough to enter the grounds first passed a fish pond decorated with animal skulls. Nearing the house, the howls of dogs and jackals chained in the kennels mingled with the cacophony of noises from the pigs, donkeys, and domestic fowl scrubbing for food in the yard. To one side, a large conservatory vibrated with the hum of bees entering and leav-

ing glass observation hives, while surrounding the entire building ran a six-
foot-deep trench that led at one end to a stout wooden door. Opening this
door revealed an underground laboratory containing a large copper vat, in
which Hunter boiled down the bodies of animals, and at times even humans, to obtain their skeletons. The great cauldron would one day play a
significant role in the most extraordinary of all his quests for specimens.

Hunter's neighbors regarded their most flamboyant resident with un-
bridled curiosity, though they hurried past the sight, and stench, of a large
whale bone, left over from one of his dissections, which lay discarded for
years in a gutter.[11] Excited village children gawped at the extraordinary wild
creatures in their incongruous pastoral setting. It has even been plausibly
suggested that Hunter provided the model for the children's book character Dr. Dolittle, the eccentric country physician whose fascination with
natural history inspired him to become an animal doctor.[12] In his fictional
home of Puddleby, John Dolittle kept exotic animals and built up friendships with traveling circus owners, just like Hunter. Likewise, Hunter on
occasion treated animals belonging to his patients. He would ultimately be
instrumental in founding the Veterinary College of London, forerunner to
the Royal Veterinary College, when it was launched in 1791.[13]

For Hunter, his country menagerie fulfilled two essential purposes: It
allowed him to observe living animals in natural surroundings and to perform investigations on animals both dead and alive. And as his growing
London clientele brought him increasing income to indulge his passion for
collecting, he spent more and more time among the bizarre flocks and
weird beasts at his rural hideaway. He enjoyed watching his wild creatures
from the windows of his study and even played with some of them.
Listening to the eerie calls of a pair of leopards mating, he noted, "Their
noise is a mew but not so loud & hoarse & have a gutteral or hollow sound
when angry very strong." On another occasion he was delighted to see a
hawk and a pigeon living happily together in the same cage, and he wrote,
"Thus Animals not chusing to devour or hurt those that they are acquainted
with is of great service in the Animal System and the Effect is beautifull."[14]

All his life, Hunter would retain the reverence for nature he had felt as
a child. Despite performing many experiments on living dogs, he was particularly fond of his canine pets, and he admired the industry of the bees
he studied for hours in the glass hives he had specially designed for the
conservatory. He noted, obviously from painful experience, that a bee about

to sting made a different noise "from that of the wings when coming home of a fine evening loaded with farina of honey; it is then a soft contented noise." He even went so far as to compare this hum with the notes of a pianoforte, remarking that "it seemed to be the same sound with the lower A of the treble."[15] And on balmy summer evenings, as the bees returned to their hives, the cattle lowed in the fields, and the pigs snuffled in the yard, the scene seemed idyllic.

Life was not always so harmonious, however. His leopards on one occasion broke free from their chains and ran into the yard, where they attacked the dogs. The commotion was heard all around, and "the howling this produced, alarmed the whole neighbourhood," Everard Home later related.[16] Without stopping to consider the risks, Hunter ran from the house and into the fray. Grabbing one leopard by the scruff of its neck as it was about to scale the wall into the village, he yanked its mate from the middle of the dog pack with the other hand. Then he carried both animals, struggling and squealing, back to their den before realizing how close he and the rest of the neighborhood had come to peril. As soon as he had secured the beasts, Home recorded, "he was so much agitated that he was in danger of fainting." In another display of his muscular strength, Hunter enjoyed wrestling with a bull he kept. The animal, a "beautiful small bull," was a present from Queen Charlotte in return for the donation of twenty-eight preparations, which Hunter had presented to create a miniature anatomical museum at the Royal Observatory, Kew, for the edification of the young princes and princesses.[17] But he obviously underestimated his combatant, for one day the bull threw him and was about to gore him, until a quick-thinking servant drove the beast off.[18]

Yet behind all the madcap antics with dangerous animals and the musical comparisons with the hum of bees lay a serious scientific intent. Hunter was among a long line of anatomists, dating back to earliest times, who studied animal bodies. Aristotle had dissected numerous animals, including whales, and since then many anatomists had used animals, dead and alive, to further their explorations. In his milestone anatomical works, Vesalius included a helpful drawing of a custom-made table on which live beasts could be secured during experiments.[19] Among Hunter's contemporaries, the Monros, father and son, in Edinburgh and Albrecht von Haller in Switzerland were all keen vivisectionists. As with other anatomists of the time, their research was intended to illuminate human anatomy and physiology; there was little interest in the internal structure of animals for its own sake. Only in France, where Buffon and Daubenton collaborated to produce their thirty-six-volume *Histoire Naturelle,* did anyone come close to Hunter's single-minded mission to understand all natural life.[20]

At the same time, many naturalists were busy collecting and cataloging newly discovered forms of fauna and flora, either by risking their lives traveling to distant lands, like Solander and Banks, or by pottering around their own neighborhoods, like Gilbert White, who collected his specimens in Selborne, Hampshire. Generally, however, this passion for natural history was confined to describing the external appearances of known or new species and categorizing them to fit into a particular system of classification, principally from the late 1760s the binomial system invented by Linnaeus. And though many amateur naturalists built up large stores of plants and animals, these tended to be either random selections of stuffed skins and dried flowers or highly specialized collections of a particular species. The huge hoard Sir Hans Sloane left to the nation in 1753—forming the foundations of the British Museum, which opened six years later—comprised a gigantic jumble of stuffed animals, birds' eggs, dried plants, fossils, and precious stones, with little semblance of order. Other collectors owned "cabinets of curiosities" containing a confusion of natural history exhibits, ethnic artifacts, and antiquities.

John Hunter's explorations in natural history were radically different.[21] Initially, he had begun, like the Monros and Haller, examining animals largely to illuminate the human body. His early experiments injecting milk into dogs' intestines were aimed at understanding how the human lymphatic system worked; when he cracked open hens' eggs to examine the developing fetuses, he was keen to know how the human embryo grew. But his animal studies soon constituted an important pursuit in their own right. His chief aim was to discover general principles or rules governing all organic life. As he explained, "All animals must have certain general principles, or they would not be an Animal; and it is from the different combination of these principles that produce different animals."[22] His labors would make him a pioneer, though rarely recognized as such, in the discipline that would become established in the following century as biology.

The stirrings of this lifelong quest began early. When Hunter had tested the hearing of fish in Lisbon, it was not fish in particular that interested him, but the sense of hearing in all animal life. Refusing to accept other anatomists' view that fish had no ears, he quickly established that they did indeed possess internal hearing organs, thereby confirming his beliefs about the function of hearing in general. He explained, "I am still inclined to consider whatever is uncommon in the structure of this organ in fishes as only a link in the chain of varieties displayed in its formation in different animals, descending from the most perfect to the most imperfect, in a regular progression."[23] Everything, he was certain, fitted logically into a pattern.

> We may observe that in natural things nothing stands alone, every thing in Nature having a relation or connection to some other natural production or productions and that each is composed of parts common to most others but differently arranged, and therefore in every natural product there is an appearance of an affinity in some of its parts to some other natural production because it has some of its parts in its composition; and where there are the greatest number of those affinities or parts as also the closer the connection or affinity between those of one production with those of another, the nearer are those allied.[24]

But always a staunch believer in the power of visual evidence rather than the ephemeral quality of words, Hunter's aim was most clearly to be seen in his collection. The human and animal preparations that he had moved into the fine rooms in the Jermyn Street house were no haphazard medley of curiosities, but an ordered display systematically exploring organic life. Similar or analogous parts from different animals and humans were grouped together. So differently sized eyeballs bobbed in alcohol on one shelf, while variously shaped kidneys floated in jars on another; digestive organs were displayed in one section and reproductive organs in another.[25] No other collection of its kind existed. And every time Hunter dissected an animal he had not seen before, his chief aim was to discover where it would fit into nature's system—precisely how its particular organs mirrored or differed from those of similar species, how its internal structure slotted into an overarching pattern. From the 1760s onward, he began actively looking for similarities and variations among different species, the discovery of which, he believed, might explain the secrets of life itself.

No animal was too large or too small, too simple or too complex, for

Hunter's knife. One day might find him stooped over his bench searching for the brain in an earthworm, while another would see him wrestling with the stomach of an elephant. He dissected the most basic animal forms, such as polyps, and the most sophisticated, the human body. Mostly working without the aid of a microscope—for fear that it would distort reality— he observed, "The circulation of the insect is probably very slow, if we may judge of the whole class by the motion of the heart in the caterpillar. In the silk-worm, for instance, the heart beats only 34 in a minute."[26] With incredible patience and dexterity, he dissected brains in leeches, worms, and centipedes, tongues in bees, and reproductive organs in silkworms.

Large mammals, however, could be equally challenging and just as entrancing. Opening a huge sperm whale on a barge on the Thames, standing on top of its blubbery carcass as he had in 1759, he noted that "the tongue was almost like a feather-bed." With awe, he added, "The heart and aorta of the spermaceti whale appeared prodigious, being too large to be contained in a wide tub, the aorta measuring a foot in diameter. When we consider this as applied to the circulation and figure to ourselves that probably ten or fifteen gallons of blood are thrown out at one stroke, and moved with immense velocity through a tube of a foot diameter, the whole idea fills the mind with wonder."[27] Hunter's poetical descriptions of whales would later be cited as inspiration for *Moby-Dick*. And even if Dr. Johnson informed the reading public that whales were fish, Hunter had no doubts about their position in the overall plan. After comparing their organs with those of other mammals in his collection, he remarked, "I shall always keep in view their analogy to land animals." But impressed as he was by the grandeur of cetaceans, Hunter was just as fascinated by the worms he found in the whale's intestines.

While similarities among normal forms of species were a compelling interest, abnormalities were just as absorbing. Hunter was intrigued by deviations from the norm, or "monsters" in contemporary language, and he avidly collected such peculiarities. Already by 1767, his collection included the brains of a calf with two heads, a pig with two bodies, a kitten with two mouths, and a creature he described as "the Elephant-Pig or Cyclops." These he grouped together with a child born without a skull and an infant with spina bifida. He was equally bent on obtaining living examples of such abnormalities. At one point, he kept a lamb with three legs at his Earls Court farm.[28] Such oddities were a common obsession in eighteenth-century Britain. People flocked to see animals and human freaks at fairs and shows. But Hunter's interest was more than curiosity. He realized that all animal life possessed an innate ability to become altered or deformed— his double lizard tails were a perfect example—and he was gradually coming to believe that this propensity for change might explain how life first developed.

As well as examining and collecting dead animals, Hunter devoted hours at Earls Court to experiments on living creatures. His interminable trials investigated every bodily function, including temperature, circulation, respiration, generation, and digestion, in every possible animal. In one series of experiments, he measured the body temperature of dogs, hens, fish, slugs, a rabbit, a carp, a viper, an ox, and "sleeping"—hibernating— dormice, as well as humans.[29] Initially, he used a standard mercury thermometer for these trials. Since its invention by Gabriel Fahrenheit in 1714, such instruments had been readily available from specialist instrument makers in London. But finding this too clumsy for his purposes, he commissioned his friend Jesse Ramsden, the royal instrument maker, to make him a smaller, more accurate model, just six inches long and less than one-sixth of an inch wide, with a sliding scale made of ivory. With this, he concluded that all warm-blooded animals have their own regular body heat, which varies when sleeping or awake.

Other experiments begun at Earls Court in the 1760s were designed to discover how bones grow. Having caught a chicken in the yard, he drilled two holes in one leg and fixed a piece of lead shot into each.[30] After a while, he killed the hen, cut off its leg, and measured the distance between the two pieces of shot. While the leg had grown more than three inches, the gap was exactly the same. His trial suggested, correctly, that limb bones grow by accumulating material at the outer ends, rather than universally throughout their structure. Next, he repeated the experiment on a piglet. When it was slaughtered at full size, the results replicated those with the hen. Still, this failed adequately to explain how bone tissue grows and develops. For this, Hunter selected two more pigs and fed them with madder root, traditionally used as a vegetable dye, for a fortnight. He was not the first to use the plant for such experiments. A surgeon named John Belchier had reported to the Royal Society in 1736 that eating madder stained pigs' bones red. A few years later, a French researcher, Henri-Louis du Hamel du Monceau, had used this effect to establish that bones grow widthways like trees, by the addition of layer upon layer, as well as lengthways in layers added at the ends, although his conclusions were still hotly disputed. Hunter killed one of his two pigs at the end of the fortnight, while the other was fed for a further two weeks on a normal diet before being killed for comparison. Dissection of the bones of this second pig revealed red and white strata in certain parts, confirming du Hamel's theory that bone grows layer upon layer. Other parts of the bone remained white, however, suggesting that old bone matter had been absorbed. Hunter named this simultaneous absorbing and depositing of matter the "modelling process."

Though the spacious grounds at Earls Court allowed Hunter to observe numerous animals over long periods, other people's creatures were equally rewarding to watch. Hunter described a zebra that had been imported to England by Lord Clive, the hero of conquests in India, who had fallen from grace since returning home. After attempting in vain to persuade the female zebra to mate with an ass, Clive had ordered the donkey to be painted with black and white stripes; suitably embellished, "she received him very readily," Hunter recorded.[31] His observations would later be cited by Charles Darwin as part of the explanation of his theory of sexual selection in *The Descent of Man* (1871).

Generation was an abiding interest, not only in the pregnant human corpses Hunter had dissected in Covent Garden but in all manner of animals, too. One winter in the late 1760s, Hunter caught six male house sparrows and compared the size of their testes as the mating season progressed. Preserving their tiny bodies to show the tenfold increase in testicle size from the dimension of a pinhead to those of a marble, he recorded with admiration, "If we compare their size in January with what it is in April, it hardly appears possible that such a wonderful change could have taken place during so short a period."[32]

But in the late summer of 1768, with Cook, Banks, and Solander heading for the South Pacific, Hunter had his sights set on rather more exotic creatures than house sparrows. He would have to be patient. It would be three years before the adventurers returned to England with their treasures.

Meanwhile, Hunter's collecting passion was proving an incessant drain on his resources. While some specimens and animals were gifts, he expended large sums procuring the anatomical curiosities he craved from dealers, circus owners, and auctioneers. At the same time, his expanding

Earls Court establishment required an increasing number of staff to tend its animal population. There was his future wife to consider, too: As an assiduous society hostess, she would certainly expect generous household support when she settled in the house on Jermyn Street. Although Hunter was making his way as an up-and-coming surgeon, his income from patients' fees was still insufficient to keep pace with his escalating expenses. On several occasions, he mortgaged land at Earls Court to raise funds for his ventures.

Unlike William, who was building a vast collection of coins, books, and paintings as a solid investment, John had not the slightest inclination to generate a profit. All he earned, he spent. And he regarded his routine work, pandering to the whims of malingering rich clients, principally as a way of fueling his collecting habit. On one occasion, he wearily told an assistant, "Well . . . I must go and earn this damned guinea, or I shall be sure to want it tomorrow."[33] When one of Hunter's friends, the physician George Baker, was asked by a mutual acquaintance whether the anatomist would appreciate some form of edible delicacy as a birthday present, Baker replied, "He cares not what he eats or drinks and I am sure that a curious case or some anatomical curiosity would be more agreeable to him than all the wine and all the venison in the country."[34] It was flesh for his dissecting table, not meat for the dining table, alcohol for his preparation bottles, not wine for his palate, that Hunter desired. Nevertheless, his expensive lifestyle and impending marriage made it imperative that he augment his earnings. And so, when a vacancy arose at St. George's Hospital in November, it was an opportunity Hunter could not afford to miss.

On December 9, 1768, the boardroom of St. George's Hospital was crammed full. More than 160 governors had turned up for the meeting to elect a new full-time surgeon. Bishops, lords, dukes, and other pillars of Georgian society took their seats to cast their votes for one of the two candidates: the junior surgeon David Bayford, who had dutifully served his apprenticeship at St. George's, and John Hunter, the controversial maverick, who had not treated a patient at the hospital since his brief stint as a house surgeon twelve years earlier. As far as the three current members of the surgical staff were concerned, there was no contest: Bayford was a St. George's man, schooled in the St. George's traditions, and the position was morally his. Although three of the four physicians on the staff may have been equivocal, there is no doubt that the fourth member, Donald Monro, older brother of the Hunters' bitter rival Alexander Monro, Jr., would have rallied to Bayford, too.

But the clinical staff had not reckoned with John Hunter—or rather, with his influential brother, William. Since becoming governors themselves, both brothers had diligently attended board meetings during 1768.[35] Through these gatherings, and William's contacts with the rich and powerful in Georgian society, the brothers had become friendly with fellow governors, such as the actor David Garrick and the artist Joshua Reynolds, both of whom made sure to turn out for the St. George's vote. This was no casual favor from Reynolds, for on the same day he had three crucial meetings with fellow artists in order to prepare for an audience with George III the next day to establish the Royal Academy of Arts.[36] The kindness was not wasted. Hunter garnered 114 votes, to Bayford's 42.

At forty, having gained his surgical diploma only five months earlier and never having completed a traditional hospital training, but with more anatomical experience than the rest of the hospital's staff put together, Hunter had finally secured his first proper surgeon's job. It was the beginning of a steady rise in his prospects. A staff surgeon's job brought no salary—all hospital appointments were honorary—but it still promised a substantial boost in income. Not only could hospital surgeons earn fees from teaching apprentices and house pupils; their private practice benefited from their leap in status. Moreover, Hunter now had access to a ready pool of patients on which to practice his art and perform his experiments. His patients at St. George's were quickly recruited as human guinea pigs in his tireless research program.

Conflict was inevitable. Hunter's dogged determination to question accepted doctrines, his fascination with innovation and experiment, and his commitment to founding surgical practice on sound scientific principles were anathema to his fellow surgeons at St. George's. For his part, Hunter was never one to employ tact or diplomacy to win friends and influence colleagues. When offered resistance, like the fierce little bull he kept at Earls Court, he was more likely to put his head down and charge straight at his opponents.

For the moment at least, an uneasy peace prevailed. Hunter attended the hospital to admit emergency patients on his allotted day, completed his ward rounds according to the agreed rota, and patched up casualties in the operating theater as best he could. And if the surgeons at St. George's kept their own counsel for the time being, Hunter's surgical skills were certainly in demand elsewhere in the capital. On October 21 of the following year, he was called out urgently to help perform a rare and exceedingly risky emergency operation.[37] Twenty-three-year-old Martha Rhodes had gone into labor at her home in Holborn, but at just four four and with severe deformities in her spine and hips, the delivery was going badly. The midwife had quickly discerned that her patient's contorted frame would make a natural delivery exceptionally difficult. In accordance with common practice when births were progressing poorly, she had sent for a medical man, William Cooper, for advice. Equally at a loss as to how to help, Cooper called in a bevy of colleagues for a bedside conference.

Altogether, six physicians and five surgeons crowded around Martha's bed in the cramped and dark room. An examination confirmed what the midwife had already suspected: Martha's pelvis was too narrow for her to give birth naturally. Surgeons and physicians were unanimous. The only chance of survival for both mother and baby was a cesarean section, and poor Martha "cheerfully consented" to the operation. Henry Thomson, a surgeon at the London Hospital, agreed to perform the surgery. John Hunter, with his unsurpassed knowledge of the anatomy of the pregnant womb, was his assistant. None of the medical men gathered around Martha's bed had ever performed or witnessed such a perilous procedure before, and when Thomson hurriedly skimmed through the surgical texts he kept at home, he found very little help there, either. All would have heard reports of successful cesarean operations, but these were vague, unauthenticated stories, almost in the realm of folklore, and certainly lacked any useful technical advice.

Cesarean operations were undoubtedly carried out in ancient times, but usually posthumously, when women died in labor.[38] Although the term was widely believed to have emanated from the manner of Julius Caesar's birth, the Roman emperor had certainly not been born by this method, for his mother was still alive many years after his birth. On the extremely rare occasions when the operation was performed on living women, few, if any, survived. Since then, accounts of cesarean sections had circulated, with varying reports of success. In 1500, a Swiss farm laborer had reputedly performed a cesarean on his own wife, who survived to give birth to five more children. In Ireland in 1738, a midwife had reportedly performed a successful operation on a woman who had been in labor for twelve days; the midwife had held the wound together with her hands while a neighbor ran to fetch needle and thread for the stitches. There were even stories of women who, in desperation, had cut open their own wombs during difficult labors. But there had been no accounts, not even folktales, of surgeons performing successful cesarean sections in Britain.

Thomson and Hunter were undeterred. Dosed with opium, Martha Rhodes was placed on a table, her head on a pillow, as Thomson took up his knife. With physicians and surgeons craning to see, he made a six-inch-long incision in her belly, then opened the womb to reveal the baby curled inside. One of the physicians yanked the child out by its feet, whereupon it "cried as heartily as children commonly do." Now Hunter pushed the woman's intestines out of the way while Thomson swiftly stitched the wound. Throughout the grisly procedure, Martha barely complained. As Cooper later recorded, "She behaved with surprizing fortitude during the whole process, lost very little blood, and seemed to be most of all sensible to pain when the needles were passing through the peritonaeum, especially on the right side of the wound."

Afterward, however, things took a turn for the worse. Martha declined and died five hours later, most probably from internal bleeding. Her child, who had seemed initially healthy, had severe brain damage and died two days later. Hardly able to refuse the persuasive appeals of eleven eminent medical men, the grieving family consented to an autopsy and Martha was dissected in the presence of another large audience, this time including William Hunter, who reported details to the Royal Society. John Hunter would perform the operation for a second time, in 1774, with a degree more success.[39] This time, the baby survived, though the mother still expired. It would be several more decades before a cesarean section was performed by a surgeon in Britain—by James Barlow, in Blackburn in 1793, assisted by one of the Hunter brothers' early pupils, Charles White—after which the woman survived. Even then the operation was carried out without the aid of either pain relief or antiseptics.

Demand for Hunter's presence at postmortems continued unabated. In December of 1769, Hunter was called out by a friend, the physician John Pringle, to dissect a sixteen-year-old youth who had died of a fever.[40] After watching Hunter saw open the skull of his former patient, Pringle was gratified to observe that the boy's brain appeared diseased, just as he had diagnosed. But as Hunter continued to remove and examine the youth's organs with his usual thoroughness, Pringle was baffled to see that the boy's stomach had a huge hole at one end. Convinced that his patient had not suffered any digestive complaint, Pringle questioned his friend on the organ's appearance. Hunter had seen the same effect many times in his twenty-one years of dissecting bodies. At first, he, too, had assumed the gaping holes in the stomach wall were connected to the cause of death, but having witnessed a similar phenomenon in those who had died a violent death but had previously been in good health, he had since surmised that the damage was due to the solvent power of the stomach itself.

Although the French naturalist René Réaumur had discovered gastric acid in animals' intestines in 1752, intense debate over how digestion occurred continued. Rival anatomists argued over whether the process was caused by heat, muscle pulverization, or a chemical reaction.

Hunter was in no such doubt. Pressed by Pringle to commit his thoughts to paper in 1772, in only his second communication to the Royal Society, he presented views that were lucid and decisive. Pointing out the clinical importance of distinguishing appearances in the dead body as a result of disease from those due to natural postmortem changes, he explained that after death the digestive powers of the stomach acted on the body just as they had previously done on food. "These appearances throw considerable light on the principles of digestion," he informed the fellows; "they show that it is not mechanical power, nor contraction of the stomach, nor heat, but something secreted in the coats of the stomach, which is thrown into this cavity."[41] This "gastric juice," as he called it, was an acid, "a little saltish or brackish to the taste," he added.

Hunter's shelves were filling up fast. Before long, the youth's preserved stomach was joined by the aorta of an old campaigner, Maj. Gen. Robert Armiger. At sixty-eight, the experienced soldier had married a woman of forty, and after an evening's celebrations in March 1770, he had taken her home to consummate the relationship. But the wedding-night excitement was all too much for the veteran of heavy battles, and within half an hour of retiring to bed, he suddenly died. Called in to dissect the body, Hunter confirmed the general had suffered a heart attack, and he confided in his casebooks, "It is not known whether or not he was taken ill in the Act of Consummation."[42] Hunter's interest was not prurient; he simply knew that obtaining the body of someone who had died in the act of copulation was a rare occurrence indeed and might feasibly shed light on the physiology of generation. Accordingly, Hunter dissected the general's sexual organs, noting that his penis was "very large, & almost half erected" and that the sperm-carrying vessels were "full of semen."

While one day saw Hunter laboring over a naked corpse with a smile on its face, another found him examining the carcasses of peculiar-looking beasts newly arrived on home shores. One such was the nilgai, or "nyl-ghau," which had first been brought to Britain from India in 1767. A pair of the bull-like creatures with their distinctively blue coats had been sent to Lord Clive as a present. The following year, a second pair was donated to Queen Charlotte, another enthusiastic menagerie owner, who was happy to oblige when her faithful male midwife, William Hunter, asked to borrow the animals. William kept the beasts in the stables at his new home in Great Windmill Street. He even had one of the curious animals, with its goatishly small head and tiny horns stuck incongruously on its robust, muscular body, painted by one of his artist friends, George Stubbs, who was fast becoming famous for his animal pictures.[43]

As he watched the gentle creatures, which licked his hand when he fed them, William suspected the nilgai might be a hitherto-unknown species.[44] He knew this question could only be settled by one person, his brother John, and only then by dissecting the dead animal. So when, shortly afterward and rather conveniently, one of the pair died, William obtained permission from Queen Charlotte for John to work on the beast. Just from examining its teeth and digestive system alone, John was convinced the nilgai was indeed a distinct species. While William reported the findings to the Royal Society, adding another paper to his name, John skipped home to Jermyn Street with his booty—the skeleton of the creature for his collection.

Hunter now knew that before long even more exotic creatures would be in his hands, for news of the *Endeavour*'s arrival in Java in October 1770, after circling the islands of New Zealand and charting the western coast of New Holland, as Australia was then known, had reached London. "The Naturalists write with great spirits," Pringle reported to Haller, "& say they are loaded with new plants & other natural productions."[45] Waiting in eager anticipation of the new species he hoped to examine, that same month Hunter found a fellow enthusiast to share his passion for the natural world.

Edward Jenner, a vicar's son from Berkeley, Gloucestershire, enrolled as a pupil at St. George's Hospital that October and immediately took up residence as Hunter's first house pupil.[46] Well mannered, impeccably dressed, and gifted in poetry and music, the twenty-one-year-old Jenner might have seemed more suitable a companion for Hunter's talented young fiancée than for his brash, bombastic teacher. But in spite of the age gap and their divergent cultural interests, Hunter and Jenner were kindred spirits. Having lost his father, like Hunter, as a child, Jenner had spent his spare time as a youth hunting, as Hunter had, for wildlife in the surrounding countryside. After having been apprenticed to a local surgeon since the age of thirteen, Jenner came to London to complete his medical education. As Hunter's pupil, he walked the wards at St. George's, assisted as his "dresser" in the operating theater, and accompanied him on visits to his wealthy clients in their West End homes. But his most productive and pleasurable hours were spent at Hunter's side in the Jermyn Street house they now shared, and at Hunter's Earls Court retreat, exploring human corpses, examining the bodies of rare animals, and experimenting on living creatures.

From the start, the inquiring young student revered and adored his charismatic mentor, who fostered his interest in natural history and taught him to think for himself. It was Hunter's doctrine—of observation, experiment, and application—that Jenner would faithfully follow when, nearly thirty years later, in 1796, he tested the smallpox vaccine. Not only would Jenner's innovation save millions of lives; it would become the only medical therapy ever to eradicate a disease completely from the face of the earth.[47] Hunter was equally enchanted by his industrious pupil, who shared his love of nature and novelty. Jenner would always be not just his first but also his favorite disciple. So, as the *Endeavour* finally approached the English Channel at the end of its three-year world tour, both Jenner and Hunter were waiting impatiently to gain first sight of its haul of natural treasures.

They were not alone. Since hearing that Cook's ship was on its way home from Java, every naturalist in Europe had scanned the newspapers for details of the daring expedition's discoveries. When at last the *Endeavour* anchored off Deal on July 12, 1771, nobody was disappointed. The ship's holds were crammed to bursting with a bounty of seeds, dried plants, bottled marine creatures, and preserved animals of varieties never seen in Europe before. Chests of gold plundered from Spanish merchant ships could scarcely have aroused more excitement. It was a collector's dream. The adventurers had brought back no fewer than fourteen hundred new plant species and more than a thousand new species of animals, including a handful of mammals, more than a hundred birds, over 240 fish, and assorted mollusks, insects, and marine creatures.[48] It would take twelve years to classify the haul.

Hastening to welcome his friends, Hunter immediately recommended that his talented new pupil help Banks and Solander catalog their finds. Most were destined for Banks's own collection, with a few choice items to go to the British Museum, but there were generous pickings for Hunter, too. Among the preserved new animals he added to his collection were a sea pen, a simple aquatic animal, from Rio de Janeiro; sharks' eggs and eels from the South Seas; a mole rat and a zorilla, or striped polecat, from the Cape of Good Hope; and part of a giant squid.[49] But most astonishing of all was the peculiar gray animal that Banks had named, with his limited understanding of native dialects, a "kangooroo," or "kangaru." The ship's crew had first sighted the curious creature—"an animal as large as a greyhound of a mouse colour and very swift"—while repairing their vessel on the east coast of New Holland on June 22, 1770.[50] Banks saw the animal for the first time a few days later. The following month, the ship's second lieutenant, John Gore, shot a large male, and Banks recorded in his journal, "To compare it to any European animal would be impossible as it has not the least resemblance to any one I have seen." The next day, the kangaroo was served up for dinner and "proved excellent meat." It was enough to make a comparative anatomist weep; the kangaroo was one of many exotic creatures the crew devoured in their hunger for fresh meat. A second kangaroo fell victim to Gore's shotgun a few days later, and on July 29, one of Banks's greyhounds caught a small female. The flesh having been long since consumed, Banks had brought back the skins and skulls of the extraordinary creatures. Back in London, he commissioned George Stubbs to paint a picture of the marsupial using a stuffed skin and a little imagination. It was the first portrayal of a kangaroo seen by Western eyes. He gave his friend John Hunter one of the skulls.[51]

It was not much to go on, but Hunter still hoped he could tell something about the strange animal Banks had described by examining its teeth and jaws and comparing them with species he had already dissected. He was confounded and had to admit that "the teeth did not accord with those of any one single class of animals I was acquainted with, therefore I was obliged to wait with patience till I could get a whole."

It would be almost two more decades before Hunter could fulfill his ambition, when John White, surgeon general to the first British colony in New South Wales, sent back a treasure trove of preserved animals in 1788. As well as various birds, fish, and insects, a dingo, a kangaroo rat, and several possums, it included several complete kangaroos. The task of describing the mammals for White's subsequent book, *Journal of a Voyage to New South Wales,* published in 1790, fell to Hunter.[52]

At last, Hunter was able to explore the entire anatomy of the kangaroo, to examine the female marsupial's unusual double reproductive organs and appreciate its peculiar method of nurturing its young in pouches. He was certain that the species Banks and Cook had discovered were unique to the antipodean islands, noting, "They are, upon the whole, like no other that we yet know of," and stressed that their anatomy showed particular adaptations to their environment. This realization that there were species in parts of the world unlike those anywhere else provided yet another piece to slot into the big puzzle of how life on earth had developed. His descriptions of kangaroos and other creatures from Cook's first voyage would eventually appear in his extensive work of comparative anatomy, *Observations on the Animal Oeconomy* (1786); he dedicated the book to Banks.

The summer of 1771 was a busy one. Jenner and Solander were immersed in cataloging the *Endeavour*'s haul and Hunter was finally going ahead with his marriage. The fact that his three-year experiment on venereal disease had ended the previous year may or may not have been the spur. No doubt his new hospital post and the two hundred pounds he received in 1771 for his treatise on teeth helped to boost his suitability for marriage.[53] And strangely, after waiting seven years, he was suddenly in a great hurry to arrange the wedding. A marriage license was obtained on July 21. That evening, John dashed off a hurried note to his brother William: "To morrow morning, at eight o clock at St James's Church I enter into the Holy State of Matrimony. As that is a ceremony which you are not particularly fond of, I will not make a point of having your company there."[54] William, who believed wedlock was incompatible with a career devoted to anatomy—he had dissolved his partnership with his then assistant, William Hewson, the previous year chiefly on the grounds that he had married— did not attend the ceremony.

John Hunter and Anne Home were married on Monday, July 22, 1771, with her parents as witnesses; he was forty-three and she was twenty-nine.[55] Having returned to London just over a week earlier, Banks and Cook were home in time to enjoy the ceremony. According to one tale, the travelers not only attended but gave the couple a wedding present of some hickory wood that they had cut down in New South Wales.[56] John had it made into a set of dining chairs for his "Anny." Sparing a few days from Hunter's busy schedule, the couple spent a honeymoon "out of Town," most probably at Earls Court.

One old family friend who had been unable to attend the wedding was Tobias Smollett. The long-suffering writer was ailing in his adopted home of Italy that summer, and on September 17, two months after the publication of his most acclaimed novel, *The Expedition of Humphrey Clinker,* he died. Earlier in the year, in a moment of foreboding, he had promised John his body after death to display in his collection, declaring in a letter from Livorno, "You shall receive my poor carcase in a box, after I am dead, to be placed among your rarities."[57] Yet despite his offer, Smollett escaped dissection. When he died, his wife was either unwilling or unable to ship his corpse from Italy.

There would be plenty more bodies, both human and animal, to fill Smollett's place. And as Anne looked out of the bedroom window of her new country home at Earls Court on the rare and varied creatures grazing her English country garden, she could have been in no doubt of the extraordinary life that lay ahead. Returning to their town house on Jermyn Street, with its elegant rooms filled with preserved organs, heads, bones, and fetuses, it was abundantly clear that she had married an extraordinary man.

CHAPTER 11

The Electric Eel's Peculiar Organs

∽〰〰〰∾

London

1772

John Hunter's bachelor days, rattling around a large, empty house, were over. His home was constantly busy, his doors never closed, his neighbors' peace shattered. Throughout the day, patients arrived for private consultations, pupils turned up for dissection practice, colleagues dropped by to discuss cases, visitors called for dinner, and servants scurried in and out on errands. In the evenings, coaches and sedan chairs drew up outside to disgorge guests for the musical and literary parties gregarious Anne hosted. And even in the middle of the night, the neighbors were awakened by the bumps of hampers being delivered to the basement. Every space in the house was pressed into use. As well as the Hunters' private rooms, there were servants' quarters, lodgings for the house pupils, a room for dissecting, and several rooms on the first floor dedicated to Hunter's anatomical collection.

In June 1772 came another addition to the household, when Anne gave birth to the couple's first child. The healthy red-haired boy was named John Banks Hunter, in recognition of his father's explorer friend, although he was soon variously nicknamed "Jock" or "Jack," just as his father had been. As if the house was not already crammed to the rafters, in the autumn Hunter took in an assistant, nineteen-year-old William Lynn, and at the same time Anne's younger brother Everard moved in with the family. Having attended Westminster School, the sixteen-year-old gave up a place at Cambridge to accept Hunter's offer of an apprenticeship.[1] For the moment at least, young Everard was grateful, declaring himself "ambitious to tread the paths of science under so able a master." But while Edward

Jenner, Hunter's favorite pupil, would refer to his teacher as "the dear man," and William Lynn later called his master "Glorious John," Everard Home's lasting tribute to his mentor would be significantly less noble.[2]

The house on Jermyn Street, in the heart of London's West End, was both a hive of domestic and social activity and a center for intellectual and scientific discourse. Despite their divergent interests, Anne ran the two households smoothly and hosted her soirées, which were acclaimed for their jovial atmosphere.[3] Though she published no more while her husband was alive, she continued to write poetry and shared her efforts with family and friends. Alongside the Bluestocking ladies, she counted among her male friends the prolific letter writer Horace Walpole and, later, the composer Joseph Haydn.

Although Hunter rarely participated in his wife's gatherings—literary debates were decidedly not his idea of fun—he usually took time out from his studies to greet her guests as they arrived.[4] But they were not always made welcome. Arriving home one evening, only to find his drawing room filled with dancing guests, he stormed into the room and announced, "I knew nothing of this kick-up, and I ought to have been informed beforehand; but as I am now returned home to study, I hope the present company will retire."[5] Sheepishly, Anne's friends shuffled out. Yet though he was certainly more comfortable in the company of his male, scientifically minded Royal Society chums, he did accompany Anne to the theater, concerts, and house parties. After all, what he lacked in cultural accomplishment, he could make up for in diverting conversation. Dr. Johnson's close friend Hester Thrale was struck by one of Hunter's revelations at one social engagement in the 1770s. She noted in her diary, "The heart of a frog will not cease to beat *says John Hunter* for four hours after it has been torne from the Body of the Animal," adding dolefully, "Poor Creature."[6] Lord Holland, another member of the Hunters' social circle, remarked, "John Hunter was neither polished in his manner nor refined in his expression, but from originality of thought and earnestness of mind he was extremely agreeable in conversation."[7]

Just as John usually indulged Anne in her literary pursuits, so she made no attempt to curtail her husband's passion for experiment. Certainly it must have been irksome at times trying to ensure her guests did not bump into the Resurrectionists on the stairs or stumble into a room where a strange breed of animal was being dissected. It was awkward, too, having friends to stay at their Earls Court country house—where she introduced refinements such as painted panels on the doors[8]—only to be interrupted by the shrieks of animals suffering her husband's experiments. And in both homes, she had to struggle to mask the smell of decay and alcoholic preservative clinging to the furnishings. Yet she respected her husband's work, and their relationship proved both a harmonious and successful alliance. Within the first five years of their marriage, Anne gave birth to four children, although only two—John Banks and Agnes—would survive infancy. For all his surgical expertise, there was little even Hunter could do to protect his offspring from the virulent childhood diseases that abounded.

Among the guests arriving at Jermyn Street to visit the new family was James Beattie, the Scottish poet and moral philosopher, who stayed to supper at Jermyn Street with his wife, Mary. "Mr Hunter showed us his Anatomical curiosities, which are very numerous and well arranged," Beattie recorded in his diary. "He seems to be very profoundly skilled in Comparative anatomy."[9] Another party of guests was rather less enamored of Hunter's anatomical collection. George Cartwright, a trader and explorer, brought back a group of Inuit people, some of the first "eskimoes" ever to set foot in England, on his return from northeast Canada at the end of 1772.[10] Keen to learn about their lives and environment, Hunter asked Cartwright to bring his party to dinner at Jermyn Street. After enjoying the meal, the head of the family, a priest named Attiock, wandered into a room within Hunter's backyard, only to find himself face-to-face with a glass case full of human bones. Rushing back in a state of shock, he demanded to know whether these were the bones of other Inuit guests who had been killed and eaten by Hunter. Laughing at their confusion, and not entirely truthfully, Hunter assured the party that the bones had all belonged to executed criminals, who had been dissected for the greater good of science. The visitors left rather hurriedly, and according to Cartwright, "Attiock's nerves had received too great a shock to enable him to resume his usual tranquillity until he found himself safe in my house again." Yet the family recovered their confidence in Hunter sufficiently to allow him to commission portraits of all five members.

With guests to entertain, a young family to support, private patients to visit, house pupils to teach, and hospital duties to fulfill, as well as his relentless program of research to pursue, Hunter had more than enough to occupy his days. Yet inevitably, given his boundless zeal for novelty, he now embarked on a new project. Since his childhood days, when he had skipped lessons to explore country life, he had been bent on self-education. As an autodidact, he excelled. And so in the autumn of 1772, he resolved to teach others what he had learned—or, more precisely, to teach them to learn for themselves, too. The mission he embarked upon, to provide the foundations of a medical education for London's growing body of surgical pupils, might have seemed impossible to fault. Yet it set him on a headlong collision course with his colleagues at St. George's.

Since Hunter had taken up his staff surgeon's post at the hospital in 1768, would-be surgeons had flocked to enroll as pupils under him. His growing reputation as a radical thinker was a magnet to the new generation of youngsters keen to enter surgery. In the "age of experiments," when revelations in natural history, chemistry, and anatomy were constant topics of conversation in newspapers and coffeehouses, upcoming surgeons sought out the hospital's most adventurous surgeon as their role model and mentor. More than 120 youngsters signed up as pupils at St. George's between 1768 and 1772, and by far the majority enrolled with Hunter. In 1771 alone, more than two-thirds of the pupils signed under his name, each staying at the hospital between six months and one year.[11] Yet no matter that Hunter topped the students' popularity polls year after year, the pupils' fees were divided equally among the four surgeons. Since the sum total was relatively small—only £141 in 1772, for example—for the moment Hunter turned a blind eye to this incongruity.

With his house pupils, his hospital students, and his apprentice, a veritable posse trailed behind Hunter as he conducted his ward rounds, performed operations, and undertook postmortems. But though this offered valuable practical experience in the traditional manner, it was not, in his view, an adequate medical education. Hunter aimed to nurture questioning, inquisitive, skeptical young surgeons who were capable of thinking for themselves, and for this he needed to convince them of his novel doctrine. Shortly after joining St. George's, Hunter had attempted to persuade his colleagues to give lectures on surgery free of charge. A hospital, he later explained to the governors, should be not only a charitable institution offering aid to the poor—indeed, even a place where surgeons gained experience before trying their luck on wealthier clients—but also a center for educating the surgeons of the future.[12]

At a time when none of London's six general hospitals offered an all-around medical education, this was a futuristic vision. Hunter's proposal amounted effectively to creating at St. George's both a medical school and a research center. "My motive," he later explained, "was in the first place to serve the Hospital and in the second to diffuse the knowledge of the art that all might be partakers of it." And in a clear dig at his colleagues, he added, "This indeed is the highest office in which a surgeon can be employed; for when considered as a man qualified only to dress a sore, or perform a common operation, and perhaps not all of those that may be reckoned common, he cannot be esteemed an ornament to his profession."

The idea was given short shrift by his three surgical colleagues. Despite several overtures—he even called a meeting with them to discuss how the pupils' education might be improved—none would consent to his proposal. Although his elder colleagues, Caesar Hawkins and William Bromfield, had staged anatomy lectures in their youth, they were either too busy or too lethargic to bother now, while John Gunning, his declared enemy, dismissed the idea with disdain. Disgruntled at the lack of attention, even those pupils signed up with Bromfield, Hawkins, and Gunning now turned to Hunter for extracurricular direction. According to Hunter, "They all wished to be dressers under me and at last complained to me and even threatened to recommend their young acquaintances to other hospitals if more general attention was not paid to them." Although Hunter passed on their comments to the governors, and some of the students even complained anonymously to the board, it was to no avail.

Rebuffed by his colleagues but egged on by the students, Hunter resolved to go ahead alone. In the autumn of 1772, he invited pupils from St. George's, regardless of which surgeon they had enrolled with, to a series of lectures at his own house free of charge.[13] A few years later, he threw the lectures open to a wider audience, advertising his course in London newspapers and charging a modest four guineas for almost a hundred lectures running on alternate evenings from October until April. The move split London's surgical fraternity down the middle. One half acclaimed Hunter as a genius and his lectures as inspired; the other half condemned him as a charlatan and his lectures as incomprehensible.

Jessé Foot, who stood in the establishment camp, described the lectures as "a sort of a skirmishing course—something new, and which could not be compared—consisting of surgical, physiological, and comparative anatomy branches,—and so mixing them together, as either to confound or illustrate each other."[14] In fact, Foot had contemplated attending the course himself. He called at Jermyn Street for a syllabus, then changed his mind, on the grounds that the "terms were high" and the "design was not liberal." Hunter's St. George's colleagues were equally dismissive. "What his lectures were we cannot tell," they later remarked snidely; "we have understood that they were ingenious but physiological rather than chirurgical."[15] Yet for Henry Cline, a twenty-four-year-old newly qualified surgeon at St. Thomas' Hospital, attending Hunter's lectures would prove cathartic. He later declared, "Having heard Mr Hunter's lectures on the subject of disease, I found them so far superior to every thing I had conceived or heard before, that there seemed no comparison between the mind of the man who delivered them, and all the individuals whether ancient or modern who had ever gone before him."[16] There were many more who were similarly moved.

The schism was hardly surprising. As both his devotees and his enemies made plain, Hunter's lectures were unlike anything ever staged in London, or indeed Europe, before. There were plenty of private lectures on offer. William Hunter's lectures in anatomy were immensely popular, Percivall Pott's course on surgery was highly regarded, and there were reputable lessons in midwifery, too. But essentially these all offered practical training. Hunter offered something fundamentally different. Although he titled his lectures "The Principles and Practice of Surgery," he barely mentioned operations; in fact, he encouraged his pupils to avoid operations altogether unless absolutely necessary. And while he was undoubtedly London's most experienced anatomist, he scarcely discussed anatomy, either. Instead, he offered his pupils a comprehensive introduction to the workings of the human body—its physiology—in order to teach them how the body works when healthy and how it reacts when diseased, along with many of his radical views on treatment. His course was based on the belief that any surgeon planning to treat disease and injury should first understand basic principles of health. "They are to the surgeon what the first principles of the mathematics are to the practical geometrician, without the knowledge of which a man can neither be a philosopher nor a Surgeon," he explained in a magazine article describing the course.[17]

Although few practitioners still seriously believed wholesale in the ancient Greek ideas of humoral balance, the language and the practices remained. The ubiquitous therapies of bloodletting, purging, and vomiting all emanated from the notion of letting out unwanted humors. Hunter, in contrast, wanted to sweep away all past superstitions and unproven doctrines and begin from absolute basics. Most important, he aimed to provide his disciples with a grounding in life sciences—despite how inadequately they were understood—in order that they should learn to reason for themselves. "I do not intend to give my lectures as a regular course," he explained, "but rather to explain what appear to me to be the principles of the art, so as thereby to fit my pupils to act as occasion may require, from comparing and reasoning on known principles."[18]

It was a revolutionary concept. The idea of thinking for themselves was shocking, uncomfortable, even distasteful to some, but liberating and inspirational to others. And so, as pupils made their way to Jermyn Street for the first lessons in October 1772, there were wide variations in expectations.

Squashed together, elbow-to-elbow, in one of the cramped rooms in Hunter's house, the students waited with their paper and quills at the ready for their nervous-looking teacher to begin. Always anxious about public speaking, Hunter knocked back thirty drops of laudanum, opium mixed with alcohol, before starting each lecture.[19] As he shuffled his notes, the class became impatient, so that "the more humorous and lively part of the audience would be tittering, the more sober and unexcitable quietly dosing into a nap," one pupil, John Abernethy, recalled.[20] Beginning slowly and hesitatingly in his gruff Scottish drawl, Hunter immediately confounded his audience by telling them not to take notes, or, if they did, to burn them later.[21] It was not that he was concerned they would produce pirate copies of his lectures; he merely believed that as a perpetual student himself, his views were constantly changing. When one of the pupils, sufficiently intrigued to attend a second course, accused him of altering his views from one year to the next, Hunter calmly retorted, "Very likely I did. I hope I grow wiser every year."[22] When asked by another pupil whether he had put a particular viewpoint in writing, Hunter replied, "Never ask me what I have said, or what I have written; but if you will ask me what my present opinions are, I will tell you."[23]

But Hunter's revolutionary opinions were not always easy to determine. Even his most devoted pupils would admit that Hunter experienced "great difficulty in communicating what he knew" and that most of his students "acknowledged the difficulty they found in comprehending him."[24] Yet they would also note that "his wit, or more particularly his archness, was always well directed" and that Hunter relished taking questions at the end of each session. Notes of Hunter's lectures, transcribed by numerous students despite his exhortations, testify to his dry sense of humor and informal lecturing style, for he sprinkled his discussions with anecdotes and caustic asides. And while even he acknowledged he could sometimes seem "a little abstruse," there was every reason that his views were difficult to elucidate. As he always warned his pupils in the introductory lesson, "Many of my ideas, and the arrangement of my subject, are new, and consequently my terms become in part new."[25] Quite simply, the language required to explain some of Hunter's doctrines did not yet exist.

Some of the novices, who, reasonably enough, had enrolled for the lectures to pick up some tips on surgery from the celebrated surgeon, were further bemused as he launched into his course. Far from initiating them into the secrets of successful surgery, Hunter began by discussing the very nature of organic life.[26] This was no easy task, since any useful studies in geology and biology were still many decades ahead. Hunter began by outlining the differences between inanimate matter—rocks, minerals, earth—and living matter—animals, plants, vegetables—before moving on to grapple with the thorny issue of what distinguished living animals and humans from dead ones. Still struggling to explain that particular puzzle himself, he proposed that life, or his "living principle," was "something superadded."

Having discussed organic life in general, he moved on to humans. In a series of lectures stretching over the winter, he described the various functions of the human body in its natural or healthy state, discussing muscle action, nutrition, respiration, circulation, bone growth, and sleep. Understanding these normal functions was an essential grounding for any would-be surgeon, he insisted.

Only after this lengthy foundation in physiology did Hunter discuss the effects of disease and injury, or pathology, in humans. First, he explored the signs of disease, such as inflammation and fever, before describing specific illnesses. Discussing a range of conditions, including cancer and scrofula, broken limbs and hernias, venereal disease and infectious fevers, he illustrated his opinions with beautifully prepared specimens from his collection, as well as case histories gleaned from his twenty years' experience as a surgeon. When describing inflammation, he passed around his preparation of a human tooth implanted in a cockerel's comb, and a maggot buried in the skin of a reindeer. The treatment of aneurysms, a pet topic, was illuminated by a preparation of a man's chest with a bulging aorta, while the bones of a lion were pressed into service when describing bone growth. Meanwhile, his growing acclaim as a surgeon furnished him with entertaining anecdotes about aristocrats and celebrities with which to amuse his students. He could always raise a laugh with his tale of the recently deceased novelist Laurence Sterne, who had asserted a person only died when they resolved not to live. When Sterne attempted to prove his conviction by leaping out of bed during his final illness, said Hunter, "death, which soon followed, shewed his mistake."[27]

But if the fidgeting, dozing, secretly scribbling pupils now believed they were at last coming to the meaty topic of practical surgery, they were to be disappointed. Hunter described almost as many occasions when he had resolved not to operate as when he had gone ahead. This was not to say that he avoided operations, even extremely risky ones, when convinced they were necessary, as he had recently proved by assisting at the attempted cesarean delivery. He outlined sufficient details of mastectomies, castrations, amputations, and trepanning for even the most bloodthirsty young pupil, although even then it was theory rather than practice that dominated. In cases of cancer, he recommended radical surgery, knowing from postmortems that "leaving the least part of the cancer is equal to leaving the whole." He even advised examining the removed lump to ensure as much of the cancerous tissue as possible had been excised.[28] When patients died, Hunter was disarmingly candid in admitting his errors. Relating a trepanning operation that went amiss, he freely confessed, "I think it is probable I killed him." Describing another operation, he declared, "I acted like a blockhead."[29] Medical error would always be a fact of life in surgery. While other surgeons balked at the very idea of confessing their blunders, for Hunter it was vital to learn from his mistakes.

His conservative approach extended to medical care in general. Hunter did not allow the traditional hierarchy to stop him from pronouncing on "physick" as well as surgery. Many of his closest friends were physicians and his advice on general medical matters was increasingly in demand. He firmly advised using the three favorite Georgian remedies—purging, vomiting, and bloodletting—with caution. Warning that purgative medicines could severely weaken a patient, he urged, "A single purge has been known to produce death in dropsy," and he recommended extreme care when administering medicines to induce vomiting.[30] Bloodletting, too, should be sparingly used: "Bleeding, however, is a remedy of so much importance that it should be employed in all cases with great caution; yet not more than appears really necessary."[31] In truth, however, bloodletting was impossible for surgeons to eschew completely, since Georgian patients routinely demanded the procedure. Accordingly, Hunter did bleed some patients. In the summer of 1772, just before his first course began, he had bled the socialite Lady Mary Coke, on the recommendation of her physician. "'Tis not that I am very ill," she noted in her diary, "but Mr Fox, who I now consult, advised me to be blooded in my foot."[32] Still, Hunter was one of the first to suggest that bloodletting was not just largely ineffectual but potentially dangerous, as well. Even fifty years later, his nineteenth-century editors would add a footnote to his works, insisting that bloodletting and mercury, "our chief curatives," were still valuable treatments.

For all his iconoclastic views, Hunter was careful to shield his pupils from overt trouble. While expounding his controversial views on gunshot wounds, he warned, "If you were examined at Surgeon's Hall how you would treat a gun shot wound, you would do well not to mention my doctrine, but to say according to the old plan, I would open the edges of the wound." Ruefully, he added, "They do not ask the reasons of things there."[33] Hunter's aim was that the young surgeons attending his lectures would always "ask the reasons of things." He wanted them to take nothing for granted, to subject every common superstition and unproven therapy to scrutiny, to question every step they took. Essentially, he aimed to equip them to elevate surgery to the rank of a science.

But not everybody was ready for such dramatic change. The popularity of his lecture course would wax and wane over the next two decades. While sometimes fifty pupils would cram in to hear him speak, at other times only twelve turned up. Indeed, one lecture was reportedly so scantily attended that Hunter had a skeleton brought in so that he could begin his lesson with the customary address, "Gentlemen."[34]

Though some dozed, some left confused, and others shunned the course altogether, there were many for whom Hunter's revolutionary view of medicine was life-changing. Henry Cline, who strolled along to hear Hunter's views chiefly out of curiosity, was one of them. He later recalled, "When I heard this Man, I said to myself, *This is all day-light*. I felt that what I had previously been taught was comparatively nothing. I felt that I was now enabled to judge of what my experience and Observation had taught me; and thought I might, like Mr Hunter, venture to Think for myself."[35] Others included Astley Cooper, a pupil at Guy's Hospital, who was undeterred by Hunter's detractors: "The surgeons of Hunter's day thought of him as a mere imaginative speculator, and anyone who believed in him a blockhead and black sheep in the profession."[36] He was so impressed, he attended year after year. John Abernethy, an apprentice from St. Bartholomew's, was similarly affected. "I believe him to be the author of a great and important revolution in medical science," he declared, "of this I am certain, that his works produced a complete revolution in my mind."[37]

Despite his faltering speech, his obscure language, and his unconventional syllabus, John Hunter won the minds and hearts of countless pupils who sat in his lecture room. They imbibed his doctrines, emulated his practices, and preached his creed wherever they worked. Much more than his own writings, it was Hunter's disciples who would spread his message to future generations of surgeons and apply his principles to future practice. In all, including almost five hundred young men who followed him at St. George's, Hunter would teach an estimated one thousand pupils.[38] Many would become towering figures in the powerful nineteenth-century teaching hospitals. Others would take his message overseas to Ireland, the Continent, and the United States. They formed an unstoppable army of evangelists. But their dedication to Hunter's standard was bound to broaden already wide divisions and stoke up already passionate jealousies. For Hunter's colleagues at St. George's, this magnetic attraction to youngsters from all over the globe, not least their own pupils and apprentices, would create an unbridgeable rift. According to Foot, Hunter was embroiled in "continual war" at the hospital.[39] Hunter himself would say, "I know, I am but a pigmy in knowledge, yet I feel as a giant, when compared with these men."[40]

For Edward Jenner, Hunter's most loyal devotee, the cut and thrust of competitive city life was all too much. Having declined Cook's invitation to join his second round-the-world expedition as a naturalist on the *Resolution* the previous year, Jenner packed his bags in 1773 and opted for a quiet life as a country doctor in his native Gloucestershire. He would not be allowed to remain idle, however. From the moment Jenner left London, Hunter wrote him a stream of chatty letters, besieging his beloved pupil with demands to send animal specimens and to conduct experiments. In return, Jenner pressed his teacher for advice on treating patients and on his own research. The pair would exchange letters and gifts in an affectionate, lively correspondence spanning two decades. Perhaps more than anything else he was to write, Hunter's scrawled, misspelled, hurried but always fond letters to his favorite protégé reveal the tireless, erratic, ever-inquiring mind of the irrepressible anatomist and experimenter, who was always on the lookout for a fresh specimen to dissect or investigation to launch. It was plain, as Hunter remarked in one letter, that, "I do not know anyone I wd sooner write to than you."[41]

Jenner's departure left a large dent in the Hunter household. He was missed by Anne, who shared his passion for poetry and music, as well as by John. But for Anne's young brother Everard, Jenner's departure was the golden opportunity he had been waiting for—to fill the shoes of the favored pupil. This was no easy task. Jenner understood his teacher's blunt ways well enough not to take offense when Hunter criticized his "damned clumsy fingers" for breaking one of his precious thermometers. The sensitive and self-important Everard bristled when Hunter snapped that he would never be able to make a decent preparation, since "his fingers were all thumbs."[42] As Hunter's assistant, tied to his master's remorseless schedule, Home often felt the lash of his brother-in-law's sharp tongue. He would later complain, "His temper was very warm and impatient, readily provoked, and when irritated, not easily soothed."[43] Others, too, would testify to Hunter's fiery disposition. Foot accused him of the "bitterest utterings of swearing," and even Lord Holland, who was a friend, admitted that he was "apt to be positive, dogmatic and angry."[44]

Hunter could prove a formidable opponent to anyone who crossed him, and his outspoken views often aroused strong passions in others. Yet he also evoked intense devotion. Though one of Hunter's long-suffering assistants, the artist William Bell, testified to his boss's foul temper, it was said that he "absolutely idolised" Hunter.[45] There were plenty more assistants, pupils, and associates who would defend Hunter's conviviality, hospitality, and kindness. Plainly, he prompted strong feelings both ways. But while some whom he offended went so far as to threaten him physically, Everard Home merely stored up the supposed slights for future revenge.

Now that he was forty-five, however, Hunter suddenly found himself the victim of his own irritability. One morning in 1773, soon after breakfast, he was struck by a violent abdominal pain, which his usual thirty drops of laudanum did nothing to relieve. Scrupulously recording his own symptoms, just as he would for a patient, he found, to his alarm, that he could feel no pulse, his breathing had stopped, and his countenance looked, in his expert opinion, "like that of a corpse."[46] Later describing the episode to his pupils, for no opportunity for learning was to be missed, he said, "I cast my eyes on a looking-glass, and observed my countenance pale, my lips white, and I had the appearance of a dead man looking at himself." Through sheer power of will, he forced himself to breathe. After taking brandy, madeira, and ginger as stimulants, he was back at work by 2:00 P.M. Ironically, while Hunter seems not to have recognized his complaint at the time as an attack of angina, only the previous year he had himself dissected a man who had died from the condition. Hunter's notes of the postmortem were included with a paper published in 1772 by the physician William Heberden, which gave one of the first accurate descriptions of angina.[47]

Conceivably, Hunter's first angina attack was the result of his self-inflicted syphilis reaching the tertiary stage. During this final stage, which can begin anywhere from three to twenty-five years after initial infection, heart problems can occur. Equally, the angina, which would afflict him for the rest of his life, may well have been prompted by the continual strain of his hectic lifestyle.

With Anne pregnant for a third time—their second child, Mary-Ann, born in December 1773, was a "weakly" infant—it was a substantial relief when she gave birth to a healthy boy in November 1774. The couple called the infant James, after John's favorite brother—there would never be a William—but at home, he was known as "Jemmy." Regardless of the fact that his father pronounced him "a fine boy," Jemmy was to live for only three months. Anne would denote the month he died as "Black February" in a poem written seven years after his death.[48] His sister, Mary-Ann, would follow him to the grave at the age of two. Hunter was plainly fond of his children, enjoyed playing with them, and was proud of their achievements. Nevertheless, he seemed less excited that November by the delivery of his second son than by the arrival of some particularly curious marine animals from South America.

Electricity was an absorbing preoccupation in eighteenth-century Britain, providing a lively topic of discussion at Royal Society debates and in coffeehouse conversations. Benjamin Franklin, the American politician and businessman, had helped stir up interest in the phenomenon in 1751 with a paper to the Royal Society, in which he argued that lightning was an electrical force. The following year, he demonstrated the fact with a perilous experiment that entailed conducting lightning down a kite string with a key attached.[49] The society's fellows initially scoffed at Franklin's conclusions, but they soon had to accept the veracity of his findings. The invention, in the 1740s, and rapid popularity of the Leyden jar—a metal-coated glass container that could store electricity to produce electrical charges on demand—made the properties of electricity the subject of both intense scientific speculation and wild fantasy. Physicians and quacks alike proclaimed the benefits of electrical therapy. In 1772, the preacher John Wesley advertised his "Electrical Machine" for curing all manner of ailments, and James Graham, the Scottish physician turned quack, would claim similar success for his electrical bath.[50] So when electric eels were shipped to London from Surinam in November 1774, the news created a sensation.

John Hunter was one of the first to hear. The strange fish—they were not really eels, as Hunter quickly observed—had first caused a stir when a showman had demonstrated their powerful electrical charges in South Carolina (a large eel can produce a charge of up to six hundred volts, sufficient to injure or even kill an adult). The spectacle was witnessed by Alexander Garden, the naturalist who had sent the greater siren, and he had persuaded a British sailor to take five of the creatures to England, principally for the interest of the Royal Society. Unfortunately, all five had died before reaching London, although one that remained alive when the ship docked at Falmouth had favored several people with its famed shock, luckily without fatalities. Four of the eels, however, had been preserved in spirit and were taken immediately to Hunter. He was ecstatic. "Mr John Hunter danced a jig when he saw them," said his friend Daniel Solander in a letter to fellow enthusiast John Ellis, "they are so compleat and well preserved."[51]

Hunter was already familiar with the electrical powers of certain animals. Two years earlier, he had dissected some torpedo fish—electric rays—at the request of John Walsh, a Royal Society fellow and MP, who had obtained several of the creatures in the waters off La Rochelle, France. Encouraged by Franklin, Walsh had become obsessed by the ability of the flat fish to produce weak electric shocks. At La Rochelle, he performed a series of experiments linking circles of as many as eight people to the wings of the fish in order to form an electric circuit. Writing excitedly to Franklin, Walsh said, "The effect of the Torpedo is absolutely electrical." Having brought several dead specimens back to England, he asked "the ingenious Mr John Hunter" to investigate their unusual anatomy.[52]

Peeling away the skin on each wing to reveal the two electrical organs,

Hunter carefully investigated the hundreds of columns of tightly stacked disks, which together concentrated naturally occurring electricity (the structure of the torpedo fish's electric organs would later inspire Alessandro Volta to create the first battery). At the same time, Hunter noted the extraordinary volume and size of the nerves connected to the electrical organs, and he presciently remarked, "How far this may be connected with the power of the nerves in general, or how far it may lead to an explanation of their operations, time and future discoveries alone can fully determine." By electrifying frogs' legs in 1792, the Italian Luigi Galvani would demonstrate how electrical impulses prompt the nerves to generate muscle movement, simultaneously inspiring literary fascination with electricity in people such as Mary Shelley, who went on to write *Frankenstein.*

Hunter's acutely observed anatomical observations on the torpedo fish were presented to the Royal Society on July 1, 1773. At the meeting, Hunter passed around preserved specimens of a male and female fish with their organs dissected to reveal their remarkable structure. Yet the torpedo discoveries left much unexplained. Many still doubted that their weak charge, which failed to produce a spark like the Leyden jar, was genuine electricity.

The electric eel, it was hoped, would provide the answers. The volunteers in Cornwall could already testify to the powerful shock a single eel could deliver. The excitement that ran through the Royal Society was hardly less charged. Solander immediately began soliciting subscriptions for a scheme to bring back more live specimens from Surinam; he wrote to Banks to confirm the plan. Rushing to town to see the preserved eels for himself, John Walsh promptly ponied up sixty guineas for three of the specimens "so they may soon be examined and dissected by John Hunter." Evidently, Solander was having difficulties restraining Hunter and Walsh, for he warned Banks, "If you don't come to town soon Walsh & Hunter seem to be bent upon beginning with opening one at least at the beginning of next week."[53] Banks failed to make it in time. Unable to wait a moment longer, Hunter began dissecting the eels on November 15, and he quickly discovered their "peculiar organs," which exhibited a structure similar to that of the torpedo but extended along much of the fish's body. Again noting the unusual size of the nerves supplying the electric organs, he furnished the Royal Society with a meticulous description the following May, while retaining two of the fish for his museum.[54]

That same summer of 1775, the first live electric eels arrived in

London and parties to test how many people could feel their electrical charge at once became a popular craze. Walsh invited more than forty Royal Society members to his house to witness the experience firsthand. He made the shock pass through a chain of twenty-seven people and even produced the elusive "spark," which convinced remaining skeptics that the force was indeed authentic electricity. John Pringle, who had become president of the Royal Society in 1772, witnessed a second display with an even larger chain. "The sparks were vivid & repeated," he wrote to tell Haller, "& were seen, the last time I attended, by about 70 people at once; and in another experiment the shock was conveyed through the bodies of the same number of people, when they joined hands to hands to form a circle."[55]

No doubt fired by so much popular excitement with animal anatomy, Hunter now conceived a scheme to create a school devoted to the study of natural history. He asked Jenner to become a partner in the project, but the country-loving doctor declined. The idea was soon forgotten in the flurry of continual industry.

Hunter had more than enough work to keep him occupied. That same month saw Cook's return from his second voyage, the explorer having brought back a further haul of natural treasures. On August 14, Solander visited the ship with a group of women friends. The party was shown a preserved Maori head, minus a portion that had been "boiled and eat" on board the ship by two New Zealanders after the crew had purchased the head from a native hunting party.[56] The exhibit and its gory history had "made the Ladies sick," Solander recorded. He knew exactly who would appreciate it. The next day, Solander took Hunter down to Deptford to claim the trophy, which was promptly added to his collection. So intense had his researches in anatomy and natural history now become that Hunter employed a young surgeon and talented artist, William Bell, offering him a ten-year contract to churn out drawings of the preparations he made, as well as to assist him in the dissecting room.

Despite the bitter loss of young Jemmy, it had been a busy and fruitful year, one that combined a happy family life with a successful work program. Hunter reflected his satisfaction with his lot in a newsy letter that November to his brother-in-law, the Reverend James Baillie, who had married his sister Dorothea in 1757 and had just been appointed professor of divinity at Glasgow.[57] "As for my self," said Hunter, after congratulating Baillie on his appointment, "with respect to my family, I can only yet say, that I am happy in a wife; but my children are too young to form any judgement of." With his Anny content to see Hunter pleasing himself, he concluded, "I must continue to be one of the happiest men living."

His happiness was almost complete when, in the new year, London's most extraordinary surgeon was appointed surgeon extraordinary to George III. Although purely an honorary position, the post carried significant status. Hunter was not expected to treat the king, though his no-nonsense manner would have appealed to the plain-living monarch far more than the preening surgeons and physicians who would torture him with bloodletting, blisters, and purges during his incapacitating spells of illness. Hunter bought a new coach—"the handsomest that was at court"—to receive his honor in January 1776.[58]

The royal seal of approval meant Hunter's professional opinion was increasingly in demand. His advice was now sought by more and more well-known Georgian figures. When David Hume, the Scottish philosopher and historian who had made an enemy of the kirk with his skeptical writings, fell ill with abdominal pains in 1775, he was surrounded by the most eminent physicians of the day. None of them could offer a satisfactory explanation or effective remedy for Hume's obvious decline. Though they were happy to posit elaborate theories and propose assorted therapies, nobody was prepared to examine the patient in order to determine what might be causing his suffering. It was only when Hunter met the ailing philosopher and performed a physical examination that Hume finally discovered the true cause of his illness.

Hunter had arrived in Bath in June 1776 on a social visit and had immediately inquired after Hume; they were related by marriage, since Anne's father, Robert Home, was Hume's cousin—the spellings Home and Hume were interchangeable and both names were pronounced "Hume." On being invited by the local physician, Dr. Gusthart, to examine Hume, Hunter laid his hands on the suffering man's abdomen and could plainly feel a tumor, which he suspected was cancerous, in the liver. Since abdominal surgery was out of the question, there was no hope that an operation could save him. Although Hunter immediately conveyed his diagnosis to Gusthart, even then the philosopher was kept in the dark for several more days. Hume's relief when his physicians finally confessed Hunter's verdict, no matter how bleak, was palpable in the letters he wrote home. Relating the news to his brother, John, Hume declared, "This Fact, not drawn by reasoning, but obvious to the Senses, and perceived by the greatest anatomist in Europe, must be admitted as unquestionable, and will alone account for my Situation."[59] The verdict of Hunter's fingers, to Hume at least, carried more weight than the hypothetical musings of all the country's best physicians. Despite his physicians' dogged persistence with their meddling medication, the philosopher died in Scotland, of suspected liver cancer, two months later.

The outcome was more cheerful when Hunter treated another well-known writer, the diarist William Hickey, that same summer. Hickey had traveled to Margate on holiday in August but had burned his foot when he fell asleep by the fire and one of his companions attached a burning taper to his boot as a prank.[60] By the time Hickey woke up, in extreme pain, the foot looked so badly injured that a local surgeon feared he would never walk again. A messenger was dispatched urgently to London to summon Hunter for a second opinion. Racing to the coast in order to examine Hickey's foot, Hunter "at once declared no ill consequence would arise." And after taking Hunter's advice to keep his leg propped up, with a cold poultice applied to the burn, Hickey was back on his feet within days.

In the light of his royal honor, his controversial lectures, his senior hospital post, and the recognition of his skills among the country's top physicians, John Hunter had become a household name. And in characteristic Georgian style, his celebrity was acknowledged with a satirical poem, composed by a radical Scottish journalist, James Perry. Entitled rather perversely *The Torpedo: A Poem to the Electric Eel*—plainly Perry was no naturalist—the lengthy piece of doggerel, published as a pamphlet in 1777, was addressed to "Mr John Hunter, surgeon."[61] Using the vogue for electric eels as an excuse to mention the various scandals concerning the country's nobility, the poem graphically captured many of Hunter's best-known ventures to date. It began:

> *O Thou! Whose microscopic eye*
> *Can every living thing decry,*
> *And search Dame Nature's womb!*
> *Whose power can raise the lifeless clay,*
> *Drag the pale spectres into day,*
> *And starve the hungry Tomb!*

Now, John Hunter's powers to "raise the lifeless clay" were about to be tested to the full.

CHAPTER 12

The Chaplain's Neck

London
June 1777

On Sunday, June 22, 1777, the Reverend William Dodd was not delivering his customary sermon. Instead, he sat in his dismal cell in Newgate Prison, composing a desperate appeal to Samuel Johnson in a last-minute effort to save his life. Dodd, a foolish and foolhardy clergyman who enjoyed the good life as much as good works, was sentenced to hang in five days' time.[1]

Having been ordained a priest some twenty years earlier, Dodd had devoted himself more to physical than spiritual pleasures. He threw himself into the social whirl of Georgian society, published a novel, and dressed in the dandiest of fashions, earning himself the nickname "the Macaroni Parson." The characterful curate was a generous benefactor to numerous charities, not least the Magdalen House for "fallen women," where he was made official chaplain. Although his sermons were popular—he had been appointed a chaplain to the king in 1763—ultimately his extravagant lifestyle led to his downfall. With debts mounting, he staged an amateurish attempt to forge a bond for a hefty £4,200—upward of £250,000 in today's terms—in the name of his erstwhile patron, Lord Chesterfield. After the discovery of his fraud, he was found guilty of forgery at the Old Bailey in February 1777, and in May, he was sentenced to hang.

Dodd's plight immediately provoked an eruption of popular feeling, resulting in a demand for a reprieve. A petition with 23,000 signatures pleading for a royal pardon was presented to George III, and Samuel Johnson, who had met Dodd once briefly, agreed to lend his powers of persuasion to the cause. The illustrious writer penned an impassioned speech, which

Dodd addressed to the Old Bailey, as well as an eloquent sermon, which he delivered to his fellow prisoners at Newgate. After Dodd's last appeal for help on the Sunday before his expected execution, Johnson again came to his aid, this time by ghostwriting a letter to the king.

Despite Johnson's well-chosen words, media sympathy, and mass public support, all was in vain. No royal pardon arrived, and on June 27, crowds turned out to watch as Dodd was transported to Tyburn. "On this occasion there was perhaps the greatest concourse of people ever drawn together by a like spectacle," reported *Gentleman's Magazine*. "From Newgate to the place of execution the streets were thronged, and never were seen so many weeping eyes."[2] Yet even as Dodd felt the noose around his neck, he had not lost hope. If Johnson's talents could not prevent him from dying, he fully believed John Hunter's skills would bring him back from the dead.

Not far from the sobbing crowds, at an undertaker's parlor in Goodge Street, Hunter was waiting with a number of medical friends to receive the curate's body. A bed had been prepared, a fire had been lighted, medicines were lined up, and a pair of bellows was placed at the ready. As the minutes ticked by after Dodd was presumed to have swung, the waiting men listened anxiously for the rattle of hooves that would signal the arrival of the coach bearing his lifeless body. So long as the delay could be kept to a minimum, Hunter was confident his plan to revive the unfortunate curate had every chance of success.

Hunter's optimism was by no means misplaced. He knew, from his numerous dissections of Tyburn Tree corpses, that most convicts who had been hung died from a long and slow process of asphyxiation, rather than from a swift and irredeemable broken neck. It was widely known that such convicts sometimes returned to life, occasionally on the dissecting table.[3] In Oxford, a servant named Anne Greene had revived under an anatomy knife after being hung unjustly for murder in 1650. More recently, in 1740, a seventeen-year-old thief named William Duell swung on the gallows for half an hour before he was delivered to the barber-surgeons. Just as the gathered surgeons were about to slice open his chest, the youth emitted a groan and sat up. Having cheated death once, his sentence was commuted to banishment to a penal colony.

Aside from these well-publicized cases of spontaneous recovery, Hunter was convinced from both his experiments and his theories that reviving a person after hanging was perfectly feasible. He had devoted years to attempting to understand precisely what constituted life, where life emanated from, and what caused life to end. Having initially proposed that blood contained a "vital principle" that distinguished dead from living matter, more recently he had suggested that every particle of the body contained some kind of life force. But however imperfectly he was able to explain this "living principle," Hunter's chief aim was to re-create it. The surgeons' equivalent of the philosophers' stone—defeating mortality, and even securing eternal life—was the goal that drove him.

Hunter's early efforts to freeze animals and bring them back to life had been one attempt to effect this power over death, and these were still favorite pursuits. The exceptionally cold winter of 1775–1776 had afforded him ample opportunities to return to his freezing experiments. When he observed that his cockerels at Earls Court had lost the jagged edges to their combs that winter, a servant explained that they often fell off during a hard frost. Hunter immediately replicated the situation by freezing the comb of a rooster. To his delight, the tissue grew back within a month.[4] In January 1777, a few months before Dodd's hanging, he was at work again with salt and ice, freezing a rabbit's ear for an hour before bringing it back to full-blooded life. At the same time, Hunter was busy investigating how long an animal's heart could beat after being removed from its body. It was in June, the very month that Dodd was transported to Tyburn, that he chirpily informed Mrs. Thrale of the ability of a frog's heart to continue beating for four hours after "death." Perhaps the enterprising surgeon really could "raise the lifeless clay."

Having devoted so much time and energy to such matters, Hunter was the obvious person to approach when a group of philanthropists resolved to draw up advice designed to save people who drowned. William Hawes, an apothecary in the Strand, first raised the idea of a society dedicated to rescuing victims of drowning.[5] Joining forces with a physician, Thomas Cogan, the pair founded the Humane Society—later the Royal Humane Society—in 1774. The charity promptly offered rewards of up to four guineas to anyone who succeeded in restoring life to any person "taken out of the water for dead" within thirty miles of London. Well-meaning surgeons and physicians living near the Thames offered their aid free to help revive those who were rescued. But this was easier said than done. With their unerring lack of imagination, physicians recommended bloodletting as the best method of reviving a victim. If this failed, ingesting tobacco vapors, usually by enema, was the fallback. Needless to say, neither method had met with much success. So in 1776, Hawes asked Hunter to develop a more factually based regime for resuscitation.

Hunter was eager to oblige, not only preparing detailed directions on attempted resuscitation for the charity but presenting his ideas to the Royal Society in March the same year.[6] In these, he advanced the view that a person who drowned should not automatically be considered dead but "that only a suspension of the actions of life has taken place." Indeed, this approach might be applied to anyone who suffered a violent death without irreparable injury to vital organs, he argued. And though he had had no opportunities so far to prove his views by experiment, he firmly believed it was possible to bring people back to life after drowning so long as their rescuers acted quickly—at least within an hour—and followed set guidelines.

Naturally, Hunter dismissed bloodletting and tobacco enemas, along with purging or vomiting remedies, as more likely to "depress life" than to restore it. Instead, the first aim of any rescuer should be to throw air into the victim's lungs, he argued, correctly determining that restoring breathing was the most effective route to revival. For this purpose, he recommended a pair of double bellows, "such as are commonly used in throwing fumes of tobacco up the anus"—though it is to be hoped not the same ones—for pumping air into a person's mouth or nose. Perhaps, he suggested, the newly discovered "dephlogisticated air"—oxygen—which Joseph Priestley had described in 1775 might prove even more effective. In addition to inflating a victim's lungs, he recommended holding stimulating vapors under the nose, warming the person slowly in a bed, and rubbing the body with essential oils.

If all else failed, Hunter suggested attempting to restart the heart with electric shocks. "Electricity has been known to be of service, and should be tried when other methods have failed," he advised. "It is probably the only method we have of immediately stimulating the heart." In fact, Benjamin Franklin had first suggested that electricity might be used to revive people who were apparently dead, although he had never put his theory into practice. But Hunter may well have been referring to a case from 1774, when a three-year-old girl who fell from a first-story window was revived with electric shocks to her chest, probably from a Leyden jar, in the first recorded example of successful defibrillation.[7] It would be nearly two centuries before this remarkable achievement became routine practice.

Finally, Hunter proposed that two people should work in tandem to effect resuscitation, and at every attempted rescue "an accurate journal" should be kept of the methods used and the degree of success. As ever, he believed that continual reassessment of practice was the route to improvement.

Hunter's recommendations to act quickly, concentrate on restoring breathing, and apply artificial respiration would become the cornerstones of standard resuscitation practice, although simple mouth-to-mouth resuscitation would eventually be adopted in 1959 as the most successful method. His recommendation to use electric shock treatment, or defibrillation, to restart the heart or regulate its rhythm would only become widely adhered to in the 1950s. But though effective methods of rescue would take time to introduce, the society's laudable aims attracted numerous supporters. Among them was the Reverend Dodd, who made a donation in 1776, the same year in which Hunter produced his guidelines. Now that his fortunes had turned, Dodd had the chance to make an even greater contribution to the charity's cause. In death, he represented the first opportunity for Hunter to test out his theories.

The events that ensued in the undertaker's parlor in Goodge Street would remain a secret for nearly two decades. Hunter would never refer in writing to what may well have been his most remarkable experiment. Yet almost as soon as the hapless Dodd swung from the Tyburn gibbet, speculation about his fate began. "Experiments were said to have been tried to bring Dr Dodd to life," reported *Gentleman's Magazine,* "according to the instructions formerly published by Dr [sic] Hunter, but without effect. He hung an hour, and it was full forty minutes before he was put into a hearse."[8] Fascination with Dodd's fate continued. Later that year, a London magazine reported an Irishman's claim that he had dined with Dodd in Dunkirk shortly after his supposed execution.[9] Then twenty years later, interest was revitalized in *Gentleman's Magazine* with a reader's query about the reported revival. This elicited a flurry of replies, which repeated the stories of Hunter's experiment. According to one account, Hunter and his helpers had tried to resuscitate the curate in a hot bath, and believed they would have succeeded had the crowd not delayed their efforts by half an hour. After it proved clear their plan had failed, the correspondent continued, the body was interred at St. Laurence's Church in Cowley, west London.[10]

The obsession with Dodd did not end there. In 1794, a Scottish newspaper suggested Dodd was alive and well and living in Glasgow, seventeen years after being brought back to life by John Hunter. Relating an account of his dramatic resurrection, the anonymous writer revealed, "When he was turned off, he felt a sudden impulse of pain at first, by his body whirling round very swiftly, he was soon deprived of all sensation, and afterwards remained totally senseless, until he found himself in bed, surrounded by Doctor C, Mr H, Mr D and Mr W, whom he perceived to be in tears, which may be considered as an effusion of joy at his recovery, of which they at one time despaired."[11] According to this version, Dodd's body had been conveyed to the undertaker's house, where Hunter and his three medical friends were waiting ("Mr H" was plainly Hunter; "Dr C" could well have been the cofounder of the society, Thomas Cogan). As soon as Dodd's body was bundled out of the hearse and into the house, Hunter and "Mr D" stripped the corpse and rubbed the skin vigorously for two hours before at last they saw a sign of breathing. Dodd's skin then broke out in a sweat, a groan emerged, and the curate sat up. Now fully restored to "sound health" but in "melancholy spirits" at his enforced absence from his native country, Dodd was living at the house of a friend, or so the writer claimed. So, did the life-loving curate breathe again after swinging at Tyburn? Had Hunter really defeated death?

Hunter never committed to paper the events that followed Dodd's hanging..But he did disclose details of the attempted revival in the privacy of his weekly coffeehouse club. Charles Hutton, professor of mathematics at the Royal Military Academy, Woolwich, and a Royal Society fellow, later recalled the evening, shortly after Dodd's execution, when Hunter was persuaded by his friends to relate the story.[12] It was true, Hunter confided to his rapt audience in the noisy coffeehouse, that he and several other Royal Society fellows had concocted a scheme to procure Dodd's body in order to attempt an experiment to bring him back to life. They had duly waited at the undertaker's in readiness for the arrival of the hearse containing his corpse. But the delays in transporting the body from Tyburn meant that by the time it arrived, the waiting medical men had all but given up hope, he told his listeners. Even so, Hunter and his friends had "tried all the means in their power for the reanimation," but after laboring for a considerable time, they were forced to give up. The chaplain's body remained cold and lifeless. The experiment, Hunter dolefully admitted, had "entirely failed."

Perhaps Hunter might have had more chance of success had he been in a more robust state of health himself when laboring to restore life in Dodd. In April, he had been struck down by a sudden dizziness, which confined him to bed for ten days—a considerable period of enforced rest for someone used to working a nineteen-hour day—and he was still unwell in early May. Although he was sufficiently fit in June to spend two hours trying to engender life into Dodd's corpse, by August he was persuaded by Anne to repair to Bath for the quintessentially fashionable Georgian remedy of sampling the spa waters.

Considering his declared skepticism when it came to the efficacy of mineral waters, it is unlikely Hunter believed that the Bath sojourn would confer any real benefit. In lectures, he cynically referred to physicians sending their patients "to die at Bath or Bristol" once their patients could "take no more physic or the physician can obtain no more fees."[13] His cynicism was justified. The Bath waters did little to help, and by the autumn he was considered extremely ill.[14] Suffering from angina again, Hunter believed that his life was in danger, as did his family.

Though Hunter may not have recognized the symptoms—or more likely deluded himself into dismissing them—he was all too aware of his own mortality by now. Such was his concern for the future that he decided to lose no time in putting his anatomy and natural history collection into better order. He knew it would provide the only source of income for his wife and two young children after his death. Before leaving for Bath, he had instructed Everard Home and his assistant William Bell to start producing a catalog of all the contents. But, according to Home, "his impatience to return to town made him come back before he was well."

Hunter was anxious, as always, not to miss the start of the winter season of lectures. He was giving his course this year at his brother's lecture theater in Great Windmill Street, just as he had the previous year. Among the new pupils was William Hamilton, the nineteen-year-old son of Thomas Hamilton—the friend who had traveled with John from Lanarkshire in 1748, now a professor of anatomy in Glasgow—who arrived in London in December 1777. Like many of the students, he had enrolled for both the brothers' courses, as well as paying his fee to use their shared dissecting room.

Despite a few difficulties obtaining research material—"bodies are vastly scarce at present," the lad wrote home to his father—young William Hamilton enjoyed his tuition.[15] "I am vastly pleased with both Mr. Hunters lectures," he assured his mother; "they are worth coming to London to attend, besides the benefit one derives from dissecting." Everard Home, working as an assistant in the shared dissecting room, had helped him to inject the veins of a leg, he added. But once the lecturing season was over, in April 1778, it was time for both Hamilton and Home to move on.

Now twenty-two, Home had completed his six-year apprenticeship with John, as well as gained his diploma from Surgeons' Hall.[16] But since his brother-in-law had neither the financial resources nor, seemingly, the inclination to continue supporting him, the young surgeon now had to make his own way in the world. No doubt he had expected Hunter to take him on as his assistant or even partner, but there was never the kind of affection or admiration between the snobbish young man and his plain-dealing brother-in-law that existed between Hunter and Jenner, or between Hunter and many other pupils, for that matter. It was another slight Home would store up for the future. Forced to fend for himself, Home joined the navy and took a position as staff surgeon in Jamaica, where he busied himself pitting snakes, tarantulas, and rats against one another in a series of experiments on venom.

Relations between the Hunter brothers had been sufficiently cordial for them to work under the same roof again for two years—they had even jointly dissected one of the king's elephants, which had died in 1776[17]—but the fraternal harmony was short-lived and was about to be crushed forever. In the autumn of 1778, John abruptly moved his lecture course from William's school to a hired room in the Haymarket. He would never return to his brother's premises again. There was no explicit reason for the shift, but there were plenty of sources of conflict. Though they had shared premises, pupils, and cadavers for two winters in a successful working arrangement reminiscent of their productive youth, they shared little in the way of temperament, interests, or beliefs. With the brothers now having reached the ages of fifty and sixty, respectively, and both revered in their different spheres, the balance of power had shifted seismically since the pliant young Jack had labored to fulfill William's demanding work program in Covent Garden. Now the younger brother had at least as strong a reputation in the London medical world as his older sibling, as well as a devoted following of pupils who acclaimed his novel philosophies. Moreover, he was at the center of a close-knit coterie of fellow enthusiasts and enjoyed an exalted place within the scientific community. Their priorities were completely at odds, too. While John plowed all he earned into his anatomical collection, so that he could barely afford to look after his own family, let alone keep his brother-in-law in work, William had accumulated substantial wealth through shrewd investment in books, coins, and art, so that he could comfortably manage to support his sister and her young family—albeit with his usual parsimony—after her husband died in 1778.

Both strongly opinionated and resolutely obstinate, the pair had clashed several times before. By early 1779, when their nephew, Matthew Baillie, came to stay with William, the brothers were no longer on speaking terms. "At this time too I waited upon Mr and Mrs Hunter, and they received me kindly," Matthew later recalled, "although there was a disagreement between him and his brother, Dr Hunter."[18] Seemingly, they had squabbled over that old bugbear, the ownership of preparations. John had invited William to see a particularly interesting organ he had removed from a soldier at a postmortem. Immediately coveting the specimen for his own collection, William had coolly taken the preparation home and refused to give it back.[19] For John, this was the ultimate injury. He was content to share his patients, his pupils, and his knowledge with his older brother, but no longer was he willing to surrender his beloved preparations. The theft of this one insignificant piece of flesh raked up the long history of William's appropriation of past preparations—and, more important, past discoveries—throughout their tempestuous relationship.

The storm had been rumbling for years. Lightning finally struck at the beginning of 1780, when the London medical world was stunned and the normally cordial atmosphere of the scientific community shattered as John Hunter turned on his own brother at a meeting of the Royal Society.

The gathering on January 27 began innocuously enough, and Banks, who had been elected president two years earlier, when Sir John Pringle resigned, anticipated a quiet evening.[20] The usual recommendations for elections were made, followed by a discussion of plans to strike a medal in memory of Captain Cook, who had been tragically killed the previous year by natives in Hawaii while on his third voyage. Then the business moved to the presentation of papers. The first was the conclusion of a paper started the previous week, describing the birth of a child with congenital smallpox; it was read by John Hunter. There was little controversy there, but Hunter had a surprise up his sleeve. He now rose to deliver a second paper, "On the Structure of the Placenta," which he nonchalantly announced would describe the separate blood supplies of the mother and fetus.

Just minutes after John had begun in his usual stilted manner, it became startlingly apparent that Hunter was accusing his own brother—a fellow member of the society and the queen's own physician—of outright plagiarism. Describing for the first time the discovery that he and Colin Mackenzie had made in 1754, he declared, "I mean to exhibit my claim to a discovery of no small importance."[21] For although William had already outlined the circulation of the placenta in his classic book, *The Anatomy of the Human Gravid Uterus,* published six years previously, he had done so, alleged John, "without mentioning the mode of discovery." This mode of discovery, made by himself and Mackenzie—who had since died—was now meticulously detailed by John, to the shock and embarrassment of the aghast members. He concluded by declaring that he now considered himself "as having a just claim to the discovery of the structure of the placenta and its communication with the uterus." The atmosphere in the room as John sat down was as electric as one of John Walsh's eel parties.

Stung by this surprise ambush, William wrote to the society a peevish but sanctimonious letter, which was read at the next meeting, held on February 3.[22] In it, William blustered that the discovery in question had already been published as his own, that he had long taught the discovery as his, and that he had always paid tribute to his brother's help as "an excellent assistant." Tellingly, at no point did he insist that he had actually made the discovery in question, nor did he dispute his brother's right to precedence, although he declared himself willing to let the Royal Society act as arbiter in deciding the merits of "Mr Hunter's claim." At the same time, William plaintively reminded the fellows that it was he who had first taught his upstart brother the skills of anatomy, "and put him into the very best situation that I could for becoming what the society has, for some time, known him to be."

As the dispute descended into outright acrimony, John now sent an impassioned and, for him, extremely eloquent letter to Banks, reiterating his claim with redoubled zeal. The response was accordingly read to the increasingly embarrassed fellows on February 17.[23] John explained that he had been impelled to reply, for fear that "silence on my part after his charge may be interpreted by my Enemies into an acknowledgement that I have intentionally claimed to myself a discovery in reality his due." Restating the details of his discovery with Mackenzie, he proclaimed, "I am as tenacious as he is to Anatomical discovery, and I flatter myself as tenacious also of truth." And yet he was willing to forfeit this claim should his brother produce evidence to show he had himself made the discovery in question prior to 1754. Extending a somewhat prickly olive branch, he also declared himself ready to accept joint credit for the discovery—"and abolish all remembrance of the succession of time"—should William refuse to accept his version of events.

Either battered into submission or infuriated into silence, William never responded. John had had the last word and thus had set right twenty-six years of injustice—or so he thought, for neither brother was aware that the crucial finding had really first been demonstrated by the Dutch anatomist Wilhelm Noortwyk in 1743. But if John could now regard himself as the victor in his wrangle with William, the price of permanently estranging his own brother was a heavy one to pay. Wisely, the Royal Society declined to umpire the brothers' unseemly public dispute, and John's paper was never published in the *Philosophical Transactions,* although he would later include a revised version in his *Observations on the Animal Oeconomy* in 1786.

Yet there were other, more fundamental reasons that William and John had reached such a cataclysmic rift by this stage in their careers. John's radical and outspoken views on the nature and even origins of life had grown increasingly at variance with William's more conventional and more God-fearing outlook. For pupils who had attended their joint lectures, there was a stark contrast between William's eloquent exposition on the divine creation of the human body and his younger brother's faltering and increasingly irreligious utterings. William's fulsome tribute to the mysterious works of God was entirely in keeping with the contemporary worldview. He may even have sought to caution the students against imbibing the iconoclastic views of his less restrained brother when he warned them, "The man who is really an Anatomist, yet does not see and feel what I have endeavoured to express in words, whatever he may be in other respects, must certainly labour under a dead palsey, in one part of his mind."[24]

Although William well knew that his brother's mind was far from dead, he was aware that John's lectures on physiology were running perilously close to countering accepted religious dogma. For not only did John never once refer to the traditionally acknowledged role of "the Creator" in forming all life—nature was always his preferred candidate for this wondrous achievement—he also suggested that life had originally appeared spontaneously out of inorganic material, rather than over a six-day period in the Garden of Eden. He told his pupils, without equivocation, "Animal and vegetable matter has certainly arisen out of the matter of the globe, for we find it returning to it again."[25] Although John had married in church and enjoyed friendships with several clergymen, he certainly entertained views that bordered on heresy. In one unpublished manuscript, he even wrote, "All innovation on established systems that depend more on a belief, than real knowledge (such as Religion) arise rather from a weakness of mind than a fault in the system."[26] Even the philosophy behind his lecture course—spurring on students to question established doctrines and daring them to think for themselves—was an encouragement to heterodoxy. William made no attempt to conceal his disapproval, condemning his brother's lecture course with the comment "though it may strike weak, uncultivated minds, yet by men of finer intellects it is construed with great propriety as an indignity thrown out against the Great Author of All Things."[27] William still might have hoped to protect his pupils from the antireligious influence of his fraternal lecturer, but he had doubtless realized that he was unlikely to deter his brother from a course that was steering him directly toward a head-on collision with the Georgian establishment.

The brothers' breach would never be mended. Shortly afterward, William complained bitterly to Dorothy that "I have lived to have my affections much disturbed by ingratitude."[28] But if John regretted his assault on William, he never once referred to their argument. When he later obtained a copy of a memoir on William published after his death, John scribbled comments in the margins. Many of the remarks were laudatory, some of them dry, some of them comical, but none of them spiteful. He made no mention of his own unresolved dispute and would only comment, "He perhaps did not make sufficient allowance for the natural frailty of human nature."[29]

With *Hunter's many* business interests forever escalating, at last even he had to admit that his Jermyn Street house was far too small for his busy household, tireless research pursuits, and burgeoning collection. Two remarkable and uncommonly large additions to his natural history specimens in 1780 and 1781 made continuing in the house just about intolerable. Even obliging Anne's patience was severely tested.

The first of these arrivals was a creature that had attained almost mythical status in eighteenth-century Europe. Giraffes, as the animals were sometimes known, from the Italian term, seemed more fabulous even than unicorns and centaurs. But in 1779, the traveler William Paterson had journeyed into uncharted parts of southern Africa on an expedition funded by the keen amateur botanist Lady Strathmore. It was during this journey that he shot and killed a large male giraffe.[30] When the trophy was brought home in 1780, it was the first giraffe ever seen in England. Lady Strathmore generously donated its skin and much of its skeleton to Hunter. Despite its lack of internal organs, Hunter was elated with his prize and made sure nothing was wasted. Having preserved several of the bones, he dissected the elastic ligaments in the neck in order to understand how its neck muscles worked, and then had the entire animal stuffed. A rather amateurish watercolor of the animal as it might have looked alive was probably painted by Hunter himself.[31] Unfortunately, the astonishing stature of the stuffed beast, estimated to have measured as much as eighteen feet, made its accommodation rather difficult. With the rooms housing his collection already bursting at the seams, Hunter was forced to hack off the giraffe's legs and stand it in his entrance hall.[32] The sight presented a dramatic welcome to visitors and patients.

The following year, another exciting find created further logistical problems for the household. In September 1781, fishermen captured a bottle-nosed whale that had ventured up the Thames beyond London Bridge. The carcass, measuring almost twenty-six feet long, was bought by an "oil-man"—presumably a dealer in whale oil—for the hefty sum of seventy pounds. According to Daniel Solander, relating the latest gossip to Joseph Banks, its buyer recouped his outlay by exhibiting the "stinking whale."[33] Hunter was among the first to view the animal. He had already dissected an orca of about the same length in 1759, and a second in 1772, as well as a bottle-nosed dolphin he had persuaded Jenner to send. Despite the buoyant whaling industry, complete whales in fresh condition were still a rarity and live ones impossible to transport. At some point, Hunter, in the hope of obtaining useful specimens, had even paid a surgeon to board a whaler set for Greenland, "but the only return I received for this expense was a piece of whale's skin, with some small animals sticking upon it," he lamented.[34] So the opportunity of investigating an entire whale of a new species was too good to miss, and Hunter quickly persuaded its owner to let him dissect the creature. Helped by a crew of pupils and assistants, Hunter stood on top of the great animal, wielding his largest knives. The scene created a stir in the press, one newspaper reporting that the whale's vena cava—the largest vein in the body—"was capable of containing a child of a year old." Hunter carefully noted its two tusklike teeth in the lower jaw, while in its stomach he found "the beaks of some hundreds of cuttle-fish."

But finding a place to display its remarkable skeleton was a rather weightier problem. By now, not only was Hunter's Jermyn Street house impossibly crowded, so that "the whole suite of the best rooms in his house were occupied by his preparations," according to Home, but also the lease was fast running out, too.[35] Hunter needed somewhere to display his extensive collection in its unique order, and he hoped that, in time, such a place could be opened as a museum, where the public could view this collection. With this in mind, Hunter now searched not only for a fitting venue but also for a dramatic new exhibit that could take its place as the crowning attraction of the museum. He was not to be disappointed on either score.

CHAPTER 13

The Giant's Bones

London

April 1782

News that the giant was lumbering his way toward London reached the capital long before he did. Ever since making the short sea crossing from his home in Ireland to Scotland, Charles Byrne had become a national celebrity, and his majestic-sounding stage name was on everybody's lips. So tall was the "Irish giant," according to accounts that preceded him, that in Edinburgh he had lighted his pipe from one of the lamps on North Bridge without even standing on tiptoe.[1] Reported to be as tall as eight four, Byrne drew crowds wherever he went. As he crossed the border to journey south, appreciative audiences queued up to witness his astonishing stature, hear his booming voice, and shake hands with the polite young man. Byrne was used to the attention, for he had caused a stir ever since he was an infant.

Born in a hamlet near the border of County Derry and County Tyrone in 1761, Byrne had soon achieved local acclaim. Although not an unusually large baby, he quickly outgrew his playmates and pretty soon towered over adults, too. As news of the peculiarly tall boy spread through the countryside, wild myths developed, such as the tale that he owed his extraordinary height to his parents having conceived him on top of a haystack.[2] With the infant's growth showing no signs of slowing down, a sharp entrepreneur from a neighboring village, Joe Vance, convinced Byrne's parents that the boy could become their ticket to a fortune. Acting as his agent, Vance exhibited the youth at country fairs and village greens as a freak curiosity. The boy more than outdid his manager's expectations, attracting excited onlookers wherever he went, and Vance was soon convinced he could pull in even bigger crowds and more money by taking his charge to Britain and possibly even the Continent. After touring Scotland and northern England, the pair steadily journeyed south toward the capital in early 1782.

John Hunter was not the only person in eighteenth-century London with an interest in "monsters." Each autumn, crowds flocked to Bartholomew Fair in London's East End to view the assortment of strange exhibits, both animal and human. In between times, curiosities were shown at taverns and lodging houses. Londoners thronged to gape at not only the wild beasts on show throughout the year but also at unusual or deformed animals, such as a cow with two heads, which was on show in early 1782. A young physician, Sylas Neville, was among the spectators in February. "Saw a monster of the brute creation," he recorded in his diary, "a heifer with two heads, four horns, four eyes, four ears, four nostrils through all of which it breathes, eats & chews cud with both mouths."[3] Neville revealed that "an eminent surgeon & anatomist" had paid its owner to receive the carcass when the cow died.

More fascinating even than double-headed cows were the human freaks and oddities that had been exhibited in London for as long as anyone could remember. Incited by folktales of men with tails, centaurs, and mermen, Londoners queued up to see people born with deformed or extra limbs or with appalling medical conditions. Since there was no state help for vulnerable people, appearing in such shows was the only source of income for most people with deformities, and many traveled to London expressly for the purpose. One boy, born without legs, came from Austria to show himself at the Eagle and Child in Fleet Street in 1714, while a boy with one head and two bodies was exhibited at Bartholomew Fair at around the same time.[4] Conjoined twins were a regular sight in the capital, and a man who was "covered all over his body with large scales"—he may have suffered from the skin disorder ichthyosis—was on show in 1758.[5] And in case the novelty of watching static human exhibits should pall, many professional freaks learned tricks to keep their audiences entertained. One man without arms, described as "the eighth great wonder of the world," could beat a drum, play a trumpet, fire a pistol, and embroider with his feet.

Dwarfs were popular attractions in the capital, too, and often took pains to accentuate their almost mythological status. Corsican-born Madame Teresa, known as the "Corsican Fairy," arrived in London in 1773 before traveling to Dublin, where, immaculately fitted in a miniature suit to resemble a little mannequin, she posed for a portrait. The painting was acquired by John Hunter, and he would follow her career with particular interest.[6] Yet while dwarfs could generally pull in a mesmerized crowd, Londoners always had a soft spot for giants.

Even more than dwarfs, giants were the stuff of legends.[7] Simultaneously evoking wonder and fear, exaggerated tales of their proportions and feats became mingled with folklore and fairy tales. Even in the early eighteenth century, it was widely supposed that extraordinarily large bones discovered in caves across Europe had belonged to gigantic human ancestors. Many believed the human race had been gradually shrinking since creation. In 1718, a French academic calculated that Adam must have been 134 feet tall, while Eve had measured 128 feet.[8] Extremely tall men had come to exhibit themselves in London for centuries. But when Charles Byrne arrived on April 11, 1782, everyone agreed he was the tallest giant ever to have walked the capital's streets.

Although he hardly needed to broadcast his presence, since he stood a full two feet taller than the average Londoner, an advertisement in the *Morning Herald* two weeks later proclaimed his arrival:

> IRISH GIANT. To be seen this, and every day this week, in his large elegant room, at the Cane-shop next door to late Cox's Museum, Spring Gdns, Mr. Byrne, the surprising Irish Giant, who is allowed to be the tallest man in the world, his height is eight feet two inche [sic], and in full proportion accordingly, only 21 years of age. His stay will not be long in London, as he proposes shortly to visit the Continent.[9]

Although in reality Byrne was an only slightly less remarkable height— seven eight—the undeniably surprising giant captivated London overnight. Residing at his apartment in Spring Gardens, near Charing Cross, Byrne entertained audiences from 11:00 A.M. to 3:00 P.M. and from 5:00 P.M. to 8:00 P.M. six days a week for a fee of two shillings sixpence per person. Overcoming their initial apprehension, audiences were bewitched by the gentle giant, who dressed in an elegant frock coat, waistcoat, knee breeches, silk stockings, frilled cuffs and collar, topped by a three-cornered hat, and who spoke so politely and displayed such genteel manners. Byrne's large, square jaw, wide forehead, and slightly stooped shoulders enhanced his placid demeanor.

He was soon the talk of the town. While newspapers published flattering reports of his physique, gossip columnists speculated on his love life; one article suggested an assignation with a female dwarf, known as the "Bird of Paradise," who was on show simultaneously in town.[10] Within weeks of his arrival, Byrne had met the king and queen at Kew, been fêted by members of the nobility, and been presented before the Royal Society, its members anxious to assess his prodigious stature and proffer theories about his peculiar condition. He even became the hero of the summer's pantomime, called "Harlequin Teague or the Giant's Causeway," which ran for nearly a month at the Haymarket Theatre, to rapturous applause.[11]

One visitor who was particularly moved by his encounter with the giant was Count Joseph Boruwlaski, the diminutive "Polish dwarf," who had arrived in London from the Continent only weeks before Byrne. He had been taken under the wing of the undisputed doyenne of fashionable Georgian society, Georgiana, the duchess of Devonshire.[12] Having been patronized by royal families across Europe—although he had no true claim to his own assumed aristocratic title—Boruwlaski was disinclined to exhibit himself as simply another human freak. He had taken lodgings at 55 Jermyn Street, almost opposite Hunter's house, through the generosity of Georgiana, who had likewise paid a tailor to make him a suit "embroidered with gems and silver." Artfully, he managed to survive on the goodwill of benefactors, supplemented by giving occasional concerts.

Ever eager to observe abnormalities, whether in animal or human form, Hunter was quick to make Boruwlaski's acquaintance. Finding him a lively and intelligent man who lamented being forced to make a living out of his unusual condition, both John and Anne would become subscribers to the count's memoirs when he published them in 1788. Hunter was fascinated by Boruwlaski's medical condition. He introduced the count to friends, invited him to meet his pupils at Sunday-evening scientific gatherings, and commissioned a portrait of the diminutive man.[13]

It was shortly after arriving in London that Boruwlaski was escorted by his patrons, the duke and duchess of Devonshire, to join the crowds at one of the Irish giant's shows. "He was eight feet three or four inches high," recorded Boruwlaski. "His shape was very well proportioned, his physiognomy agreeable; and what is very uncommon in men of this sort, his strength was equal to his size: he was at that time two-and-twenty." As soon as the audience gawping at Byrne realized that the tiny dwarf was in the same room, all clamored to see the two side by side. Boruwlaski was pushed to the front, and as the towering giant reached down to greet the miniature man, the delighted onlookers were no doubt reminded of Gulliver's encounter with the Lilliputians. "Our surprise was, I think, equal," Boruwlaski noted in his memoirs; "the giant remained a moment speechless, viewing me with looks of astonishment; then stooping very low to present me his hand, which would easily have contained a dozen like mine, he made me a very polite compliment."[14]

Members of the Royal Society were equally impressed when they gathered at their private viewing to admire Byrne's "stupendous" height and "admirable symmetry."[15] An interest in giants had long been a feature of Royal Society inquiry. The early eighteenth-century president Sir Hans Sloane had examined some of the huge skeletons discovered across Europe, as well as a collection of large bones exhibited in London as a "giant's hand." He had concluded that the skeletons belonged to elephants and whales transported to Europe by the great Flood, and he realized that the "giant's hand" was a whale's fin.[16] But the existence of extraordinarily tall men like Byrne continued to flummox the natural philosophers. Certainly, few of the fellows could have failed to be curious about the youth towering over their heads, but none was more captivated than John Hunter. From the moment he set eyes upon Byrne's fabulous figure, he knew that he had to possess his body.

With no time for superstitions or folklore, Hunter was intent on explaining the existence of giants. For him, as for his contemporaries, such human curiosities fitted into the category of "monsters." Even the modern term *teratology,* describing the study of anomalies in animal life, is a derivation from the Greek word for monster. Hunter's fascination with "monsters," of course, went back to his earliest studies in anatomy. Over the years, he had observed, dissected, and collected an extensive number of human and animal oddities, and it was well known among the body snatchers, animal breeders, and even butchers he cultivated that he was always in the market for rare or unexpected finds. In the summer of 1782, not long after Byrne had arrived in town, Hunter petitioned a close acquaintance, the politician the earl of Shelburne, to help find a job for a poor butcher who had frequently brought him "curious parts of animals."[17] Hunter's hoard of abnormalities had grown so large, it now formed a separate collection in its own right. This motley selection of bizarre finds was undoubtedly the stuff of nightmares. Along with his two-tailed lizards and two-headed cows, visitors granted the dubious privilege of looking around could see malformed human fetuses, a duck with a foot on its head, a cow with an extra leg, a pig with two bodies, and several double-headed snakes. But Hunter's penchant for peculiarities was not simply an obsession with collecting the grotesque, for he regarded such freaks of nature—they would similarly absorb Charles Darwin—as a key to the origins and development of life.

Given the general belief that nature was essentially unchangeable and unchanging, human freaks such as Byrne presented something of a difficulty. If nature could produce such wild variations in individual beings, was it not conceivable such changes might occur in whole species, too? To Hunter, the production of "monsters" was clear proof of nature's ability to generate change.

Just three years earlier, he had presented a controversial paper to the Royal Society on hermaphroditism, the rare condition in which male and female sexual organs occur in a single animal. His farsighted study, "An Account of the Free-Martin," published in 1779, described three hermaphrodite cattle he had examined both alive and dead.[18] After dissecting the curious creatures, Hunter was the first to delineate their dual male and female reproductive organs. Based on this research, he concluded that every animal, and every part of every animal, possessed an innate propensity to malformation. Such a view steered perilously close to challenging the orthodox doctrine that all species had been created in a fixed and never-changing state. The following year, in a paper discussing some female pheasants that developed male plumage, he had gone further. Here he had argued that the ability to produce abnormalities seemed to be present in the embryonic state and he declared that "each part of each species seems to have its monstrous form originally impressed upon it."[19] He spelled out this contentious idea with startling clarity for the time:

> Every deviation from that original form and structure which gives the distinguishing character to the productions of Nature, may not improperly be called monstrous. According to this acceptation of the term, the variety of monsters will be almost infinite; and, as far as my knowledge has extended, there is not a species of animal, nay, there is not a single part of an animal body, which is not subject to an extraordinary forma-

tion. Neither does this appear to be a matter of mere chance; for it may be observed that every species has a disposition to deviate from Nature in a manner peculiar to itself.

So by 1780 at least, Hunter had established to his own satisfaction that nature could produce fluctuations in form and had speculated that this propensity existed in some inherent congenital state. He was aware, too, from his contacts with livestock breeders, that such change was a regular and progressive process, noting that in eye pigmentation "varieties are every day produced in colour, shape, size, and disposition." He then added unequivocally, "It certainly may be laid down as one of the principles or laws of Nature to deviate under certain circumstances."[20] It was but a small step from here to propose that not only individuals but even whole species could change their form, as Hunter was gradually coming to realize.

At the same time, Hunter appreciated that certain kinds of abnormality were inherited. "I have seen three [cases of] Spina Bifida in the children of one family," he noted in one unpublished paper; "I have seen two hair lips [sic] in the children of the same parents."[21] This hereditary principle applied to animals, too, as Hunter explained: "A cow was brought to London for a show who had a supernumerary Leg upon the Shoulder which is a very common monstrosity but the curious circumstance was, she had a calf with the same Monstrosity." And he concluded, "How far an animal . . . which is a species of Monster, is endowed with the propagation of these peculiarities to its offspring is not yet determined but there are many circumstances which would make us suspect that such a principle often takes place." The hereditary principle was clearly at work in producing exceptional stature, as well. The physician Dr. David Pitcairn, who was one of Hunter's friends, had told him that the two tallest men he had ever seen were twins. One of them was six seven and his brother six five. Yet while undoubtedly tall, the twins were hardly remarkable compared to Charles Byrne. Hunter's hunger for the unusual was insatiable.

In truth, there was another motivation, beyond the laudable advancement of science, for Hunter's determination to secure Byrne's body. Although he was now earning an estimated five thousand pounds a year— around £300,000 in modern terms—from his private practice and students' fees, Hunter was still invariably strapped for cash.[22] Now fifty-four, his tawny hair touched with gray, forced to wear spectacles for close work, and increasingly plagued with the symptoms of angina, Hunter was all too conscious of his own mortality. His only legacy, the sole future income for Anne and their two young children, would be his writings and his museum. He was keenly aware of his investment in the collection and eager in 1782 to promote its importance. In an article in *European Magazine* detailing his ventures that year, Hunter revealed that he had already expended ten thousand pounds on purchasing and maintaining his preparations, including more than two thousand pounds on buying dead animals alone.[23] He was still searching desperately for larger premises for the household and the collection. It was imperative to find a new home for the cramped exhibits and create a unique museum that would explain his lifetime's work. But he wanted a magnet to attract visitors, too. The skeleton of a giant, who was even now drawing rapt crowds, would form the perfect centerpiece.

With his all-consuming conviction in the importance of his researches, coupled with his success in collecting every manner of treasure from around the globe, Hunter naïvely believed his latest ambition should present little obstacle. He hardly expected the giant himself to pose any objection. Assuming a satisfactory transaction could be cordially agreed upon, he took steps immediately to put an expert on the case. The man he recruited to help him in his mission was John Howison.

A shadowy figure, Howison was evidently of immense service to both the Hunter brothers.[24] At one time, he had purchased some coins on behalf of William, and he had also taken notes in both brothers' lectures. Yet it was bodies—not coins—that Howison was most skilled at procuring. Howison was, by all accounts, a general factotum, a jack-of-all-trades who was willing to perform whatever underhanded and dirty tricks were necessary for the smooth running of the brothers' separate anatomy businesses. Despite the brothers' irreconcilable rift, Howison was still at the bidding of John Hunter in 1782. It was for him that Howison would help to pull off the most daring assignment of all.[25]

He would not have long to wait. As Hunter watched the giant parade in front of delighted audiences, his astute observation skills told him that Byrne was not a well man. Symptoms of Byrne's condition, the result of overproduction of growth hormone caused by a benign tumor on the pituitary gland, now known as childhood-onset acromegaly, or gigantism, were all too apparent.[26] By now, he would have suffered from painful joints, excessive sweating, and headaches, while the placid nature noted by onlookers could well have been a sign of slowed intelligence. Although the true cause of Byrne's extreme height was a mystery to Hunter—it would be many more years before the cause of his abnormality could be understood and treated—he was certainly aware that the untreated condition generally spelled an early death. Giants had a reputation for short lives, and Byrne, as he freely declared in his advertisements, was already twenty-two. Hunter knew he had only to bide his time.

As the summer faded, so, too, did London's mania over the Irish giant. Even the "tallest man in the world" could hold the attention of fickle Londoners for only a few short months. By early autumn, queues at Byrne's shows had dwindled, the curtain on "Harlequin Teague" had dropped for the last time, and the city's residents were scouting for new diversions to excite their imaginations. There was no shortage of amusements to rival Byrne for their attention and disposable income. Idle Londoners could, for example, witness acrobats, a camel, and daring horse antics at Mr. Astley's Amphitheatre Riding School on Westminster Bridge. In Piccadilly, the eccentric Mr. Katterfelto offered the chance to see "insects" in water, beer, milk, and blood through his "New Improved and greatly admired SOLAR MICROSCOPE." And in Spring Gardens, a stone's throw from Byrne's own show, the versatile Mr. Breslaw thrilled crowds with his mind reading and magic tricks, in which he commanded "A FRESH EGG TO DANCE upon a Stick, in the middle of the Room, by itself," to the accompaniment of the violin and mandolin.[27]

Faced with such energetic competition, by early autumn Byrne was forced to move his show to another apartment, probably a smaller and cheaper one, above a sweet shop in Charing Cross. By November, he had not only moved again, to a room at the Hampshire Hog in Piccadilly, but had dropped ticket prices for children and servants to one shilling.[28] More worryingly still, there were rival claims to his title. On display at his former haunt in Spring Gardens was the "wonderful GIGANTIC CHILD," who at eighteen months old had already attained a height of three feet.[29] There were even rumors of another "Irish giant," reputedly taller than Byrne, who was considering making his debut in London. Aged twenty-two, like Byrne, and, like him, sometimes adopting the stage name "O'Brien" to suggest descent from the ancient king of Ireland Brian Boru, Patrick Cotter had arrived in England in 1779 and won acclaim in Bristol. Described in some advertisements as being eight three and three-quarters, he plainly had his eye on Byrne's crown, as evidenced by the claim "The Giant is upwards of Four Inches taller than the noted Burn."[30]

With revenues falling, rival attractions stealing his audiences, and his health now rapidly declining, Byrne turned to drink. Racked with pain and befuddled by alcohol, even now he considered leaving London for new audiences on the Continent. His failing health put paid to that. By spring 1783, it was all too clear that Byrne's brief life on earth, as transitory as his short spell as a national celebrity, was coming to an end. With no hope of treatment, the growing tumor in his brain was causing terminal damage. He may also have contracted consumption. Byrne knew his days were numbered. But death itself was no longer his greatest fear. The Irish giant's worst dread was the anatomists, and especially the best-known anatomist of them all: John Hunter.

Having maintained his close watch on Byrne's health and whereabouts throughout the winter, with the aid of the giant's helpful advertisements relating each change of address, Hunter hoped that his vigilance was about to be rewarded. Like a leopard circling his prey, he closed in for the kill. Intent on beating any other anatomists to the prize, Hunter set his spy Howison to watch Byrne's every move. In truth, this was not the most challenging of tasks. Byrne towered head and shoulders above every other London pedestrian, so it was simplicity itself for Howison to stalk his quarry through the West End streets as the giant stumbled from tavern to tavern. When Byrne made one last change of address, to an apartment at 12 Cockspur Street, and a final price reduction to one shilling for all comers, Howison even took rooms a few doors away.[31]

It may well have been at this point that Hunter made Byrne an offer he felt sure he could not refuse for the promise of his corpse after death. It was an ill-judged mistake, for Byrne was horrified at the proposal. Although members of well-to-do society were increasingly accepting of the need for postmortems to determine cause of death, such enlightened attitudes had yet to permeate the general populace. To a poorly educated country boy, the notion of being cut open after death and his body quite possibly displayed like that of a common criminal, with the very real fear that this might deny him entry to the promised land on Judgment Day, was simply abhorrent. The idea that investigating his corpse might aid understanding of his medical condition and possibly help future victims meant little to Byrne. Fully cognizant of Hunter's quest, and knowing he was being stalked day and night by Hunter's accomplice, Byrne sought desperately for a way to outwit the anatomist. As Howison laid siege to the apartment on Cockspur Street and Hunter waited feverishly for news not five minutes away on Jermyn Street, Byrne made plans to foil them both.

When an urgent message arrived at Jermyn Street at the end of March 1783, however, it was not Byrne who was dying, but William, Hunter's brother. Having been taken ill in early March, William had risen from his sickbed on the twentieth to deliver the introductory lecture for his spring course.[32] Before he could reach the end of his lesson, he collapsed and had to be carried to bed. His physician, Charles Combe, determined that William had suffered a stroke, for which the usual ineffectual therapies were applied. It was Matthew Baillie, the brothers' nephew, now working in the Great Windmill Street school, who summoned Uncle John. Although they had not exchanged friendly words since their bitter rift, and there was no deathbed reconciliation between the brothers, John eased William's last days by fitting him with a catheter. Visiting daily, he dispassionately recorded details of his brother's fading life in his casebooks.[33] William died, aged sixty-five, on March 30. With his last breath, he told Combe, "If I had strength enough to hold a pen I would write how easy and pleasant a thing it is to die."[34]

William's funeral was held on April 5 and he was buried in the vault of St. James's Church, Piccadilly; he had spent almost forty years dissecting corpses, but there would be no autopsy of his body nor any risk his corpse might be stolen.[35] Led by Matthew Baillie, his mourners included the most eminent physicians and surgeons of the day. But his own brother was absent. In his will, William left a fortune in priceless books, coins, medals, paintings, and anatomical preparations.[36] The entire collection, including the vast majority of John's painstaking work carried out at the Covent Garden school, was bequeathed to Matthew, on condition it revert after thirty years to Glasgow University. Ultimately, it would form the university's Hunterian Museum and the Hunterian Art Gallery. Matthew was also left twenty thousand pounds, a half share with Cruikshank, William's last business partner, in the Great Windmill Street school, and the family farmhouse—John Hunter's own birthplace—at Long Calderwood. To his sister Dorothy, William left a measly one hundred pounds, although his nieces, Agnes and Joanna, each received two thousand pounds, and there were token gifts of twenty pounds for various friends and associates. To

John, his only brother, his best pupil, and his most illustrious collaborator, William left nothing. As his obituary in *Gentleman's Magazine* flatly recorded, "His brother, Mr. John Hunter, the surgeon, on account of some differences between them, is not named in the will."

If John felt bitter at being so decisively and so publicly cut out of William's will, he did not show it, although there was no doubt that the money would have been useful. But at the end of his last lecture of the spring term, without having mentioned his brother's death throughout the course, Hunter suddenly hesitated as the pupils made ready to leave. "Here Mr Hunter seemed to finish," remembered one of the students, Joseph Adams, "yet to have more to say; at length endeavouring to appear as if he had just recollected something, he began 'Ho! Gentlemen, one thing more: I need not remind you of the loss you all know anatomy has sustained!' He was obliged to pause, and turned his face from his hearers."[37] Recovering his composure, but with his eyes filled with tears, Hunter paid tribute to his brother's lasting contribution to anatomy. According to Adams, "The scene was so truly pathetic, that a general sympathy pervaded the whole class; and every one, though all had been preparing to leave the place, stood or sat motionless and silent for some minutes." The following year, Matthew Baillie would hand the deeds for Long Calderwood to his uncle John, having told his lawyer, "I think I shall nearly feel as much satisfaction in delivering over the deeds to J. Hunter as if a sum of the same value was to be given to myself."[38] Hunter gladly accepted the family home as his due, later remarking that "the paternal estate of Long Calderwood came to his brother, the Dr. having no right to dispose of it."[39]

Having obtained no portion of William's fortune, and despite the fact that he was earning a sizable annual income, John was forced to raise a mortgage on his Earls Court land in the spring of 1783 in order to buy a new home.[40] He had at last discovered a suitable house, which was about to be vacated by the American artist John Singleton Copley, in fashionable Leicester Fields, soon to become known as Leicester Square. Having been laid out in the 1670s, the square had become a favorite residence for aristocrats, artists, writers, and other professionals. It remained a popular address throughout the eighteenth century. William Hogarth had lived at number 30 until his death in 1764; Sir Joshua Reynolds worked from studios at number 47.[41]

The elegant and roomy four-story town house at number 28 was an ideal new home for Hunter, his family, and his large retinue of servants.[42] With its graceful front door opening into a large hall, there was ample accommodation for Hunter's study, an afternoon bedroom where he could take his naps, and a parlor to receive patients. From here, the main staircase wound up to a spacious drawing room with four tall windows overlooking the square, providing a perfect venue for Anne's soirées. Above that, there were two more floors of rooms, while beneath the house were subterranean stables for Hunter's coach and horses. Even so, there was still insufficient room for all of Hunter's needs, so he bought the house behind, at 13 Castle Street—later Charing Cross Road—as well as the land between. Immediately, he set builders to work erecting a huge two-story structure bridging the two houses in order to create custom-built premises for a lecture theater, a grand parlor, or *"conversazione* room," and, of course, his treasured collection. With his usual lack of financial acumen, Hunter spent three thousand pounds for a lease of just twenty-four years on the Leicester Square house. The building work would cost a further three thousand pounds.[43]

It was while he was in the throes of moving his household into Leicester Square at the end of May—the museum would have to wait another two years for the builders to finish their work—that Hunter finally received word from Howison. At his room in Cockspur Street, the Irish giant was dying. His thunderous voice no longer boomed; his enormous chest labored over every breath. Only twenty-two years old but dependent on alcohol and almost destitute since his life's savings of £770 had been stolen in a Haymarket tavern in April, Byrne had gathered his few friends together to extract a binding last promise. Fully aware that the moment he died Hunter would come for his body, Byrne had concocted a plan designed to thwart him. As his friends leaned forward to hear his dying words, Byrne made them vow to seal his body in a lead coffin, ship it to the middle of the English Channel, and plunge it to the bottom of the sea, far from the reach of even the most resourceful anatomist.

On Sunday, June 1, Charles Byrne died. The event was immediately reported in the London newspapers:

> Cockspur Street, Charing Cross, aged 22, Charles Byrne, the famous Irish giant, whose death is said to have been precipitated by excessive drinking, to which he was always addicted, but more particularly since

his late loss of almost all of his property, which he had simply invested in a single bank note of £700 [sic]. In his last moments (it has been said) he requested that his ponderous remains might be thrown into the sea, in order that his bones might be placed far out of the reach of the chirurgical fraternity.[44]

Hunter's ambition was no secret. Almost all of London knew of his obsession with obtaining the giant's corpse. And there were others in the running, too, for a few days later, the *Morning Herald* reported, "The whole tribe of surgeons put in a claim for the poor departed Irish giant, and surround his house, just as Greenland harpooners would an enormous whale. One of them has gone so far as to have a niche made for himself in the giant's coffin, in order to his being ready at hand, on the 'witching time of night, when church-yards yawn.' "[45]

Stories of the scramble to obtain the giant's corpse continued to fill the newspaper columns. A few days later, another report proclaimed, "So anxious are the Surgeons to have possession of the Irish Giant, that they have offered a ransom of 800 g to the undertaker. The same being rejected, they are determined to approach the church-yard by regular works and terrier-like, unearth him!"[46] But while the anatomists vied for poor Byrne's body, his friends at least were true to their word. Having obtained an oversize coffin and secured their tall friend's body inside, they kept watch over the corpse for four days. In the interval, they invited spectators to view the colossal casket, advertised as a remarkable eight feet four inches long, for a fee of two shillings sixpence—at least they would reap some profit for their pains.[47] On June 5, in accordance with Byrne's wishes, they shouldered their heavy burden and transported their load to Margate, where they chartered a boat and tipped the great coffin into the sea. But whatever it was the grieving comrades consigned to the deep, it was not the body of the Irish giant.

No sooner had the newspapers reported Byrne's disposal than rumors began to circulate. First, *Gentleman's Magazine* described the report of the sea burial as "a tub thrown out to the whale"—or, put plainly, a decoy. Then *British Magazine* dismissed the burial as "a pure fiction." Finally, *Annual Reporter Chronicle* charged, "The giant expressed an earnest desire that his ponderous remains might be sunk out at sea; but if such were his wish, it was never fulfilled, as Mr Hunter obtained his body before interment of any kind had taken place."[48] Terrified of reprisals from Byrne's bereaved friends, Hunter would never refer explicitly to his daring venture to seize the giant's corpse, but seize it he certainly did. Over the ensuing years, details of his remarkable body hunt gradually surfaced, handed down by word of mouth.[49] The story that emerged was almost more remarkable than the tall tales dreamed up by the newspapers.

The truth was that as soon as Hunter heard from Howison that the giant had taken his last breath, he sprang into action and lost no time in seeking out the undertaker who was charged with carrying out Byrne's last request. It was a relatively simple matter to bribe the man to procure the body. But Hunter was obviously not prepared for the colossal sum he would have to cough up to secure his prize. According to one version, the undertaker first agreed on fifty pounds but quickly realized he could extract considerably more from the desperate anatomist. Step by step, he ratcheted up the price, until Hunter was forced to agree to five hundred pounds for his booty. It was a colossal sum.[50] With no ready cash, as usual, and already in debt over his house purchase, Hunter had to scrape together the money from one of his animal-dealer friends, Pidcock, who owned a menagerie in the Strand. Clinching the deal was one thing; perpetrating the daring theft was quite another. With Byrne's friends watching the corpse day and night, it was plainly going to prove difficult to filch the body from under their noses. So together, the undertaker and Hunter cooked up an ingenious plot to hoodwink the mourners.

As Byrne's trusty companions bore his weighty coffin toward the Kent coast, accompanied by the duplicitous undertaker, they stopped at points along the way to partake of refreshment. With the nomadic wake becoming steadily merrier, no doubt encouraged by Hunter's accomplice, eventually the inebriated band were inveigled to call at a particular tavern, where the crafty undertaker had made prior arrangements. Persuaded to deposit the coffin in a neighboring barn, the revelers continued their boisterous wake. In blissful ignorance, they caroused inside while the undertaker's accomplices deftly unscrewed the coffin lid and swapped Byrne's body for a stash of paving stones. Byrne's friends took up their load once more, staggered on to the coast near Margate, and paid their last respects as they pitched a coffin full of stones into the brine.

Meanwhile, the giant's corpse was whisked back to London, hidden under straw in a cart and delivered to Hunter's new house on Castle Street under the cover of darkness. Hunter himself then drove the huge carcass, wrapped only in its funeral shroud, westward to his Earls Court retreat before daybreak. Trundling the body into his underground laboratory, Hunter was still so afraid of discovery that he abandoned his usual meticulous dissection, his collector's zeal for possession of the spectacular specimen for once overcoming his surgeon's curiosity in investigating the rare condition. Instead, he hurriedly chopped the Goliath into pieces, threw the chunks into his immense copper vat, and boiled the lot down into a jumble of gigantic bones. After skimming the fat out of the cauldron, Hunter deftly reassembled the pile of bones to create Byrne's awesome skeleton.

Even then, he was forced to shield his prize from public scrutiny, most likely keeping the towering skeleton secreted at his Earls Court house. As newspaper columnists had already pointed the finger at the notorious anatomist, he must also have kept the existence of the giant's skeleton concealed from even his closest friends and colleagues. That summer, however, he allowed a cryptic hint to emerge in a letter to Jenner: "I hope to see you in London about two years hence, when I shall be able to show you something."[51] It would be a full four years before Hunter confided his sinister secret to his dear friend Banks. "I lately got a tall man," he wrote in 1787, "but at the time could make no particular observations. I hope next summer to be able to show you him."[52]

For the time being, then, the giant's bones remained out of sight. But there were other human body parts, belonging to people just as famous in their time as the Irish giant, to keep Hunter satisfied. Though Hunter still relied on the grave robbers for regular supplies of body parts, more and more of the bottles containing diseased and damaged organs that wound up on his shelves were now clearly labeled with the names of respectable, wealthy, and consenting Georgians. As demand for his services at postmortems had increased, so the named body parts had grown. Privileged friends invited to tour the collection could now peek at the internal organs of assorted former pillars of Georgian society. Among them were the cancerous bladder that had once belonged to the Reverend Mr. Vivian, erstwhile vicar of St. Martin's-in-the-Fields, and the injected kidney of Lady Beauchamp, who had died of a fever. Visitors could also peruse the thickened arteries of Lt. General Thomas Desaguliers, who had been treated by Hunter at the siege of Belle-Ile and then met him again on his deathbed;

and even the crumbling thighbones that had once held aloft the Honorable Frederick Cornwallis, archbishop of Canterbury, until his death in 1783.[53]

By now, Anne was well used to her husband's macabre collecting passion and the unappealing exhibits preserved in his collection. But even she must have quailed at the idea of internal parts of her own family joining this melancholy hall of fame. Robert Home, Anne's father and Hunter's former army colleague, had been ill since the previous summer, after suffering a stroke and losing the use of one side of his body. Having almost recovered by July 1784, under the careful attention of his son-in-law, he was struck down again, becoming completely blind and entirely losing his reasoning. "The total loss of sight, with almost the intire loss of Memory produced a very curious effect," Hunter recorded in his casebooks; "he lost entirely the rememberance of Light, and did not annex any Idea to Light."[54] When Home died shortly after, Hunter performed an autopsy on the former surgeon. Quite probably, he was aided by Everard Home, helping to cut open his own father; he had just returned from navy duties in the West Indies and had resumed his role as Hunter's assistant. Having sawed open his father-in-law's skull, Hunter paid particular attention to the brain, looking for signs of the fatal stroke. Inevitably enough, a slice of Anne's father's kidney found its way into Hunter's collection.

CHAPTER 14

The Poet's Foot

Leicester Square, London
April 1785

The genteel tranquillity that the residents of Leicester Square had once enjoyed had been rudely shattered. For two years, they had endured the continual din and disruption of builders, carpenters, and glaziers laboring in the southeast corner of their fashionable square. But as the workmen finally moved out of number 28 in the spring of 1785, the commotion only intensified. Carriages and pedestrians were brought to an abrupt halt and the distinguished gentlefolk arriving to sit for their portraits at Sir Joshua Reynolds's famous studio could only stop and stare. Surrounded by crates, the square's newest inhabitant supervised the arrival of a seemingly interminable line of stuffed animals, skeletons, skulls, and bottles filled with dubious-looking contents. Bearded and beginning to gray, the eccentric new neighbor directed his assistants as they carried a legless stuffed giraffe, two elephant jaws, bones belonging to lions, tigers, and other exotic beasts, and the skull of a large whale through the door of his new home.

John Hunter had spent an anxious spring finalizing plans to transfer his anatomy and natural history collection into its new home. "I am imployed as much as a thousand bees," he told Jenner in a more than usually hurried letter that April. "I am building moving xc. I wish this summer was well over."[1] Meanwhile, he complained to his pupils that all the fees he earned from patients were immediately paid out in bills to carpenters and bricklayers.[2] But at last the work was finished and the remarkable building complete. Between the smart four-story town house fronting Leicester Square and the inconspicuous, dowdy-looking house at its rear, facing Castle

Street, stretched a spectacular brick and glass structure providing a lecture theater, grand reception room, and a custom-built museum. Accommodating Hunter's myriad businesses as surgeon, anatomist, teacher, and researcher while also fostering his continuing connections with London's underworld, the dual-fronted house would later inspire Robert Louis Stevenson when writing his horror story, *The Strange Case of Dr. Jekyll and Mr. Hyde.*[3] Although the plot for the story came to Stevenson in a dream, he is said to have based Dr. Jekyll's house—the setting for the melodramatic transformation from good to evil—on Hunter's Leicester Square home. In the Gothic tale, written in 1886, when the house was still a familiar London landmark, the honest Dr. Jekyll had bought his house from "the heirs of a celebrated surgeon." Stevenson described the visitors who entered the doctor's home being led across a yard toward a lecture theater "once crowded with eager students" and a dissecting room at the rear. It was from the "old dissecting-room door," which opened onto a dingy thoroughfare at the rear of the house, that the grim-faced Mr. Hyde emerged to commit his murderous deeds.

Visitors to John Hunter's new home were familiar with much the same layout.[4] The elegant front door of number 28 opened into a lobby, where Mrs. Hunter kept her sedan chair, and a parlor, where patients could contemplate Hogarth's cautionary cartoons on venereal disease while waiting to see the surgeon. A corridor led to Hunter's study and his "afternoon bedroom," where he took after-dinner naps. Reaching the back door, guests crossed a gravel yard to reach a great glass door, stretching more than twelve feet high, which opened into the remarkable new structure.

Three stories high and measuring twenty-eight feet wide by fifty-one feet long, the ground floor accommodated a grand reception, or *conversazione,* room, lined with paintings by fashionable artists. A door from the *conversazione* room led into a spacious amphitheater, where Hunter delivered his lectures and demonstrated his preparations on an oval slate table to pupils seated on semicircular benches. Ascending to the first floor, visitors emerged into a vast room spanning the entire top two stories of the new structure. Illuminated by skylights and circled by a gallery, it was here, in his first custom-built museum, that Hunter now anxiously directed the arrangement of his thousands of preparations.

Descending again, visitors passed a sunken basement yard protected by a glass roof. Here, Hunter displayed his prized skull of a large bottle-nosed whale, which was too big even for the museum. For the intrepid few who ventured on, a door led into the rear house, the dingy 13 Castle Street. In this cramped, comfortless accommodation, Hunter's four or five house pupils were squeezed into quarters, along with a housekeeper and a dissecting-room attendant. Most other rooms were given over to overflow space for the collection and a printing press, while the attic housed Hunter's dissecting room, as far as possible from the general household. Finally exiting by the plain street door, just as Mr. Hyde would do in Stevenson's story, visitors emerged into the grimy and busy thoroughfare of Castle Street. Beside the door, a ramp led down to subterranean stables, where Hunter and his wife each kept a coach and horses. And above this sloping drive hung a wooden drawbridge that could be lowered to allow mysterious cargoes to enter at dead of night and, just as swiftly, raised to prevent entry, should that precaution prove necessary.

Welded together, the two contrasting houses suited Hunter's purposes admirably. Through the graceful doorway of number 28 walked Hunter's patients and Mrs. Hunter's guests. At the other end of the building, Hunter's pupils stole quietly in for dissection practice and were awakened in the middle of the night by carcasses being dragged up the back stairs and by, as one student put it, "the Resurrection Men swearing most terribly."[5] The unique layout allowed Hunter to range freely from one house—and one world—to the other. He was the Jekyll and Hyde of the Georgian period, offering his patients a dramatic cure one moment and dragging them off to his dissecting bench the next.

Now settled in the heart of London's West End and relieved to have found an appropriate home for his prized collection, Hunter embarked on his most industrious period yet. At the peak of his intellectual powers, at the height of his popularity with patients and pupils, he would treat many of the era's most famous personalities, contribute a stream of papers to the Royal Society, publish his best-known works, and receive international recognition. Yet the move to Leicester Square also marked the beginning of a serious decline in health. The smallest irritation could trigger another attack of angina, but, as uncompromising as ever, Hunter would indulge in increasingly bitter confrontations.

Overseeing the transfer of his collection to the new museum took its toll. Having returned from his naval duties in Jamaica the year before, Home found his brother-in-law "much altered in his looks," so that he "gave the idea of having grown much older than could be accounted for from the number of years which had elapsed."[6] Fully aware of the growing seriousness of his illness, Hunter told his friend Lord Holland that his heart would one day "kill him going upstairs or in a passion." His life, he told others, was "in the hands of any rascal who chose to annoy and tease him."[7] Yet even at the height of his illness he could be "roused," he confided to Jenner, by the prospect of earning two guineas.[8]

Quite possibly, it was the worry and exertion of the move that triggered a serious attack in April, when Hunter fainted away so completely that Home thought him dead.[9] Despite his skepticism about the efficacy of many contemporary remedies, Hunter resorted to the customary purges and emetics, submitted to being bled, cupped, and blistered, bathed his feet in water mixed with mustard, wore worsted stockings in bed, and even applied electricity to his arm. After all, there were no effective alternatives. Much as he suspected, they provided no relief. The following month, on May 20, he suffered another attack, prompted, he later confided, by the fear that he might have contracted rabies at a postmortem. The physician David Pitcairn was the first at Hunter's bedside. When the surgeon's condition failed to improve, the next day a whole troop of physicians arrived. David's uncle, William Pitcairn; George Baker, the king's physician; and Richard Warren, physician to the Prince of Wales, all marched through the Leicester Square door. The last two would shortly be summoned to treat George III for his mysterious bouts of mental illness. And in a practice run for the tortures they would perpetrate on the king, they now subjected Hunter to a battery of almost every therapy in their medicine bags over a period of nine days. After administering copious doses of rhubarb and senna as laxatives, aromatic spices as stimulants, and opium as a sedative, the physicians raised a large blister on Hunter's back in an effort to "draw out" his illness.

Although the frightening spasms continued, the patient was sufficiently "roused" to resume his professional role in July, when his expert opinion was sought by one of his oldest friends. Benjamin Franklin had endured increasing discomfort from a bladder stone since 1782. In the midst of negotiating the Treaty of Paris in France that autumn, he had found traveling by carriage particularly painful. But desperate at the age of seventy-nine to avoid the agony and risk of a lithotomy, Franklin wrote from France in July 1785 to a friend, Benjamin Vaughan, in London, asking him to seek the advice of the capital's most eminent medical men on the best course of action. Enclosing a full account of his ailment, in the third person, Franklin insisted that the pain was bearable most of the time and declared, "Thus if it does not grow worse, it is a *tolerable* Malady, and may be supported for the short time he has the Chance of living. And he would chuse to bear with it rather than have Recourse to dangerous or nauseous Remedies."[10]

Vaughan submitted Franklin's case to five of London's best-known practitioners. Hunter was the only surgeon—he was often now regarded on an equal level with his colleagues in physick. In a reply signed by Hunter, the consultants agreed that Franklin was wise to avoid surgery at his advanced age. "What we shall advise therefore, will be such things as may either prevent an increase of the disease or palliate the most pressing Symptoms," they counseled.[11] Given the risks of the operation for anyone, let alone a man of Franklin's advanced years, it was eminently sensible advice. And if none of the medications would offer the politician much succor, at least the opium they recommended gave him some respite from the pain. Leaving France to return home that same month, Franklin continued to play a prominent role in American politics for several more years. Finally confined to bed, and almost completely dependent on opium, he died in 1790, at the age of eighty-four.

Though Hunter enjoyed a relatively restful summer, journeying between spas at Tunbridge Wells and Bath in an effort to relieve his illness, his health would never completely recover. His output of work, however, only increased. Still surviving on four hours' sleep a night, and his usual hour's nap every afternoon, Hunter was invariably at his dissecting bench by daybreak. One young surgeon keen to become a pupil at St. George's was told to present himself to the surgeon the following day at 5:00 A.M. Anxiously arriving at the unlikely hour, he discovered Hunter already hard at work. "I found him in his Museum," he recalled, "busily engaged in the dissection of insects."[12] Hunter had generally completed several hours' investigations before the first patients were shown in through the Leicester Square door at 9:00 A.M. By now, the jam of carriages blocking the square and the clamor of waiting patients had created a veritable cacophony guaranteed to wake any late-rising neighbors. As the patients poured in, there was scarcely room to accommodate them all while they waited to present their ailments.

Now regarded as one of the most fashionable surgeons in London—

only Pott could boast greater popularity—Hunter treated people from every walk of life. He had scant regard for their position in the social hierarchy, however. One of his trusty animal dealers was always encouraged to jump the queue and take a position ahead of the affronted nobility. "When I called, if the house was full of patients, and carriages waited at the door, I was always admitted," recalled Gough, who kept a menagerie in Holborn Hill. "You (said Mr Hunter) have no time to spare, as you live by it. Most of these can wait, as they have little to do when they go home."[13] Likewise, he treated artists, writers, and curates without fee and often took pity on other impoverished patients. When asked by one tradesman to perform a serious operation on his wife, Hunter fixed the fee at his usual twenty guineas; it was another two months before the couple asked him to go ahead with the procedure and presented the money. On learning that the delay had been due to the couple's difficulty in raising the not insignificant sum—equal to more than one thousand pounds today—Hunter promptly sent back nineteen guineas. He kept the twentieth, he explained, "that they might not be hurt with the idea of too great obligation."[14]

There was no lessening of demand at the back door, either. Enthusiastic young surgeons rushed to the Castle Street entrance to enroll for the autumn 1785 lectures, which began on October 10.[15] After hanging their hats on the pegs ranged in the lobby, they signed the pupils' register, which was kept on a desk beside the door to the lecture theater. James Parkinson, who would later publish the first description of the "shaking palsy" that later bore his name, was one of the students in 1785; his notes would form one of the most comprehensive records of Hunter's lectures.[16] Wright Post, a young American surgeon, attended the course at about the same time. Later appointed professor of surgery and then professor of anatomy at Columbia University in New York, he would import to the United States Hunter's method of operating for aneurysms.

Over the next few years, many of Hunter's most famous pupils, destined to become eminent nineteenth-century surgeons in both Britain and abroad, would attend his lectures. Among them were Astley Cooper, who emulated Hunter's close relations with the body snatchers, as well as his approach to surgery—he would boast that he could obtain any corpse he wished—and John Abernethy, who shared Hunter's brusque bedside manner but was similarly idolized by his students. Both pupils in 1786–1787, they would become two of Hunter's most devoted disciples. Abernethy later declared that Hunter had made "surgery a science."[17] Cooper, who attended the course year after year, would fondly remember relieving the tedium of the walk back from Leicester Square "by discussing Mr Hunter's opinions" with a fellow student. Fired by the ideas he heard, Cooper converted the sitting room at his lodgings into his own dissecting room and practiced Hunter's experimental operations on stray dogs.[18]

Another of Hunter's American pupils was a young Philadelphian, Philip Syng Physick, who would arrive in London in January 1789. When Physick's father asked the prospective tutor to list the books his son would read, Hunter simply led the way to his dissecting room, where several open cadavers lay, and declared, "These are the books your son will learn under my direction; the others are fit for very little." Suitably impressed, Physick moved into Hunter's home as a house pupil, enrolled as one of his students at St. George's and, with his support, later secured a job there as house surgeon.[19] Physick stayed on with Hunter to help organize his museum before heading north for a medical degree in Edinburgh, and it was obvious that teacher and student were equally devoted. Once Physick left, Hunter praised his talents to such an extent that the envious Home was furious. For his part, Physick viewed Hunter with an admiration that "amounted to a species of veneration," according to a contemporary. More than anyone else, Physick would import the Hunterian model to the United States. After returning to Philadelphia in September 1792, he became a surgeon in Pennsylvania Hospital and later professor of surgery at the University of Pennsylvania, where he would instruct his students to "observe, deduce and record" in true Hunterian fashion.

As well as continuing his controversial course on the principles of surgery, Hunter used his purpose-built theater and dissecting rooms in 1785 to teach lessons in practical anatomy and operations in surgery. As if this was not enough to keep himself and his pupils perpetually busy, he now launched an institution devoted to furthering the students' education. Already, two years earlier, Hunter and a longtime friend, the physician George Fordyce, had founded the Society for the Improvement of Medical and Chirurgical Knowledge. Drawing together both surgeons and physicians on an equal footing, this select band met fortnightly at Old Slaughter's coffeehouse, a rival to Young Slaughter's, to report case histories and discuss advancements in the two spheres of medicine.[20] Details of important cases were published in its transactions; Hunter would himself contribute six papers. Now Hunter and Fordyce wanted to extend this invaluable opportunity for mutual improvement to London's community of medical students.

The Lyceum Medicum Londinense, which they jointly founded in 1785, would become a crucible for Hunterian thinking. Having set this up essentially as a students' forum, Hunter encouraged the pupils themselves to draw up its rules, organize its meetings, and elect its presidents.[21] But with Hunter and Fordyce as its figureheads, keeping a fatherly watch over activities, its members were inevitably molded in the unique Hunterian style of self-education, independent inquiry, and experimental research. The society's weekly Friday-evening meetings were held in Hunter's lecture theater, and each week he laid out novel or unusual preparations for the members' interest, while also allowing them to borrow books from his library.[22] Each member was expected to present a paper on a medical or surgical topic; failure to comply resulted in a fine, and ultimately expulsion. But there were carrots as well as sticks. Every year, Hunter and Fordyce awarded a gold medal, struck with their two noble profiles, for the best dissertation. At these exuberant, stimulating, and noisy meetings, so different from the stultifying atmosphere at Surgeons' Hall, topical issues were hotly debated throughout the winter. Nothing else in London approached the Lyceum in either purpose or size. Within two years, the society could boast more than 250 members, and this number almost doubled five years later. With such a large and dedicated fan club, Hunter could hardly fail to become the chief inspiration for the next generation of young surgeons.

Yet for all his unwavering popularity among medical students and moneyed patients, Hunter remained deeply disliked by his colleagues at St. George's. Once Caesar Hawkins and William Bromfield had retired—in 1774 and 1780, respectively—Hunter had again raised his pet topic of providing free lectures at the hospital. He had even managed to persuade John Gunning, his steadfast rival, to join him in staging a few lectures in the winter of 1783 and again in 1785.[23] But the initiative proved short-lived. In early 1786, Gunning would abandon his efforts at medical education almost as quickly as he had begun them, lamenting that the lectures had cost him "a great deal of time and trouble in composing them and reading them."[24] No doubt he was aware he had no chance of competing with Hunter's young fan club.

Hunter's already shaky standing with his colleagues was further destabilized by his insistence on performing operations that they held controversial, such as his novel approach to popliteal aneurysms in December 1785. As Astley Cooper put it, "At this time none of the surgeons eminent for extensive practice placed any confidence in the surgical knowledge of John Hunter, who was chiefly known as a philosopher by means of his lectures and writings; they even contended against his views, as mystifying, if not inapplicable to the treatment of disease."[25]

It mattered little to Hunter. His colleagues could only grind their teeth and mutter conspiratorially as accolades continued to be heaped on the maverick surgeon. Already honored as surgeon extraordinary to George III, Hunter had won international acclaim for his experimental investigations with his election in 1781 to the Royal Society of Gothenberg and in 1783 to the Royal Society of Medicine and Royal Academy of Surgery of Paris. Recognition for his early military services came in January 1786, when he was appointed deputy surgeon general of the army. The post, which was largely administrative, placed him second in command—the surgeon general being his old friend and former boss Robert Adair—of all of the army's surgical services.

Though these titles brought valuable prestige, it was Hunter's new printing press that assured him lasting fame. The long-awaited *Treatise on the Venereal Disease* became an overnight best-seller on its publication in March 1786, selling a thousand copies within the first twelve months, despite the malicious attack Jessé Foot immediately rushed to publish.[26] Actively encouraged by Hunter's enemies at St. George's, Foot had little impact on the success of the text. It was translated into French and German the following year, and a second English edition would be published in 1788. The printing press was kept busy. Collecting the various papers he had submitted to the Royal Society on human and animal physiology, along with nine more previously unpublished tracts, Hunter published his wide-ranging book *Observations on Certain Parts of the Animal Oeconomy* later the same year. Dedicated to Banks, his friend and patron, it gathered together Hunter's myriad investigations on every aspect of life, from his dissections of beetles, bees, and caterpillars to his discovery of the nerves of smell, the mode of descent of the testes, and his proposals for saving the drowned—all with exquisite illustrations by van Rymsdyk. Yet for all his success—he was now earning approximately six thousand pounds a year in fees—and professional recognition, the name on the plain enameled plate beside the door of number 28 remained simply John Hunter.[27]

Finally, in May 1786, Hunter's celebrity status was confirmed when the Royal Academy opened its doors for its annual summer exhibition. Alongside oils of the most famous personalities of the period, including the Prince of Wales and the actress Sarah Siddons, there hung Hunter's portrait, painted by the most fashionable society artist of the time, Sir Joshua Reynolds.[28] Seated at his desk, with a quill in one hand, his other hand supporting his chin in the customary pose of a philosopher, and elegantly clothed in the damson red velvet suit usually reserved for physicians, Hunter was surrounded by the anatomical paraphernalia that defined his life's work.

The highly revealing portrait almost didn't make it to the summer exhibition, for Hunter was a reluctant subject and had only been persuaded to sit for Sir Joshua by a mutual friend, William Sharp, an engraver, who was keen to publish prints of the celebrated surgeon. Talked into the deal, Hunter had crossed the square to Reynolds's studio to sit for the preliminary sketches in 1785. But when Anne saw Reynolds's first attempts, faithfully portraying her husband with his straggly beard and untamed hair, she was not impressed. The wild-looking anatomist with his penetrating eyes may have been a familiar sight to Hunter's patients, friends, and pupils, but the image certainly did not strike Anne as the statuesque figure she had in mind for public edification. The picture was allowed to gather dust; Anne later gave it away to an upholsterer named John Weatherall, who had fitted out the museum.[29]

Persuading Hunter to shave off his tangled facial hair was the next step. According to stories passed down, either Anne or Reynolds convinced the surgeon to have a life mask made. Naturally enough, before the wet clay could be applied to Hunter's face, the beard had to go. Hunter made himself available for further sittings in February and March 1786 for the second portrait, and it was this more dignified representation that would grace the walls of the Royal Academy. Even then, the business was far from straightforward, for Reynolds found his friend a restless and fidgety sitter. At length, Hunter apparently fell into a deep contemplation—he was probably puzzling out some enigma of natural history that was currently plaguing him—and Reynolds briskly sketched the thoughtful face, which was the one that appeared in the portrait he unveiled at the summer exhibition.

If the intellectual gaze was Reynolds's choice, the symbolic objects surrounding the anatomist were entirely Hunter's own. Each of the books and preparations in the painting was deliberately selected by Hunter to convey his most significant interests and contributions in all his various fields.[30] So the delicately injected preparation displaying the airways of the lungs—a bronchial tree—placed upside down in a bell jar, represented his unparalleled skills in anatomy and the art of making preparations. The specimen of a bone preserved in a cylindrical bottle, possibly a leg bone from an ass, showing what appears to be a bone graft, demonstrated his skills as a surgeon and his knowledge of physiology. The sketch beneath his elbow appears to be a drawing of the fibers in muscles, probably in reference to a series of lectures on the topic he had staged for the Royal Society.[31] The two closed volumes on Hunter's desk, entitled *Natural History of Vegetables* and *Natural History of Fossils,* plainly indicated his passion for natural history. The open book at his elbow, which seems to be the third in this same series, displays sketches of human and animal skulls and bones, which no doubt most visitors to the Royal Academy that year regarded as signs of Hunter's renowned expertise in comparative anatomy. The fact that these drawings plainly depict a series of human and animal skulls in a striking gradation was doubtless lost on most visitors. Likewise the significance of the system of forelimbs or hands, ranging from the single hoof of a horse through the two-toed foot of a deer to the paw of a monkey and ultimately a human hand, would have eluded most viewers. But the curiously long feet, tantalizingly glimpsed in the top right-hand corner of the painting, would certainly have sent a shudder down the spines of most spectators that summer.

Having stealthily concealed from public gaze his prized skeleton of the Irish giant for three years, at last Hunter now felt confident, or brazen, enough to allow a sly peek at the giant's destiny. What this told the idle viewer at the summer exhibition took no subtlety to unravel. The message was clear: Here was a man whose zeal for collecting curious objects knew no bounds, who would evidently go to any lengths to obtain what he desired. Yet it would be two more years before Hunter unveiled more than the giant's unusually long feet to public view.

Reynolds's portrait was revealing in other ways, too. The intelligent, freshly shaven face that appeared in the engraving Sharp would print in 1788 from the 1786 portrait was changing rapidly. Reynolds had to rework his painting in 1789, probably in order to show the evidence of the strain induced by Hunter's continuing angina.

A few years later, the roles would be reversed when Hunter was summoned in early 1792 to observe the symptoms of his artist friend, who was in the last throes of a long illness. Reynolds had lost the sight in his left eye in 1789, three years after completing his portrait of Hunter. Having all but given up painting, he became increasingly dejected and weak, but observing no outward reason for the degeneration, his physicians dismissed his condition as hypochondria. Only when the illness reached a crisis in early February 1792 did the physicians realize that Reynolds's liver was unusually swollen. By then, it was too late, and on February 23, Reynolds died. The following day, Hunter crossed the square to perform the autopsy on his old friend. After opening the body, he noted that Reynolds's liver was more than twice its normal size, its texture hard and fibrous, its color strangely yellow. Most probably, although Hunter could not have realized it, Reynolds had died of cancer, which had spread from his eye to his liver.[32]

Hunter was in the public eye in more ways than one in 1786. For that same year, he was called to wield his knife on the nation's leading statesman. Having performed a postmortem on one prime minister, the marquis of Rockingham in 1782, now Hunter was required to operate on a second, William Pitt the Younger. After assuming the premiership three years earlier, now at twenty-seven Pitt was troubled by an "encysted tumour"—probably a sebaceous cyst—on his cheek. Advised by his physicians that it should be removed, Pitt chose Hunter to perform the delicate surgery.

Hunter duly arrived at Downing Street with his pocket set of knives and proposed to tie the prime minister's hands in the customary fashion. But Pitt was having none of it. He inquired how long the operation would take, was given an estimate by Hunter of six minutes, and adamantly insisted he would not move. Fixing his eyes on the Horse Guards' clock, which he could see from the window, Pitt sat motionless until Hunter had completed the procedure, upon which he remarked: "You have exceeded your time by half a minute." Powerfully impressed by such stoicism, but no doubt also keen to flatter the young politician, Hunter declared that he had never seen "so much fortitude and courage in all his practice."[33]

But the summer brought frustration. Though Hunter was fit enough to visit his patients on foot, he could walk only slowly.[34] By October, he was forced to travel everywhere in his carriage, since he could not walk fast enough to keep warm. The prolonged illness afforded Home the chance he had been waiting for. Making himself indispensable to his former teacher, he moved into the Leicester Square house, adopted many of the surgeon's private clients when Hunter was too ill to visit them, and assumed control over much of his other day-to-day work. The heady experience of trying on the popular surgeon's mantle, directing his pupils and assistants, advising his patients, and generally preening himself in the reflected glory of the man portrayed in Reynolds's painting proved all too tempting. Although he would never match pupils such as Jenner in Hunter's estimation, Home would do his utmost to remain the surgeon's right-hand man.

Though Hunter was now reduced to sitting impatiently in London's chaotic traffic jams rather than sprinting along the pavements as he preferred, he was still performing exhausting operations lasting up to an hour, as well as postmortems, without faltering. In November, his stamina was tested to the full when the elderly Princess Amelia, George III's aunt, died after a lengthy illness. It was Pott, Hunter's former teacher, who had treated the princess during her lifetime, but it was Hunter who was called to embalm her in death.

Inspired perhaps by interest in Egyptian mummies, the art of embalming had enjoyed something of a vogue in eighteenth-century England. Most famously, the eccentric dentist Martin van Butchell had directed William Hunter to embalm his wife when she died in 1775. A friend of both brothers, having attended both of their lecture courses, van Butchell subsequently kept his preserved spouse in his living room, where guests could view her by appointment.[35] Finally, the dentist's second wife insisted, somewhat understandably, that the shriveled corpse be removed; it ended up in Surgeons' Hall. But it was Hunter's unequaled talents in preservation, as well as his experience in unwrapping Egyptian royalty, that made him the obvious candidate to preserve and wrap a Hanoverian princess.

Hunter approached the delicate task with his usual precision, noting every step of the procedure, which took a grueling three hours.[36] Assisted by Home, Hunter removed the abdominal organs, then sawed open the head and removed the brain from the skull of the seventy-five-year-old woman. These were placed in a lead urn, while the bodily cavities were packed with a mixture of aromatic herbs and flowers. The recipe for this concoction, described as "Sweets," was meticulously noted; it included marjoram, cloves, musk, and pulverized lemon. A second mixture, containing lavender, cloves, thyme, and wormwood, was used to line the coffin. At least the atmosphere was more delicately fragranced than usual in the dissecting room. Perfectly perfumed, the princess was bound in green linen, which had been steeped in a mixture of beeswax, resin, powdered verdigris, and mutton suet. There were precise measurements for the lengths of bandages required for each part of the body, and the method for waxing the fabric was similarly supplied. "Put pack thread to the corners when you dip them," Hunter explained, "and stand on a table to draw them easily out of the pan." Whether Hunter, now fifty-eight and suffering from the pain of angina, scaled the table to perform this athletic feat, or whether the petulant Home had to carry out the task, went unrecorded.

Wrapping the corpse was an equally demanding job, as Hunter recorded: "The body is wrapped up in two pieces & the face & head are covered with two pieces and afterwards rolled over with strips in every direction. The legs are then to be brought together and the two great toes tied & then all rolled up in one; the arms brought to the sides and the whole body is to be enveloped in two pieces each 7 feet long; the whole making one mass without any appearance of neck being retained." Completely bound in green cloth, the princess's body was wrapped in white and then purple silk, Hunter pointing out, "The Purple is peculiar to the Royal family." At last, mummified and cocooned, the royal aunt was laid in her coffin and the lid soldered down, but not before certain privileged visitors had viewed Hunter's handiwork. "She is already embalmed, cered, and coffined," Horace Walpole wrote to tell Lady Ossory, adding, "her body is wrapped in I do not know how many yards of crimson silk, and she, they tell me, looks like a silkworm in its outward case."[37]

Hunter's talents were no less in demand among literary royalty. James Boswell had at last settled down to the hard work of compiling his biography of Dr. Johnson, when his eldest son, Alexander, fell ill in March 1787. Having moved back to live in his beloved London the previous year, he brought his wife, Margaret, and their children down from Edinburgh to join him.[38] The family's health was not in peak condition. After renewing his acquaintances with London prostitutes, as well as with Signor Gonorrhoea, Boswell had become depressed about his chronic condition after reading Hunter's treatise on venereal disease within days of its publication. Moreover, his wife had been ill with consumption for many years—the move to smoky London was far from beneficial—and now twelve-year-old

"Sandie" had a recurrent hernia condition. Hunter, who had first studied hernias more than thirty years ago, was the obvious person to examine him. Boswell called the surgeon to the family home on March 21.[39] "Mr John Hunter visited him," Boswell recorded in his journal, "and found that Squires, the trussmaker, was right in his discovery that one of the testicles was not come down." After examining Sandie, Hunter cast a professional eye over Mrs Boswell, too, declaring "she looked greatly better." In truth, there was little he could do to ease Margaret Boswell's condition. Nearly two centuries before the advent of antibiotics, there was no effective treatment for her already advanced consumption. Hunter visited her three times more in the following year, prescribing a few medications, letting blood—although "she had a horrour of that operation," it was still universally expected—and generally offering soothing words.[40] By then, she had only another year to live.

It was Boswell's tutor, the inspired economist Adam Smith, who next sought Hunter's help, traveling down from Edinburgh in April 1787 expressly to obtain the surgeon's aid.[41] Smith, perhaps the brightest star of the Scottish Enlightenment, had long held Hunter in affectionate regard. The two had probably met when Smith attended William Hunter's lectures in the 1770s, along with historian Edward Gibbon. No doubt the economist had also been impressed by Hunter's no-nonsense treatment of his friend David Hume in 1776. Two years later, Smith sent Hunter a copy of his epic work, *The Wealth of Nations,* a year after its publication.[42] Now suffering cruelly with a bladder infection and hemorrhoids, the sixty-four-year-old professor traveled gingerly down from Scotland to stay with a friend, the Scottish MP Henry Dundas, in Wimbledon, where he was fêted by a bevy of politicians, including William Pitt. While physicians tended to Smith's bladder complaint, Hunter performed the delicate operation of removing the economist's piles. By mid-July, one of his friends was informing another that Smith had "been cut for the piles," while his bladder problem was "much mended."[43] A few days later, patently relieved of his former agony, Smith urged Dundas to recommend Hunter for two army appointments it was the government's right to bestow, entreating that "nothing is too good for our friend John."[44] Smith enjoyed a welcome reprieve from his ailments on his return to Scotland in July, but he would continue in declining health until his death in 1790.

There was little, however, to suggest the future fame, or notoriety, of the young patient Hunter was called to visit in a shabby upstairs room off Oxford Street in early 1788. George Gordon Byron was born on January 22, after a long and difficult labor. When the baby emerged, his head was still enclosed in its caul, the inner fetal membrane, while his right foot appeared unnaturally twisted.[45] Just twenty-two and virtually penniless, due to her absentee husband's debts, his mother, Catherine, called Hunter to examine the baby's foot within days of his birth. Herself a Scot, and a stranger in London, she may have heard of Hunter's reputation from her Scottish relatives, or perhaps he was recommended by the male midwives who attended her delivery. At this visit, or possibly another, Hunter inoculated the infant for smallpox, using the traditional method of jabbing the arm with some matter from another patient's pustules. He recorded the treatment in his casebooks, noting that he "inoculated Master Byron in both arms," and after an anxious few days, during which he feared the reaction was too severe, the procedure appeared to have been successful.[46]

The infant's foot was another matter. Although Mrs. Byron would later recall that Hunter declared her son's foot "would be very well in time" so long as he was fitted with a special boot designed to straighten the deformed limb, she was seemingly too hesitant or too impoverished to heed his advice. Three years later, having returned to Scotland, she asked her sister-in-law, Frances Leigh, in London to seek Hunter's help once again to obtain such a shoe. Lamenting that she could not get one made in Scotland, she remarked, "As Mr Hunter saw George when he was born I am in hopes that he will be able to give you directions for a proper shoe to be made without seeing it [the foot] again," and she added plaintively, "I am perfectly sure he would walk very well if he had a proper shoe." Whether or not Mrs. Leigh ever sought Hunter's aid, the foot would never be completely cured. Byron's deformity—traditionally described as a clubfoot, or talipes, but possibly the result of mild spina bifida—would plague him all his life. He would successfully disguise the disability with a sliding walk, but he would remain forever self-conscious of his limp. Although he was later forced to wear an iron brace, made by a renowned quack in Nottinghamshire, this failed to mend the deformity. Many years later, the eleven-year-old boy, by then Lord Byron, would be examined by Hunter's nephew, Matthew Baillie. He concluded that the foot might have been mended had a corrective boot been used in infancy, just as his uncle had suggested.[47]

Hunter continued to offer his services free to artists, and it was Thomas Gainsborough, the portrait artist who longed to devote his time to landscapes, who requested the surgeon's aid in spring 1788. Gainsborough had spotted a lump in his neck three years earlier, but since it caused no pain, he had thought little of it. Now it was so painful that he could not sleep comfortably at night, and he anxiously sought the advice of both the physician William Heberden and John Hunter. They were equally dismissive of the lump, both assigning the pain and swelling to a cold that Gainsborough appeared to be suffering from at the same time. Hunter prescribed a saltwater dressing, while Heberden could throw no more light on the ailment. "What this painfull swelling in my Neck will turn out, I am at a loss at present to guess," the artist confided to a friend in April. "Mr John Hunter found it nothing but [a] swell'd Gland, and has been most comfortable in persuading me that it will disperse by the continued application of a sea-water poultice." He added, "My neighbour Dr Heberden has no more notion of it."[48] Somewhat inexplicably, Hunter had failed to diagnose the cancerous tumor that would kill the artist that summer. Heberden, a respected physician, had been similarly remiss. Possibly both were simply shielding Gainsborough from the dismal truth.

But there was one well-known patient who would live to regret escaping John Hunter's knife. The Austrian composer Joseph Haydn would become a familiar visitor to the Hunter household during his stay in London in 1791 and 1792. It was Hunter's attractive and talented wife, Anne, however, who attracted him to Leicester Square.[49] Haydn became a regular guest at Anne's musical evenings, and he even set six of her poems to music, his Six Original Canzonettas. As their friendship grew, Anne would become regarded as Haydn's muse.

But the talented composer suffered from painful polyps in his nose. Whether Hunter simply had an altruistic desire to ease Haydn's discomfort or, conceivably, suspected the friendship with his wife went beyond purely professional interest, he determined to employ his surgical instruments on the composer. Initially, during a consultation in Hunter's study, Haydn seemed to consent to have the polyps removed. But a moment later, when he was roughly grabbed by Hunter's assistants, forced toward a chair, and saw the surgeon bearing down on him with a pair of forceps glinting, he promptly changed his mind. "I yelled and kicked and hit until I had freed myself," Haydn later recalled, "and made clear to Mr Hunter, who already had his instruments ready for the operation, that I did not want to undergo the operation." Finally, he managed to convince Hunter to put away his instruments. Haydn noted, "It seemed to me that he pitied me for not wanting to undergo the happy experience of enjoying his skill." Returning to Austria, where his polyps continued to grow and cause him pain, Haydn later wished that he had consented to the happy experience of Hunter's skills.

For all the lionization of Hunter—the kindly, good-hearted Dr. Jekyll of Leicester Square—by fellow scientists, grateful patients, and Georgian high society, deeply seated suspicion and fear of the surgeon-anatomist— the secretive, sinister Mr. Hyde who emerged from the Castle Street door—persisted among the general populace. The boom in private anatomy schools and the continual gravitation of medical students to the capital generated relentless demand for dissection material. As the Resurrectionists became more desperate, and more bold, in ransacking graveyards to supply the students' needs, so public vigilance and hostility were growing. Already this had reached such a fever pitch across the Atlantic, where former pupils of the Hunterian schools had imported the brothers' grave-robbing methods, that a fierce and bloody riot had erupted in New York.[50] It began one Sunday in April 1788, when some boys playing behind New York Hospital had spotted a medical student dissecting an arm. Their shouts quickly attracted an angry mob, which rampaged through the building, destroyed anatomy preparations, and seized several cadavers, which were promptly reburied. But the fury did not abate, despite appeals from the mayor, and the following day several surgeons and pupils had to take refuge in the jail. Full-scale battle ensued as the protesters attempted to storm the prison, but they were rebuffed by a hurriedly raised armed guard. Pelted by stones, the soldiers were ordered to fire on the crowd, which they did, killing seven people and injuring several more. Although in the wake of the so-called Doctors' Riot the surgeons involved in dissecting activities were forced to deny in sworn affidavits that they had stolen bodies, the clandestine grave robbing and the protests would continue.

Medical students and professional body snatchers were at least as busy, if generally more discreet, in London graveyards. And now that Hunter had added anatomy classes to his lectures, his need for regular supplies of fresh bodies was greater than ever. Yet while it was widely known that corpses were delivered to the Castle Street door most nights throughout the winter, Hunter stealthily kept his involvement at arm's length. In November, he was cited in a court case—the only time he would come anywhere near legal exposure—when an employee of Westminster General Dispensary, Edward Howe, was sent to prison for "fraudulently obtaining a corpse from Mr Bradford, an apothecary of Grafton Street . . . and stripping and selling it to Mr Hunter of Leicester Square."[51] While Howe festered in jail, Hunter's legal record remained untarnished.

Given the sinister world Hunter was forced to inhabit by virtue of his calling, it is little surprise that he provided inspiration not only for Robert Louis Stevenson but for the melancholy William Blake, too. In his satirical piece of prose *An Island in the Moon,* written in the 1780s, the young Blake launched a scathing attack on intellectual society, ridiculing various members of the Royal Society but saving particular venom for those involved with surgery and anatomy. The villain of his piece, Jack Tearguts, was almost certainly John Hunter. Not only did Hunter have several acquaintances in common with the poet, but Blake spent most of his life living in the same neighborhoods. His childhood home in the 1760s was minutes from Hunter's house in Golden Square; after marrying in 1782, Blake lived on Green Street, directly round the corner from Leicester Square, where he probably enjoyed a good view of both entrances to the anatomist's extraordinary home.[52] "He understands Anatomy better than any of the Ancients," Blake's protagonist says of Jack Tearguts. "He'll plunge his knife up to the hilt in a single drive, and thrust his fist in, and all in the space of a Quarter of an hour. He does not mind the crying, tho' they cry ever so. He'll swear at them & keep them down with his fist, & tell them that he'll scrape their bones if they don't lay still & be quiet."

But as Hunter now prepared to open the doors of his treasured museum to public view for the first time, even some of his friends in the scientific world feared he had gone too far this time.

CHAPTER 15

The Monkey's Skull

Leicester Square, London
1788

Anticipation among the distinguished visitors arriving at 28 Leicester Square was intense. Ushered through the house and into the new extension, they pressed impatiently forward. As well as members of the Royal Society, the Royal Antiquarian Society, and the Royal College of Physicians, the guests included writers and several important foreign visitors. Shepherded up to the first floor, they arrived at the tall doors and stepped into the great skylit room, where they gazed around, incredulous. All had heard the rumors and speculation about the famous anatomist's collection. Some of the stories detailing exotic animals he had obtained sounded more like the stuff of fiction, while the reports of human curiosities he had procured seemed equally fabulous. What they found, as they stared in wonder at the bottles filled with fleshy substances, the cases of dried organs, and the skeletons positioned around the large galleried room, and what they now heard from the lips of their illustrious host, was more surprising than any of them could have expected.

It was five years since Hunter and his young family had moved into Leicester Square, three years since he had supervised the transfer of his already large collection into its custom-built accommodation. Only now, after he and his several assistants had spent many months sorting, arranging, and cataloging the ever-growing number of exhibits, was Hunter ready to reveal his museum to the outside world. This was not just the happy conclusion of five years' planning; it was the culmination of his life's work. A small figure in the lofty room, though recently grown a little more rotund through his enforced lack of exercise, Hunter waited as his guests perused the displays, preparing to deliver his opening words.

By anyone's standards, it was a remarkable scene. None of the guests could have failed to be amazed at the sight of the huge stuffed giraffe, captured in 1779, hacked in half to fit the Jermyn Street hall, and now reunited with its legs, as it loomed above their heads. One journalist, invited to record the opening, was plainly overwhelmed, noting that "from the report of its size and other circumstances, it was hitherto much doubted by naturalists whether such an animal did really exist or not."[1] Finally convinced that the giraffe was not a work of fiction, he wrote, "In point of size, it is above eighteen feet high, with an erect neck and long feet, and in many respects partaking of the species of the common camel. From the stiffness of its joints, it can neither stoop, nor lie down; but as nature is ever provident for its creatures, it receives its food from the leaves of trees, which from its extreme height it can readily do by putting his head in among the branches."

There were more sights, equally remarkable. Staring apprehensively around, the visitors could glimpse massive bones belonging to elephants, camels, and whales; body parts from a zebra, a leopard, a pelican, and a hyena; and entire preserved bodies of rare creatures such as an aardvark fetus, a Surinam toad carrying its young in pockets on its back, and a baby crocodile frozen at the point of emerging from its egg, the umbilical cord still intact. There were scarcely credible human curiosities, too. Few could have failed to be moved by the five pitiful bodies of quintuplets, lined up stiffly in a row. Born two years earlier, four months prematurely to a young mother in Lancashire, three had been stillborn and two born alive, though they survived only briefly. The physician who delivered the babies, all girls, had preserved their emaciated bodies in spirits, tied a label to each wrist indicating their order of birth, and sent them to the Royal Society, which promptly gave them to its foremost anatomist.[2] Just as unusual was the skeleton of a man reduced to less than five feet tall by a rare disease, *myositis ossificans,* which had progressively turned his muscle into bone. Hunter had purchased the skeleton at an anatomical auction in 1783 for the considerable sum of eighty-five guineas.[3] But most stunning of all, as they raised their eyes, the visitors came face-to-face with the magnificent skeleton of Charles Byrne, the acclaimed Irish giant, grinning down grotesquely in death at the spectators he had once enthralled in life. Revealing his prize trophy to London society for the first time, Hunter claimed he had paid 130 guineas for the body. Whether this was the genuine price, colossal in itself, or whether he was simply too embarrassed to reveal the cost of five hundred pounds that he was reputed to have spent, would remain his secret.[4]

As they stared around, few of the spectators could disagree with the journalist's description of "Mr Hunter's very curious, extensive, and valuable museum."[5] By sheer volume alone, the collection was nothing less than spectacular. Ultimately, it would total nearly fourteen thousand individual items, and it was not far off that number already. Almost certainly it was the largest collection of its kind in Britain at the time. The range, too, was incredible. Eventually, it would encompass more than fourteen hundred animal and human parts preserved in spirits; over twelve hundred dried bones, skulls, and skeletons; more than six thousand pathological preparations showing the effects of injury and disease; and more than eight hundred dried plants and invertebrates, as well as assorted stuffed animals, corals, minerals and shells.[6] In all, more than five hundred separate species were represented. In addition, there were nearly three thousand fossils, comprising one of the largest such collections in the country.[7]

The accumulation of such a vast and varied hoard had cost Hunter a fortune—more than twenty thousand pounds, according to the journalist attending the open day, "beside a very accurate and industrious collection of near thirty years." But this, too, was probably a gross underestimate. Hunter at one point told his former assistant William Lynn that he had spent seventy thousand pounds on the collection; Hunter's will, written in 1793, would put the total outlay at more than ninety thousand guineas.[8] Whatever the true cost, the continual expenditure explained why, despite earning upward of five thousand pounds a year in the early 1780s and an estimated six thousand thereafter—about £360,000 in modern terms— Hunter was invariably short of money. He was never out of debt to animal dealers, friends, and money lenders, while his properties at Earls Court were forever being remortgaged in order to drum up more cash for new acquisitions. Lynn knew this better than most. Once, when Lynn was ill after wounding his hand during a postmortem, Hunter had offered to lend him two hundred pounds. When Lynn recovered and called to thank his former boss for the offer, Hunter had completely forgotten the promise, exclaiming, "I offer you money! That is droll, indeed; for I am the last person in this town to have money at command."⁹ With the considerable expense of running two large households, plus his wife's extravagant lifestyle, it was a wonder indeed that Hunter never ended up bankrupt. Nonetheless, he was quite prepared to stand by his offer to Lynn.

Hunter was himself the recipient of extraordinary generosity. The donors who had helped him to form his remarkable collection spanned every area of Georgian life. He had received a hog deer from Lord Clive, and a gibbon, a baboon, and an albino macaque from the former prime minister, the earl of Shelburne. Queen Charlotte herself, of course, had donated the carcasses of two elephants that had died in her menagerie, as well as the little bull that had almost gored him. Explorer friends had been equally forthcoming. As well as the antipodean finds from Banks, Hunter had been given several animals by Captain Constantine Phipps, later Lord Mulgrave, from his voyage toward the North Pole in 1773. Julius Griffiths, a former pupil from St. George's, had brought Hunter back an aquatic snail and the entrails of a pangolin from his expedition to Sumatra. Sadly, Griffiths noted on his return to England, the entrails had been "entirely spoiled from their long detention at the India House."¹⁰ Indeed, although Hunter never ventured beyond British shores after his brief military career, he managed to foster contacts around the globe in his efforts to secure creatures from every corner of the world. In a typical letter to an acquaintance in Africa, he urged, "If a foal camel was put into a tub of spirits I should be glad. Is it possible to get a young tame lion, or indeed any other beast or bird?"¹¹ He was never coy about his requirements. Even his servants were drawn into the mission. Hunter gave orders that when mowing the meadows, his Earls Court staff should preserve every dormouse nest and beehive they stumbled upon.¹²

Now the fruits of this tireless quest were open to public scrutiny. As the last guests bustled in, Hunter began the first guided tour—"a kind of peripatetic lecture," according to the newspaper columnist—of his startling museum. It lasted between two and three hours. But this was scarcely surprising, for the collection encapsulated Hunter's entire career. As well as the handful of preparations he had held on to from his years in Covent Garden, there were bones displaying gunshot wounds from his spell in the army, the tooth grafted into the cockerel's comb from his period studying teeth, and tissues showing venereal infections from his research into sexual diseases. There were preparations demonstrating his surgical prowess in operations at St. George's, organs harvested from innumerable autopsies, and cabinets showing results of experiments conducted at his Earls Court laboratory. It was nothing less than a record of John Hunter's life. More significantly, it was a representation of life itself. For, as Hunter made clear to his rapt entourage, the collection was no haphazard assortment of curiosities such as any number of eighteenth-century amateur enthusiasts had acquired, but a carefully ordered series of human and animal parts arranged expressly to investigate and illustrate fundamental principles about life on earth. Nothing else like Hunter's museum existed, or ever would exist.

To most outsiders, this peculiar arrangement was simply perplexing. For his friend Horace Walpole, for example, the museum was just "Mr Hunter's collection of human miseries."[13] To Hunter, its organization was entirely logical. Always scornful of the written word, he had created a museum that functioned both as a teaching aid, graphically displaying fundamental facts about anatomy and physiology to his pupils, and as a research tool, helping Hunter to investigate the general principles of life. As one contemporary report explained, "The main object which he had in forming his Museum was to illustrate as far as possible the whole subject of life, by preparations of the bodies in which these phenomena are presented."[14] No static display of inanimate objects, such as Walpole's own magpie hoard, this was a dynamic and vibrant exposition of Hunter's theories on organic life. Not only did the museum attempt to explore the physiology of all life-forms; it also aimed—as he explained to visitors on his guided tour—to reveal the connections that existed among every kind of life. It was these unorthodox views that captivated his guests, as the newspaper report made plain: "What principally attracted the attention of the *cognoscenti* was Mr Hunter's novel and curious system of natural philosophy running progressively from the lowest scale of vegetable up to animal nature."

The system was indeed novel, although entirely in keeping with all Hunter had taught and written. His collection was divided into three main sections. The first two—the greatest proportion—illustrated normal life: One part exhibited the anatomical structure of individual animals in order to display their physiology; the other elucidated the preservation or reproduction of entire species. The third section, the pathological series, showed examples of normal life gone wrong because of disease or injury, although most of these preparations were stored in the lecture theater, where they could be brought out during class to underline Hunter's teachings. In the main museum, Hunter had organized the healthy human and animal parts according to bodily systems, as he previously had at Jermyn Street. So organs of digestion or elements of the respiratory system were grouped together, running from the simplest animal structures to the most complex. But now he went further than ever, even placing vegetables and plants alongside analogous animal parts. The similarities between the tendrils of climbing plants and the prehensile tails of sea horses and chameleons were displayed in one series illustrating parts of locomotion; in the series demonstrating circulatory systems, sap was compared to blood.[15] In all, fifteen series demonstrated the principal anatomical systems—digestion, bone structure, nervous systems, and so on—through a staggering range of species. Not only did the museum present uncompromising proof that humans, animals, and even plants shared similar structures—a common makeup—which varied only in its complexity; it also demonstrated precisely how each life-form was suited to its own particular circumstances.

If his guests were in any doubt of the potentially heretical implications of this arrangement, Hunter was quick to disabuse them. Drawing up in front of a row of skulls positioned in what he considered to be an ascending order of complexity, Hunter bracketed the human species along with monkeys. Betraying the classic eighteenth-century European belief in white superiority, which would prevail for at least the next century, he had placed the skulls of Europeans through to Africans in descending order. More significantly, and more shockingly to Georgian minds, he had included a monkey—most probably a chimpanzee—in this ordered series of skulls. Sparing no religious sentiments, Hunter explained, "There is a regular and continued gradation of these from the most imperfect of the animal, to the most perfect of the human species. The most perfect human skull is the European; the most imperfect of this species is the Negro. The European, the Negro, and the Monkey form a regular series."[16] Plainly unable to resist scandalizing his guests still further, Hunter then came out with his pièce de résistance. "He also remarked," the newspaper reported, "that our first parents, Adam and Eve, were indisputably black. This is quite a new idea; but Mr Hunter observed it might be proved without difficulty." The idea that the first humans were black—and since they were created in God's image, by implication God was therefore black—was startling. Whether or not he did indeed possess proof that human beings had originated in Africa, a notion that would be confirmed only in the latter half of the twentieth century, the radical ideas he expressed at this first public viewing of his extraordinary collection plainly demonstrated that he had moved far beyond conventional eighteenth-century dogma.

It was little wonder that Hunter's museum now attracted widespread interest from across the Continent. Keen to share his insights, Hunter opened the museum twice a year—in May to aristocrats and gentlemen, in October to fellow medical practitioners and natural philosophers.[17] Fellow anatomists from the Continent hastened to see the fruits of Hunter's laborious research for themselves.[18] Yet he was choosy about whom he invited. After the storming of the Bastille on July 14 the following year, Hunter carefully scrutinized prospective visitors' allegiances. While Hunter's approach to medicine was decidedly radical, he was a staunch conservative and royalist in public life. He told an acquaintance who wanted to bring a French friend to see the museum, "If your friend is in London in October (and not a Democrate), he is welcome to see it; but I would rather see it in a blaze, like the Bastile, than show it to a Democrate, let his country be what it may."[19]

Undeterred by the political climate, from about 1792, Hunter began to commit to paper his daring and potentially dangerous thoughts on the development of life on earth. As antirevolutionary fervor spread to Britain, and sympathizers such as the chemist Joseph Priestley had to flee when their homes were gutted by furious mobs, John Hunter spent his evenings dictating blasphemous views that could potentially shake the world. For although he would never use the term *evolution*, he now set out to make plain that he believed all animals, including humans, were descended from common ancestors.

John Hunter was not the first to express controversial ideas about the origins of life, and he would certainly not be the last. But to challenge the traditional biblical story, that God had created the earth and all life over a period of six days, was still highly contentious toward the end of the eighteenth century, and would remain so for some time to come.[20] Many in Georgian society still accepted unquestioningly the precise calculation by James Ussher, the seventeenth-century archbishop of Armagh, that the world had been created on the morning of October 23, 4004 B.C. This prevailing doctrine denied the possibility of any significant time lapse between the creation of the earth and the appearance of human beings, while it also supposed that all life had been formed in a state of perfection—no creatures had since changed or died out. Where evidence contradicting this doctrine turned up, such as the discovery of bones belonging to animals that no longer appeared to exist, or fossils of marine creatures found on mountaintops, often convoluted theories were proposed to suit the traditional orthodoxy. The remains of unknown animals simply meant the creatures had not yet been discovered, while evidence of sea life in high places had been deposited there during the Flood. Such a fixed and hierarchical order was often expressed as the "chain of being"—an idea developed in ancient Greece but still widely supported in the eighteenth century—which imagined a single ladder leading neatly upward from rocks, to plants, to the simplest animals, and ultimately to humans and even divine beings.

Naturally, there were notable rebels who wrestled with this religious straitjacket. In the seventeenth century, Robert Hooke, the eccentric stalwart of the fledgling Royal Society, had raised doubts that a single flood could have caused all the earth's diverse features or that the world could have existed for only six thousand years. Only a long period of continual change by natural forces—earthquakes, volcanoes, and sea erosion—could have wrought such forms, he argued, while he maintained that fossils were the impressions of living organisms, some of which had become extinct.[21] But Hooke's was an isolated voice.

The pattern—lone mavericks contesting predominant theological opinion—was set. Although the spirit of scientific scrutiny fostered by the Enlightenment opened the door to further questions about the biblical worldview, those daring to query the accepted ideology remained firmly on the fringes of intellectual society. So James Burnett, the Scottish judge known as Lord Monboddo, was mercilessly ridiculed for suggesting in 1774 that orangutans were members of the human species.[22] Another Scot, the pioneer geologist James Hutton, was similarly the butt of society's scorn when he concluded that the earth had been formed over a vast period of time, through a gradual process of climate change, terrestrial movement, and water erosion—just as Hooke had earlier proclaimed. Yet even while denying the biblical account of the Creation and the Flood in his theories, published in the *Transactions of the Royal Society of Edinburgh* in 1788, Hutton still insisted on God's role as a divine designer.[23]

On the Continent, the origins of life were just as hotly debated. Carl Linnaeus, the Swedish botanist who had developed the system of classifying plants and animals in the middle of the century, had gradually revised

his view that all species were fixed. Proposing that new species might occasionally be produced by interbreeding, he argued that God had simply
created the head of each genus, or class of species, and this produced local varieties—effectively, new species.[24]

The Dutch anatomist Peter Camper, an early visitor to John Hunter's
museum, had likewise ranked his skulls in a hierarchy ranging from monkey through African to European, but he still insisted on a distinct gap between humans and animals.[25] Johann Friedrich Blumenbach, a young
professor of medicine at Göttingen, who visited Hunter's museum in the
early 1790s, was similarly preoccupied with collecting and ranking skulls.
He advanced the theory of a hierarchy of human "varieties," descending
from the Caucasian down to the Ethiopian.[26] Yet even though the French
polymath Pierre Louis Moreau de Maupertuis had proposed as far back as
1745 that life-forms had changed over time and that similar species shared
common ancestors, both Blumenbach and Camper stuck resolutely to the
idea of fixed, unchangeable species.

A distinctly more daring view was advocated by the French naturalist
Buffon, who enraged theological opinion in 1749 by suggesting that the
earth was 75,000 years old—a dramatic increase on the orthodox 6,000
years, if still considerably short of today's best estimate of about 4.5 billion
years. After being censured by religious authorities, Buffon had tempered
his heretical view to suggest that the earth had developed over seven
lengthy epochs, correlating with the biblical days of creation.[27] Undaunted,
however, Buffon declared that similar animals shared common ancestors,
although he stuck doggedly to the idea that species were fixed in form by
suggesting perversely that the different animals inhabiting the earth were
varieties rather than species, all descended from thirty-eight original ancestors, akin to Linnaeus's genera. Even so, Buffon's idea of a process of divergence—an original cat ancestor bringing forth lions, tigers, and domestic
cats, for example—represented a radical new approach, one that was still
heavily resisted by theologians as well as fellow naturalists.[28]

In 1792, therefore, as John Hunter began to outline his own views on
such contentious issues, there had been isolated assaults on orthodox opinion, but there was still no notion of a single common ancestor, no consen-
sus that species had changed over time, and certainly no idea of how such
a process might occur. Society in the main held firm to the biblical version
of the Creation and the Flood. Any divergence from this view was regarded

as heresy. Even the young Charles Darwin still adhered to this version when he set out on the *Beagle* in 1831. After his own faith in it waned, his theories would be furiously attacked throughout the nineteenth century for their irreligious stance.[29]

While Hunter may well have kept abreast of the assorted theories being advanced by his contemporaries, the beliefs encapsulated in his museum, and those he now put down on paper, were entirely his own, developed over a lifetime's inquiry. Hunter had no doubts, on scientific or religious grounds, that life on earth had changed over time. His extensive collection of fossils, in which he placed his trophies side by side with their modern equivalents—"the resent"—was plain proof of that.[30] The process by which such life-forms could change had become a vital question for him. He had been preoccupied with deformities and variations within species since the late 1770s. In his study of freemartins, he had even put forward the startling suggestion that animals with separate male and female genders might originally have developed from the accidental occurrence of a natural hermaphrodite with two distinct sexual organs.[31] "Is there ever, in the genera of animals that are natural hermaphrodites, a separation of the two parts forming distinct sexes?" he pondered. "If there is, it may account for the distinction of sexes ever having happened." It was clear that he believed species were mutable and capable of gradually developing—evolving—into others. He was aware, too, from his connections with animal breeders, that variations in eye color could become permanent "with respect to the propagation of the animal, becoming so far a part of its nature, as to be continued in the offspring."[32]

By the late 1780s, Hunter had developed a fervent interest in artificial breeding—not only a vibrant eighteenth-century industry but a graphic demonstration of the way in which variations and deviations are passed down. As well as attempting to breed rare animals, including opossums, at Earls Court, he had tried to crossbreed different species. At one point, he successfully mated one of his cows with a buffalo belonging to the late marquis of Rockingham.[33] This interest had first been sparked by the wolf-dog hybrid he had owned as a young man. Nearly thirty years later, he was still enthralled by the notion of interbreeding dogs and wolves, only now it was not so much experimental curiosity as an attempt to pinpoint whether the wolf and dog belonged, in his words, to the same species.[34]

In the intervening years, there had been several successful efforts by menagerie owners to mate dogs and wolves. In 1785, Hunter had been promised a puppy bred from a she-wolf and a male greyhound by the animal dealer Gough; to his chagrin, a leopard in Gough's menagerie killed Hunter's pup, along with two others. Two years later, Hunter was more fortunate, obtaining a female puppy bred from the same wolf and another dog. Now attempting himself to mate his wolf hybrid with a dog, he pressed his friend Banks to procure testimonies from aristocrat acquaintances who had achieved the same end. In the meantime, Hunter had obtained a puppy born of a female jackal that had mated with a spaniel dog on a ship returning from the East Indies in 1786. He took his trophy to Earls Court, where it mated with a terrier. Hunter was like a delighted father when the jackal hybrid produced five puppies that November. One of its offspring was sent to Jenner, who wrote back, "The little jackal-bitch you gave me is grown a fine handsome animal; but she certainly does not possess the understanding of common dogs. She is easily lost when I take her out, and is quite inattentive to a whistle."[35]

Hunter's observations on interbreeding, published in a paper to the Royal Society in 1787, had led him to conclude that the wolf, dog, fox, and jackal were indeed the same species. He based this assertion on the grounds that as well as sharing outward appearances, they could mate and produce offspring—a definition of species still commonly applied—although he overlooked the fact that most of the females had to be coerced or restrained in order to mate. Buffon, Hunter knew, had attempted the same trial—without success—in order to test his theory that wolves and dogs were merely varieties of one species, related through a common ancestor. Hunter had plainly come to the same opinion, remarking, "Here then being an absolute proof of the jackal being a dog, and the wolf being equally made out to be of the same species, it now therefore becomes a question whether the wolf is from the jackal, or the jackal from the wolf (supposing them but one origin)?" The fact that he suggested that the animals were different varieties rather than species mattered little. What was significant was an appreciation that similar life-forms were descended from a common ancestor—"one origin"—and had changed, considerably, over time. He shared Buffon's belief that the different animals had migrated from an original population, in this case from the wolf, and had changed according to environmental and climatic conditions. Unequivocally outlining his belief that similar animals shared a common ancestor, Hunter proclaimed, "To ascertain the original animal of a species, all the varieties of that species should be examined, to see how far they have the character of the genus, and what resemblance they bear to the other species of the genus."

Within Europe, only Buffon had dared to venture such heretical views. Within Britain, nobody had yet gone so far in challenging religious orthodoxy. Despite the very real risk that he could be condemned as a heretic, Hunter now determined to enlarge his controversial theories in print. In particular, he began work on two documents speculating on the nature of fossils, while at the same time compiling a plethora of essays and notes that ranged over his ideas on the entire animal kingdom.

Given his lengthy study and large collection of fossils, Hunter was the obvious person to be consulted when a hoard of fossilized bones was unearthed in a cave in Germany and sent for examination to the Royal Society in the early 1790s; he duly prepared a paper on his conclusions.[36] Studying the fossil bones, he found them curiously similar to those of a polar bear he possessed, yet almost twice the size, with the teeth in similar proportion. He concluded, as he had with the earlier mastodon bones, that the bones belonged to extinct animals akin to their modern-day counterparts. And he made plain that the process by which such life-forms had changed required "a vast series of years"; indeed, he repeated the phrase "many thousand years" several times. Though this careful terminology might just fit the orthodox estimate of the age of the earth, Hunter had no qualms about casting doubt on the story of the deluge. The remains of marine animals buried under several strata of rock indicated that the sea must have invaded the land many times, he declared, and remained there, he repeated again, for "thousands of years."

If Hunter restrained his wildest calculations in the paper for the Royal Society, he felt no such constraints in the treatise he simultaneously produced to describe his own fossil collection.[37] In the slim two-part work, Hunter allied himself with Hutton's uniformitarian theory, which held that geological features had been formed by the same progressive forces seen in modern times and that these had occurred over a vast period. Although Hunter's pioneering contributions to geology would never be fully recognized, he gave an expert explanation of the formation of the earth's key features, describing how rock strata had formed, how fossils had been made, and how the sea had produced gradual changes through erosion and deposition. He even speculated that the Thames valley had once been an arm of the "German Ocean"—the North Sea—which had covered much of the Low Countries, too. Having arranged each fossil in his collection alongside a portion of the rock strata in which it was found, Hunter observed that the same fossils were always discovered in the same strata; Hunter came to this conclusion nearly two decades before William Smith, later dubbed "the father of British geology," described rock sequences from fossil samples. Most contentiously, Hunter took these assertions to their logical conclusion and denounced the biblical version of the Flood, proclaiming that "Forty days' water overflowing the dry land could not have brought such quantities of sea-productions on its surface."[38] Supposing that the sea had changed places with the land on several occasions, he remarked, "What number of thousand of years this would take, or how often this has happened, I will not pretend to say."[39] But in speculating how long certain fossils might have existed in any singular state, he boldly proclaimed that "many retain their form for many thousand centuries."

Such a blunt denial of religious orthodoxy was too much for Hunter's Royal Society friends. In France, Buffon had been forced by the church to recant a similar view on the age of the earth; even in 1823, William Buckland would attempt to "prove" the veracity of the deluge. For Hunter to undermine the authority of the church at a time when revolutionary fever threatened the country was unthinkable. Certainly that was the staunch reaction of Maj. James Rennell, a distinguished geographer and RS fellow, who had been asked—either by Hunter or by stalwarts of the society—to peruse the treatise. Writing to Hunter after having read his document "three times," Rennell urged him to change his time line from "many thousand centuries" to "many thousand years." Rennell himself had "no quarrel with any opinions relating to the antiquity of the Globe," he insisted, yet he warned, "There are a description of persons, very numerous and very respectable in every point but their pardonable superstitions, who will dislike any mention of a specific period that ascends beyond 6,000 years." Evidently, while members of the Royal Society might entertain their doubts about the biblical story in the exalted seclusion of their cosy meetings, such heretical thoughts must not be aired in the public domain, especially at a time of social upheaval, as Rennell made plain. Questioning Hunter's denial of the Flood and emphasizing his concerns by underlining key phrases, Rennell insisted, "Again in page 8 you have rather questioned the Knowledge of a <u>Certain Person</u> [Moses]; that also is tender ground with some people, and at <u>this time</u> we must not let the Vulgar know how far we believe in the Books of Moses."[40]

Whether or not Hunter bowed to Rennell's concerns is unclear. The fact that two surviving manuscripts, both versions of the same document, refer to "thousands of years" rather than to "thousands of centuries" suggests that he did amend his treatise. According to Home, not always a reliable commentator but someone who was working closely at Hunter's side at the time, Hunter refused to modify his comments and angrily withdrew the paper. Either way, Rennell's intervention effectively suppressed Hunter's unorthodox views—on the age of the earth and the deluge at least. The Royal Society never published the treatise, and it would gather dust for almost seventy years before the Royal College of Surgeons finally hurried it into print at the end of 1859, just a month after Darwin's *On the Origin of Species* was published.

Even if he had been induced to amend or withdraw his written views on fossils to suit the Royal Society's pious misgivings, there was no way Hunter could be persuaded to reform his radical beliefs. In the voluminous notes and essays he furiously dictated each evening, he now set out his controversial theories in their full, unrepentant glory. Ranging over natural history, geology, fossils, and anatomy, Hunter finally sought to lay down the overarching principles he believed governed all life on earth. Whether or not he intended to publish these daring ideas is unclear, although, given his disregard for convention and previous publishing record, it seems likely. Ultimately, they would be published in two lengthy volumes, entitled *Essays and Observations on Natural History, Anatomy, Physiology, Psychology and Geology*, more than sixty years later. Together, they expounded the conclusions of a lifetime's research while classifying the thousands of animals Hunter had studied into families of related species.

Unequivocably, in these assorted notes, Hunter outlined his views on the origins of life. First, he made tacit acknowledgment of the biblical story of the Creation and the Flood; then he proceeded to demolish its veracity step by step. Beyond this perfunctory denial of the accepted ideology he made no further reference to God, the Bible, or any divine mode of creation. It is dubious that Hunter saw himself as a crusading atheist, unlike his erstwhile patient David Hume, but he certainly displayed the opinions of a materialist, rejecting a divine plan of the world in line with his contem-

porary Erasmus Darwin. Although Hunter conformed to church rituals—
he had grudgingly consented to be godparent to Jenner's first son in 1789
with the comment, "Rather than the brat should not be a christian I will
stand Godfather"—he had, of course, declared that any belief system de-
pending on faith rather than fact demonstrated "a weakness of mind."[41]
Some of his disciples would later seek to downplay his irreligious beliefs.

Certainly Hunter was not held back by religious sensibilities as Darwin
would be. In the notes later published as *Essays and Observations,* he insisted not only that the earth was immensely old but that its life-forms had
changed substantially. This process of alteration had been so profound,
producing so many variations in species, Hunter declared, that "it becomes
a doubt whether they were all original, or whether any one of them are original, or none of them; or, if any one be original, which that one is."[42] He
made plain, too, that he believed similar species had descended from common ancestors, asserting, "To attempt to trace any natural production to its
origin, or its first production, is ridiculous; for it goes back to that period,
if ever such existed, of which we can form no idea, viz. the beginning of
time. But, I think, we have reason to suppose there was a period in time in
which every species of natural production was the same; there then being
no variety in any species."[43] The suggestion that there had initially been "no
variety" in any species—in other words, only original species existed—
pointed to a number of common ancestors, in line with Buffon's view. Yet
the possibility that these early beings had first stemmed from a single common ancestor was certainly implied in his remark that "it will be necessary
to go back to the first or common matter of this globe, and give its general
properties; then see how far these properties are introduced into the vegetable and animal operations." Nor did he hesitate to suggest that new life-
forms had emerged—another heresy—for he remarked that tail-less, or
Manx, cats probably first arose "from a kitten being brought into the world
without any tail."[44] Equally, he reiterated his understanding that artificial
breeding showed how variations in species were passed down through generations.

Going further, Hunter was unequivocal that all living things were in-
terrelated: "We may observe that in Natural Things nothing stands alone;
that everything in Nature has a relation to or connexion with some other
natural production or productions; and that each is composed of parts common to most others but differently arranged."[45] He declared, "Every prop-

erty in man is similar to some property, either in another animal, or probably in a vegetable, or even in inanimate matter." This applied most clearly
when comparing the early embryo states of humans—"the most perfect animal"—and other creatures, as Hunter explained:

> If we were capable of following the progress of increase of the number
> of the parts of the most perfect animal, as they first formed in succes-
> sion from the very first, to its state of full perfection, we should proba-
> bly be able to compare it with some one of the incomplete animals
> themselves, of every order of animals in the creation, being at no stage
> different from some of those inferior orders. Or in other words, if we
> were to take a series of animals from the more imperfect to the perfect,
> we should probably find an imperfect animal corresponding with some
> stage of the most perfect.[46]

So, more complex animals mirrored the most simple forms in their embryonic stages, Hunter argued. Precisely the same observation, noting the similarities between the embryos of a dog and a human, would help Darwin to
appreciate the common ancestry of all life.[47] But if Darwin held back from
speculating on the place of humans until his *Descent of Man* in 1871,
Hunter had no such qualms. "The monkey in general may be said to be half
beast and half man; it may be said to be the middle stage," he declared.[48]
Only the fact that the ape possessed toes shaped like fingers differentiated
the two, he insisted. Most striking of all, Hunter laid out his views on the
origins of different species, asking, "Does not the natural gradation of animals, from one to another, lead to the original species? And does not that
mode of investigation gradually lead to the knowledge of that species? Are
we not led on to the wolf by the gradual affinity of the different varieties in
the dog? Could we not trace out the gradation in the cat, horse, cow, sheep,
fowl, etc, in a like manner?"[49]

Plainly, Hunter had not taken that crucial extra step and proposed that
all life had developed from a single original ancestor, nor worked out the
method by which such change happened, but this was a startling insight
nevertheless. Writing almost seventy years before the publication of *On the
Origin of Species,* Hunter traversed much of the territory Darwin would
later explore to formulate his theory of evolution. Ahead of all his contemporaries, with the possible exception of Buffon, he had appreciated that di-

verse animals had developed over vast periods of time from common ancestors. And if Hunter's museum was not, as one devotee would later declare,
a "museum of evolution," certainly it displayed in the most graphic way
how different life-forms were perfectly fitted to their environments.[50]

John Hunter's most controversial conclusions would never see the light
of day in his own lifetime. Just as the Royal Society had successfully kept
his radical views on the age of the earth under wraps, so his prescient theories on original species would also be concealed from public view—and by
someone he regarded as his one of his closest allies. Because his farsighted
ideas languished unread for decades, Hunter's contribution to theories of
the origins of life would never be recognized. Ironically, when his ideas
were finally rediscovered and published in 1861, two years after Darwin's
dramatic revelations shocked society, they were compiled and edited by
Richard Owen, the man who had become Darwin's most vehement critic.
It was scarcely any wonder that in his preface Owen dolefully remarked,
"Some may wish that the world had never known that Hunter thought so
differently on some subjects from what they believed, and would have desired, him to think."[51]

CHAPTER 16

The Anatomist's Heart

⚬౼ᴍᴍᴑ

Castle Street, London
February 14, 1792

To a penniless, orphaned Cornish lad, arriving in London for the first time in the winter of 1792, the clatter of horse-drawn traffic, the splendor of shop windows, and the stench from uncollected refuse seemed overwhelming. Stepping over the threshold of 13 Castle Street, even more unlikely sights and smells awaited William Clift. Having just been apprenticed to John Hunter for a period of six years as anatomical assistant, artist, and amanuensis, the slight youth was shown to his quarters in the rear house, where he unpacked four shirts and four neckties—which constituted all his belongings—before being set to work in the museum.[1] It was his seventeenth birthday, the same day as Hunter's, but there would be no celebrations for Clift. He was too busy "making little paper boxes to hold little shells and such things" to party.

Born near Bodmin, the youngest in a large and desperately poor family, Clift had briefly attended the village school, by virtue of his mother almost starving herself, before being orphaned at the age of eight. Forced to fend for himself, he found work in a nursery, where he demonstrated his talents for nurturing the blooms and for drawing. His artistic skills were his undoing, however, for he was peremptorily sacked after sketching a mischievous caricature of the owner and was forced to subsist by running errands, writing signs, and finding odd jobs. Among his regular haunts was the local priory, where Clift became a favorite with Major Gilbert and his wife, Nancy. It was the sight of Clift's striking chalk drawings, scratched onto her kitchen floor while he waited for orders, that inspired Mrs. Gilbert to recommend the "very clever boy" to her childhood friend Anne Hunter. Knowing that John Hunter had been fervently seeking an artist to replace William Bell since he had left in 1789, Mrs. Gilbert proposed young Clift. Expecting the poorly educated, unsophisticated country youth to pass muster as an anatomical artist was a gamble, but no more so than the prospect of a Scottish farm boy rising to become London's leading surgeon. Hunter agreed to take on Clift, unseen, for a six-year apprenticeship without fee. The position would bring no wages, but at least Clift could rely on board, lodging, and, crucially, training.

It was a life of drudgery from the first—long hours, cramped living quarters, no money, and little thanks. But after taking in his new surroundings—the peculiar two-fronted house, the curious museum, the effervescent household of family, pupils, servants, and even animals—young Clift was exhilarated. Sharp and perceptive despite his foreshortened education, he relayed his first impressions in excited letters to his siblings in Cornwall. "We have a great family indeed," he wrote, listing the huge retinue of kitchen staff, maids, footmen, and coach drivers squeezed into the two houses, along with four house pupils and members of the family. Although John Banks Hunter, nearly twenty now, was away studying medicine at Cambridge, planning to follow in his father's footsteps, sixteen-year-old Agnes still lived at home, attracting suitors at her mother's musical evenings. In all, Clift reckoned there were never fewer than fifty people, including the staff running the house and farm at Earls Court and the regular workmen in both establishments, all dependent on Hunter's employment.[2] With the Leicester Square basement occupied by two coaches, one each for Mr. and Mrs. Hunter, and stabling for six horses, it made for a cramped existence. But after being shown the city sights by Robert Haynes, the dissecting-room attendant—even spotting the royal family promenading in St. James's Park—Clift enthused, "I like London very well."

The chaotic city streets were enthralling; the shops selling "ready-made" clothes amazing; the glimpse of George III and the royal princesses astonishing. But most of all, Clift was captivated by his new boss, who at sixty-four was a stout figure with curling white hair springing from either side of his face. It would be several days before Hunter would lay eyes on Clift, because Hunter was laid low with another bout of angina—the spasms were increasingly common now. When the two did finally meet, it was a moment Clift would remember all his life. "He is a verry curious man," he wrote to his sister, "and plain as well for he has hair as white as snow and has never got it drest, I think there is not a bit of Pride in him and all his clothes so plain (But very rich) and I am sure you would not think he was such a Grand Gentleman."[3] Grand gentleman or not, Hunter treated his youngest, lowliest assistant with the same civility as his wealthiest patients. His master was "mild and kind in his manner," Clift told his sister, and "spoke as kindly and familiarly to his gardener or myself as to his equals or superiors."[4] Like the assistants who had gone before and the pupils who still flocked around, Clift was smitten. The honest but steely Cornish lad would become Hunter's most devoted disciple.

If Hunter's health was failing, his energy was not. Every morning at six o'clock, Clift met his master in the Castle Street dissecting room. Hunter would already be stooped over his bench, examining minute creatures through a pair of spectacles modified to provide lenses of different magnifications. While Clift painstakingly learned the skills of dissecting, making preparations and illustrating his efforts, his boss would "stand for hours motionless as a statue, except that with a pair of forceps in either hand he was picking asunder the connecting fibres of the vessels or parts, till he had unravelled the whole structure."[5] Finding his teacher "generally, though cheerfully, taciturn," several hours might pass without the two exchanging a word. At other times, slowly stretching and standing upright, Hunter would relax and entertain Clift with a "shrewd and witty" remark. Among Clift's first tasks was to help unpack and catalog a treasure trove of antipodean species donated by Banks; his own house overflowing, the Royal Society president had split his zoological collection between Hunter and the British Museum.[6] Not having had the benefit of Hunter's expert preservation techniques, however, many of the specimens were in poor condition. A black swan had rotted so much that Hunter decided it not worth stuffing and gave the bird to Clift as a first exercise in preparation making.

When Hunter reluctantly abandoned his dissecting bench to attend to his patients, Clift was kept busy with errands and jobs in the museum. There were drawing classes, too. Hunter initially enrolled Clift with a tutor at one of the local art schools, but he soon afterward took in a French refugee, Monsieur St. Aubin, who taught Clift at the house. The youth's artistic talents were quickly recognized. After Clift sketched an exotic bird from the collection, Hunter said "he was in hopes I should do very well in a little time."[7] After dinner at 4:00 P.M., when the entire household ate together, there was scant time for leisure. Each evening, except for Sunday, Clift sat with Hunter in his study, copying out notes and taking dictation from eight o'clock until midnight.

As well as making neat copies from Hunter's scrawled notes of experiments, dissections, cases, and autopsies, Clift took dictation, replying to Hunter's friends, ex-pupils, and acquaintances across the world, offering medical advice and begging specimens in return. By Clift's estimate, Hunter received no fewer than three or four thousand letters in a year. Hunter tore strips from their edges as spare paper on which to scribble memos, and Clift would copy these for inclusion in the correct places in the catalogs and other volumes. The discarded notes were used as spills to light candles or sent to the dissecting room as wastepaper to mop up blood.[8] Clift was quick to observe how scrupulously Hunter looked after his manuscripts, updating the different volumes meticulously as new information came to light, whether on a particular disease or an interesting insect. Letters from favorite pupils such as Jenner and Physick were carefully preserved, too, and sometimes brought out so that Hunter could check over observations they had sent.

On Wednesday evenings, when the upper floors vibrated to the rhythms of Mrs. Hunter's parties, Hunter would leave his studies briefly to put in an appearance and "shake their mutual friends by the hand" before resuming his relentless work.[9] When Clift was finally released for the night, he left Hunter still engrossed in his papers. "I never could understand how Mr Hunter obtained rest," Clift mused; "when I left him at midnight, it was with a lamp fresh trimmed for further study, and with the usual appointment to meet him again at six in the morning."[10] As Hunter burned the midnight oil, Clift fumbled back to the Castle Street house in darkness— candles were too expensive and too risky to trust to a mere apprentice— feeling his way "through the dead bones."[11]

If Clift's day was closely structured, life in Hunter's household was anything but routine. On Sundays, which Hunter usually spent at Earls Court, Clift often went along; sometimes he was corralled into helping the farm staff catch insects. Hunter was spending more and more time at his country retreat, almost returning to his rural roots. He liked to sleep at Earls Court whenever he could, traveling to town for patients and pupils, then returning every evening to potter in the orchards, hothouses, and conservatory. In addition to his other botanical experiments, he was investigating the movement of plants, measuring how far the leaves of sensitive plants could unbend and planting beans in rotating baskets to check whether they would still grow upward.[12] But his chief delight was still to watch his beloved bees humming in and out of their observation hives in the conservatory. He delivered the fruits of twenty years' study of bees in a paper to the Royal Society a week after Clift's arrival. Minutely describing their anatomy, even to the length of their tongues, Hunter outlined their intricately organized society, described their peculiar dance, explained their methods of generation, and even concluded, much to the incredulity of his contemporaries, that bees made the wax found in their hives. "The wax is formed by the bees themselves," he boldly asserted; "it may be called an external secretion of oil, and I have found that it is formed between each scale of the under side of the belly."[13]

New additions to Hunter's animal collection continued to arrive, both at Leicester Square and Earls Court. With each arrival, the debts accumulated. A favorite animal was Hunter's dog, a Great Dane named Lion, who was obedient to nobody but his master. "He was gentle as a Lamb," remarked Clift, "but alarmed every Stranger as he passed, by his great height and size, and romping disposition, when let loose." If any of the students let Lion off his chain, Hunter made them pay for the ensuing glazier's bill, and if they tried to take him for a walk, Lion inevitably escaped and returned "always without his Collar."[14]

Almost as alarming for the villagers as seeing the escaped Lion loping toward them was the sight of Hunter himself driving off to town in a cart pulled by three Asian buffalos, or zebus.[15] For Londoners, the sight of their most senior surgeon clenching the reins of the three beasts with huge curling horns as they ambled through the hectic streets was bizarre even by the city's eccentric standards.

The zebus were a regular sight. According to Clift, they left Earls Court every Wednesday, pulling a cart loaded with fruit and vegetables for the Leicester Square table, and were usually led by one of the farm laborers. After the produce was unloaded in Castle Street, the cart was filled with manure from the underground stables and waste material from the dissecting room. Both were used to fertilize the farm. On one Wednesday, however, the cart was driven to town by a simpleminded farmhand known as

"Scotch Willie."[16] After lugging the usual hamper full of rotting flesh and bones down from the attic dissecting room, Willie led the buffalos into the stables and repaired to the kitchen for "beef and beer," leaving the foul-smelling basket uncovered in the cart outside. Before long, some school-boys spotted the load and, knowing it usually contained apples from the farm, jumped into the cart to investigate. Their disappointment was only matched by their horror. Clift recalled, "The first object that struck their attention, instead of the Apples they expected, was the Putrid half dissected arms of a Man, green blue and yellow;—Livers, intestines and other parts."

The boys' exclamations quickly drew a crowd, which grew bigger and more unruly as aghast passersby took turns to view the contents of the cart. The all-too-plain evidence of the notorious anatomist's nocturnal raids soon threatened to turn the curious crowd into a violent mob. At last, the commotion reached Clift's ears in the house and he hurriedly summoned the house pupils, who were obliviously cutting up bodies in their blood-daubed aprons, to come down from the attic and help. Scotch Willie was likewise dragged from his lunch to fetch the buffalos and attempt to harness them in order to steer away the cart. But by now the "haloo-balloo" had grown so intense, and the beasts were becoming so agitated, that all that could be done was to shove the cart with its incriminating load down the ramp to the basement and quickly batten the doors.

With the buffalos trapped outside and the pupils barricaded behind the doors, thus protected from the baying mob, a hairdresser who had been lured out of his salon by the noise decided to make himself the hero of the hour. Egged on by the crowd, he stepped forward with his shears and cut one of the zebus free, only to find that it immediately turned on him and pinned him between its horns to the coach-house doors. As mob fury now turned to playhouse farce, the spectators whooped with delight when the buffalo charged away, seeking its freedom by running down neighboring Green Street and into Leicester Square, where it cantered around the genteel quadrangle with the luckless hairdresser "holding on at the end of the long halter." It was evening before the crowds had dispersed and the cart could be led discreetly from its subterranean hideaway in the dark. It was sent back to Earls Court, for once minus its haul of flesh.

It was soon after moving into Leicester Square that the young Clift first met Everard Home. Strutting about the house as if he owned the place—

and by now he almost felt as if he did—Home bumped into the new apprentice. Taking him for a naïve country boy with a head full of superstitions, Home leered at him and asked if he was afraid of ghosts. "I don't know, sir, I never saw one" was Clift's logical reply. "But I was never afraid of my school-fellows," he added, "though bigger boys than me."[17] It was a warning Home would fail to heed.

At the age of thirty-six, still living in his sister's house and still dependent on his brother-in-law's goodwill, Home was desperate to make a mark for himself. With his luxuriant curly blond hair, sensual thick lips and large doleful eyes, he cut a handsome figure striding in and out of Leicester Square, and he knew it: He reveled in fine clothes and preened himself in front of the students.[18] He was planning to marry a widow, the daughter of a vicar, later in the year, but he was as yet unable to ensure a secure financial future. Ever since moving back into the Hunter household in 1784, Home had worked hard to render himself indispensable to his former teacher. Talented and respected pupils had come and gone, some of them having been offered partnerships, which they declined, but Home remained, clinging firmly to his position at the center of Hunter's livelihood, biding his time. Having helped arrange Hunter's celebrated collection, he had taken over much of the surgeon's routine work at St. George's, inherited several of his wealthy clients, and, since 1790, delivered most of the lectures each winter. When younger and cleverer surgeons, such as the American Philip Syng Physick, won Hunter's warmest praises, Home had to bite his tongue. When ex-students such as Astley Cooper, John Abernethy, and Henry Cline—all now secure in their own hospital positions—clustered around Hunter's dinner table, idolizing their former tutor, Home kept his own counsel. Whether it was simply his constant proximity to Hunter or the growing strain of Hunter's angina, which dealt him painful spasms at the slightest exertion or annoyance, it seemed to Home at least that he bore the brunt of the aging surgeon's bad temper. His brother-in-law "spoke too freely" and "sometimes too harshly to his contemporaries," Home would later complain, while his temper was "readily provoked."[19] Oddly, it appeared that Hunter's temper was rarely provoked by his other ex-pupils, or by his assistants, such as Clift, who appreciated his boss's evenhanded, generous nature. All the same, Home stayed put, ensuring that Hunter depended on him more and more for day-to-day support. Having delivered Hunter's lectures for two seasons, he had grown certain that he could match the acclaimed teacher and surgeon. Ambitious and vain, he was even beginning to believe that he could outshine his former master.

In the spring of 1792, Home's chance arrived when the post of senior surgeon became vacant at St. George's. Having worked at the hospital as Hunter's assistant for five years, Home firmly believed the job should be his, assuring him a regular private clientele and a secure future. But he had not reckoned with St. George's medical establishment. More than twenty years of skirmishes and snubs had fermented into a seething, poisonous resentment directed against Hunter. All Hunter's acclaim within London society, his esteem in the scientific community, and his popularity among the city's medical students counted for nothing with his fellow surgeons at St. George's, who still clung to their antiquated practices and hierarchical education system. So when Hunter put forward his brother-in-law for the vacant position, John Gunning, his lifelong rival, immediately weighed in with his own candidate—his assistant, Thomas Keate.

With the date for the election by the board's governors fixed for May 11, the scene for a showdown was set. No doubt Home's obsequious manner, and his plodding technique, did little to bolster his case with the governors or fellow surgeons. Even so, Hunter did not intend to give up without a fight, and he made sure that his tame governors, Banks among them, turned up to cast their votes. In the end, the vote was close—102 votes for Home and 134 for Keate. Bitter from this humiliation, Home had no choice but to labor on in his junior position. Livid at his defeat, Hunter was determined to make his colleagues pay.

Hunter took the ballot result as a personal injustice. Decades of trying to introduce reforms in medical education, attempting to persuade his colleagues to present lectures for the students, and always shouldering the greatest responsibility for the hospital's pupils—not to mention the conflicts over Hunter's surgical techniques—finally boiled over. He had the Royal Society at his feet, nobility knocking at his door, physicians bowing to his advice, but he could not win over his peers at St. George's. Hunter resolved to teach his recalcitrant colleagues a lesson, and he fired off a furious missive on July 9.[20] Addressing all three senior surgeons, Gunning, Keate, and William Walker, Hunter announced he was no longer prepared to share the pupils' fees as custom decreed. Instead, he intended to pocket all the payments from the students enrolled with him—by far the majority.

It was not that he needed the money—less than one hundred pounds a year divided among four surgeons would make little impact on Hunter's debts; rather, he hoped the move would act as an "incitement" to force his fellow surgeons to take their teaching duties seriously. But Hunter left no room for compromise as he fumed, "I will not say it is a disgrace to be a surgeon to St George's Hospital; but I will say, that the surgeons have disgraced the Hospital." It was an uncompromising assault on the honor of his three colleagues, an audacious challenge to their pocketbooks and their pride, and they were suitably apoplectic. Writing back on October 4, the three opponents jointly declared their "entire disapprobation" of Hunter's proposal and referred his conduct to a general board of governors, a meeting of which was planned for February 1793.[21] They knew they had him cornered. What Foot described as the "continual war," which had rumbled on behind the scenes at St. George's for decades, had now burst savagely out into the open.

As the board meeting loomed, Hunter spent an anxious and increasingly lonely winter. Philip Syng Physick, who had been helping in the museum since completing his degree in Edinburgh, had returned to Philadelphia in September 1792. Home persuaded his fiancée to go ahead with their wedding in November, despite his failure to secure a senior hospital post, and the couple moved into a new house nearby. As Hunter had now handed him complete charge of the lecture course, he remained a regular visitor, but only Clift was left in permanent attendance, remaining at Hunter's side from dawn until midnight, diligently improving his dissection, preparation, and sketching skills while imbibing his master's teachings with a growing devotion.

Although Hunter had relinquished the lectures, his workload had scarcely diminished. The death of Percivall Pott at the end of 1788 had left Hunter unassailably the premier surgeon in London, with a commensurate increase in clientele. Patients still thronged for treatment to his Leicester Square surgery and Hunter set off in his coach daily to call at the London villas of his moneyed clients. The relentless round of animal experiments continued, too. Still intent on tracing the very first indications of life, Hunter had returned to the quest he had begun as a novice anatomist dissecting chicken embryos—only now he was slaughtering sheep. Adopting the same method, he killed eight ewes at different stages, from a few hours to eight days after they had mated with the ram; there was mutton for the dinner table at Leicester Square every day for over a week.[22] And as if Hunter did not have enough to occupy his time, he was now in supreme command of all of the British army's surgical services at home and overseas.[23]

His appointment as surgeon general of the army, awarded him in March 1790 by William Pitt the Younger, was recognition both of his un-paralleled prowess as a surgeon—not least in operating on Pitt—and a vin-dication of his controversial approach to treating gunshot wounds. But with hostilities among various European powers rumbling on and fears growing as a result of the aggression of the revolutionary regime across the Channel, the job was no comfortable sinecure. Although Hunter would never leave his London home to inspect the military hospitals in the West Indies nor supervise his charges in their regiments, his army duties entailed a gruel-ing daily grind of administration on pay, appointments, and supplies. He used the opportunity to set right the injustices that still rankled from his own military experience, jettisoning the system of patronage that perme-ated not only the services but all Georgian society in favor of a fair, consis-tent, and progressive career ladder for all surgical appointees. Hunter's new system ensured that a lowly surgeon's mate within a regiment could progress by stages to become a hospital mate, a regimental surgeon, a staff surgeon or apothecary, and ultimately even to the grand position of physi-cian or hospital purveyor. Rewarding merit, rather than bowing to privilege, was the basic tenet of his army scheme.

As winter advanced, Hunter's army duties grew daily more onerous due to revolutionary France harassing its European neighbors. When his elderly colleague Sir Clifton Wintringham could no longer cope with the pressures as physician general to the army, Hunter took over his responsibilities, too. Despite his own dubious health, he was now in command not only of the army's surgeons but all its physicians, as well. And when France executed Louis XVI on January 21, 1793, and just over a week later declared war on Britain, it was Hunter who had to supervise both surgical and medical sup-port for the armed forces heading for Holland, France, and the West Indies.

Britain's naval supremacy would ultimately ensure victory against the French revolutionaries, but the revolutionary surgeon lacked such steadfast support when finally he faced his own adversaries at St. George's on February 28. It was a one-sided engagement from the start. Hunter set out his case in a long, impassioned, and impeccably argued letter to the board, detailing the grudges and grievances that had plagued his entire hospital career.[24] His aim from the first had been to treat the sick and to educate the nation's future surgeons, he argued, yet he had met with opposition at every step, his early colleagues rejecting his moves to set up lectures and his later peers similarly blocking his efforts to improve the pupils' instruction. Reminding the board that he had taught more pupils than all three of his colleagues put together—a total of 449 since 1770 to their 284—he had concluded it was no longer just to share the pupils' proceeds equally. He implored the governors not to regard the wrangle as "a contested election between two men," even though "I may appear to stand single." Yet it was, and he did.

As Gunning, Keate, and Walker delivered their response, Hunter knew he was outflanked and outmaneuvered. In a counterblast to the board, the three surgeons launched a vicious assault on Hunter's character and reputation, not only accusing him of neglecting his responsibilities to St. George's but ridiculing his lecture course. Insisting that "the old practice of instruction"—the copycat method patronized since medieval times—was still the best form of educating young surgeons, they fired a snide shot at Hunter's proposed lectures: "If they had been practical and contained principles and rules founded upon judgement and experience, with regard to the authority of others as well as our own, they would have been highly useful: if on the contrary they had leaned to Physiology and experiment with a contempt for all other opinion but their own they would have been pernicious."[25] Even now, as Hunter's doctrines were being taught by his ex-pupils in hospitals throughout Britain and the United States, the term *physiology* was a dirty word. Knowing he could muster the board's support, Gunning persuaded the governors to set up a special committee to consider the future of medical education. John Hunter, the man who had done most to promote the cause of teaching at St. George's, who had taught more students than any other surgeon in his day, was excluded.

The committee, which met in May, agreed in full to a list of exacting rules drafted by Gunning and his allies, who had outlined the expected conduct of surgeons and pupils.[26] Hunter was not even consulted. At first sight, the regulations appeared to represent victory for Hunter: At long last, he had won his campaign for the surgeons to give lectures on surgery free of charge and in strict rotation. In reality, however, it was a calculated ruse to outwit him, for there was a string of onerous demands all the surgeons were required to fulfill. These included visiting their patients at least twice a week, supervising dressings once a week, and meeting in joint consultation every Friday. Any defaulters, the rules stipulated, would be reported to the weekly board for breach of duty. As Hunter valiantly supervised his army surgeons and his angina frequently brought him to his knees, these additional duties were plainly ruinous. There seemed little doubt that his jealous rivals were attempting to force his resignation—or worse. And there was one further demand that would have profound reverberations for Hunter. The new rules directed that all future pupils must have certificates proving "their having been bred up to the profession." In other words, only those who had completed a full apprenticeship would now be admitted. No longer would clever, talented young men from modest backgrounds, without the connections or the money to arrange a lengthy apprenticeship, such as Hunter had been, be allowed to cross the threshold of St. George's.

Hunter knew that his scheming opponents had won the battle, even if they were the last of a dying breed. He had little choice but to comply with the punishing new regime. Throughout the summer months, he dutifully attended the required meetings, delivered the requisite lectures, and tended to his needy patients with good grace. Yet all the time that his failing energies were focused on his declared adversaries plotting at St. George's and on the French foe overseas, the real enemy was always closer to home.

Despite all the conflict, in London and abroad, life at 28 Leicester Square seemed to carry on much as normal. Mid-October found Hunter making the final revisions on his major new volume, the *Treatise on the Blood, Inflammation and Gun-shot Wounds,* which was ready to go to press. It summarized the fruits of his experiences as a young army surgeon during the Seven Years War. The time-worn sixty-five-year-old surgeon general dedicated the work to George III in a fit of topical patriotic fervor. The introduction to his fossil collection, with its daring estimates of the age of the earth, was waiting for the press, too, while his considered views on the German fossil bones he had recently examined were ready for delivery to the Royal Society.

In the meantime, three new house pupils had moved their belongings into the cramped bedrooms of Castle Street. Among them was an earnest young lad, James Williams, who was finding his new quarters rather comfortless after his home in rural Worcester. "My room has two beds in it and in point of situation is not the most pleasant in the world," he wrote to his sister Mary. "The Dissecting Room with half a dozen dead bodies in it is immediately above and that in which Mr Hunter makes preparations is the next adjoining to it, so that you may conceive it to be a little perfumed."[27]

Every morning at seven o'clock, Williams and his two fellow students met their white-haired teacher in his Castle Street workroom to assist in making preparations, "and to see with what patience he does everything is astonishing," enthused Williams. Gradually, he became accustomed to his eccentric mentor, telling Mary, "Mr Hunter is a very good kind of man when you have been used to him a little tho' he has some oddities," and adding, "but thank God he has none of that pompous haughtiness by which great men make themselves disgusting."

On Wednesday, October 16, with the dissecting season poised to begin, John Hunter met his pupils in the preparations room at 7:00 A.M. as usual.[28] He appeared to be in a buoyant mood, regaling the pupils with whimsical anecdotes about children who feigned sickness to avoid school—rarely did they fool the wily old surgeon.[29] The start of the autumn term, with the prospect of an entire winter's dissecting ahead, had perhaps contributed to his jaunty manner. He was no doubt also looking forward to the return at the end of the week of Anne and his children, twenty-one-year-old John and seventeen-year-old Agnes, who had spent the past six weeks in Brighton.[30] Young John would be returning to his medical studies at Cambridge, while Agnes would soon be entertaining guests again at her mother's soirées. But their father's anticipation at their impending return may well have been equaled by the thought of the pair of golden eagles that had been delivered to Earls Court that same morning by Gough, his reliable supplier.[31] Hunter would have been looking forward to observing their antics and devising some suitable experiments.

After helping the pupils make their preparations with his usual patience, Hunter joined them for breakfast. He sauntered into the room "humming a Scotch air," remembered William Clift. "He was in very good spirits," noted James Williams, "and eat hearty as usual."[32] At noon, Hunter left the house with a full caseload of patients to visit and a meeting at the hospital to attend. Almost immediately, the butler realized he had forgotten his itinerary—the list of appointments the surgeon always carried with him—and Clift had to run after the carriage, which he overtook at Hunter's first appointment near St. James's Square, five minutes away.[33] After handing him the list, Clift stood for a moment and watched his master climb into his carriage "as well as ever I saw him in my life." His next stop was St. George's Hospital.

In reality, Hunter's bluff humor, amusing anecdotes, and Scottish melody were something of a front, for he knew that another conflict awaited him at the hospital. Two Scottish youths had recently written to ask if they could enroll as pupils with him at St. George's; like Hunter in his youth, they had not served the requisite apprenticeships. Hunter had accordingly informed them of the new requirement but had agreed, in his usual generous spirit, to press their case all the same. Accordingly, he had written to the board, supporting their enrollment as pupils, and his letter had been tabled for the forthcoming meeting. Hunter had hoped that Home would accompany him to the debate, or even go in his stead, but Home had been called away urgently to treat wounded troops arriving at Deal from the battles in France. Preparing to face his combatants alone, Hunter knew the effect such an ordeal might have on his weakened heart, for only a few days earlier he had sunk wearily into a chair and told a colleague, "One of these days, and not long first, you will hear that I have dropped down dead."[34]

The hospital staff was gathered in the ground-floor boardroom, directly beneath the operating theater where Hunter had saved so many lives. Gunning and Keate were absent, but William Walker, their fellow conspirator, was at his place, along with a junior surgeon, William Mathew, who had almost certainly been recruited to the opposing side.[35] Among the physicians, Hunter's only sure ally was his nephew, Matthew Baillie. Before long, the issue of the two Scottish youngsters came up and Hunter cleared his throat to speak in their favor.[36] But even as he began to argue their case, he was immediately contradicted by one of his colleagues. Unable to contain his temper, Hunter lashed back in fury. As the meeting threatened to erupt into an ugly debacle, Hunter suddenly stopped speaking, rose from his chair, staggered into an adjoining room, and with a loud groan sank into the arms of one of the physicians. There was nothing the combined talents of the St. George's staff could do to revive the hospital's renegade surgeon. Hunter had died, as he had lived, in rebellion, speaking his mind.

Ĵ*ohn Hunter's lifeless* body was transported back from the hospital in a sedan chair, while his empty carriage trailed behind. The surgeon's corpse was delivered, like so many corpses before, to the door of his Leicester Square home. Yet the demise of the man who had dealt daily with death seemed inconceivable, and the effect on the household was devastating. Servants and house pupils were stunned. Having known his master barely twenty months, Clift found himself "at a loss for words." Having met his teacher just weeks earlier, James Williams struggled to write a legible letter home. Naïvely, he told his sister, "He will be universally regretted even by his most inveterate Enemies."[37] With Home still patching up war wounds in Kent, Baillie rode through the night to fetch Anne and the children, who returned the following day. Like the rest of the household, they were speechless with grief. "Mrs Hunter and Miss and Mr John are almost breaking their hearts," wrote Clift. "I did not see Mr John till today morning when he burst into tears and could hardly speak when I spoke to him."[38]

Friends and acquaintances were equally shocked. "I have just heard that Dr [sic] Hunter is dead suddenly at St George's Hospital in a fit," Horace Walpole wrote to a friend. "It is such a blow to his family, as he was in such repute." His thoughts immediately turning to Anne, who was a dear friend, he told another acquaintance, "I am exceedingly grieved for the great misfortune that has happened to Mrs Hunter, and I heartily regret the very amiable Doctor."[39] Joseph Farington, who had been one of Hunter's last patients, was moved to record in his diary, "much concerned at an acct. in the newspaper of the death of John Hunter, the excellent surgeon, to whom I was greatly obliged in the course of last summer."[40] Glowing epitaphs filled the press. Applauding Hunter's unparalleled skills, *European Magazine* pronounced, "He rose to a rank in his profession scarce ever remembered, that of an acknowledged superiority over the most eminent of his rivals." *Gentleman's Magazine* declared Hunter "an honour to his profession and to his country" and appended a poem by a physician friend eulogizing his achievements. And the *Sun* charted Hunter's rise from a humble carpenter to quite simply "the first Surgeon in the world."[41] At St. George's, however, the triumphant surgeons voted not to send their condolences to Hunter's widow.[42] Before long, they would agree to pay Foot four hundred pounds to write his damning biography of their despised colleague.

On the day after Hunter's death, as his widow and children wept, the house pupils gathered in the attic dissecting room as if for their routine anatomy lesson—only this time the corpse laid out on the table in front of them belonged to the great anatomist himself. Unlike many of his fellow surgeons, who had decreed that their bodies be incarcerated in locked family vaults to escape the anatomist's knife, Hunter had expressly requested that an autopsy be performed after his death. With Baillie observing, Clift assisting, and the pupils craning to see, Everard Home, fetched back from Kent, picked up his knife and began to carve up his brother-in-law's body.

When he opened Hunter's chest, the telltale signs years of angina had wrought were plain to see in his heart and arteries.[43] If ghosts had indeed haunted the dissecting room, as Home had intimated to Clift, John Hunter would have enjoyed nothing better than to peer over the shoulders of his engrossed students and confirm his own prognosis. After recording his findings, Home sewed up the body for burial, flagrantly disregarding Hunter's request that his heart and Achilles tendon, which he had torn back in 1766, should be preserved.[44] To Hunter, it had seemed perfectly natural that his damaged heart and his repaired tendon should be pickled in spirits and then take their place on his museum shelves alongside the body parts of his numerous patients in order to help teach generations of future surgeons fundamental facts about injury and disease. In life, Hunter's heart had always been in his museum; in death, the museum would have been its most fitting resting place. Yet this neglect would seem trifling compared to the outrage Home would go on to perpetrate on his former tutor.

The funeral, on October 22, was a small, quiet ceremony at St. Martins-in-the-Fields, with only the family in attendance. Compared to Sir Joshua Reynolds's extravagant send-off a year previously, it was "a very private burying," Clift recorded. Not possessing a black suit, the apprentice was unable to join the official mourners, but, desperate to say a last farewell, he persuaded the undertaker to let him inside the vault. There, he took a final look at Hunter's black coffin. "None of our people saw me there I believe," he wrote to Elizabeth, "and I did not want them to."[45] For Clift, as for the family, life was now in turmoil. "I am afraid our house will be turned quite upside down now the wall and support is gone," he wrote.[46]

He was right: Both of the houses and all their occupants had been shaken to the foundations.

To his family, Hunter had left principally debts. They ranged from the two pounds owed to George Bailey, the "birdman" in Piccadilly, to the outstanding huge mortgage on the Earls Court property.[47] There was no money for mortgage payments, so Earl's Court House was sold, its fine furnishings auctioned, and its animals sent to Gough's menagerie. With only six years left on the Leicester Square lease, the house was rented out and the coaches, horses, paintings, library, and furniture sold. Only Long Calderwood remained in the family, but the tiny farmhouse was far too humble to provide either home or income. The entire household was therefore homeless and penniless. All the servants and helpers were dismissed, with the exception of Clift; the housekeeper, Mrs. Adams; and the dissecting-room attendant Robert Haynes, who remained in Castle Street. After the debts were settled, there was nothing left. Hunter's entire lifetime's earnings—a veritable fortune—had been plowed into his collection, worth an estimated seventy thousand pounds. He had intended the museum to be his family's inheritance, stipulating in his will that it be sold—to the nation in its entirety if possible—to provide his descendants with a livelihood. Unfortunately, the government was in no mood to invest in scientific heritage. When petitioned by Hunter's friends and pupils to drum up even twenty thousand pounds, Pitt, in the midst of war with France, exclaimed, "What! Buy preparations! Why, I have not got money enough to purchase gunpowder."[48]

The prospects for the family were bleak. With no means of continuing his education at Cambridge, John Banks Hunter had to abandon all hope of a medical career. He joined the army, where he would lead an ignominious career. With no form of support at home, Agnes had little option but to accept a marriage proposal, this only seven months after her father's death. It would be a miserable union, ending in an acrimonious separation, although she eventually found happiness in a second marriage. Neither she nor John would have children. The future was even less promising for Anne. Leaving her elegant home and servants, and abandoning her circle of literary and musical friends, she was forced at fifty-one to take a job as a ladies' chaperone, catering to the whims of the daughters of an army surgeon.[49] Although she was allowed a temporary government pension, and published her collected poems in 1802, she would never again entertain.

While the government invested in gunpowder, Hunter's extraordinary museum remained in place, lovingly tended by eighteen-year-old Clift, who spent what was left of his meager allowance from Home—coexecutor with Baillie of Hunter's will—on replenishing the alcohol in the jars. Too poor to go out, he devoted his evenings to studying Hunter's catalogs and manuscripts, feverishly working to understand his hero's radical theories and, because it was what he had been trained to do, copying as much as he could in his neat script. His art classes and anatomy training having abruptly ended, Clift hoped that Home might find him dissection work, although he shrewdly acknowledged, "I should never learn anatomy under him for he is quite a different man from Mr Hunter."[50]

Certainly, Home appeared to be the only member of Hunter's household to escape the devastation left by his death; in fact, he even seemed to derive some gain from his brother-in-law's passing. Leaving his sister, nephew, and niece impoverished, he stepped briskly into his brother-in-law's shoes, filling his vacant job at St. George's, presenting his unread papers to the Royal Society, and cementing a friendship with Banks, a relationship that helped him forge valuable connections among the aristocracy. Ultimately, he would become a close intimate and drinking companion of the future Prince Regent, later George IV. The long-suffering Clift, painstakingly preserving Hunter's priceless legacy year after year, was all but forgotten until the government finally agreed to buy the museum in 1799 for the knockdown sum of fifteen thousand pounds.[51] The entire collection, some 13,687 preparations, was placed in the custody of the Company of Surgeons, renamed the following year the Royal College of Surgeons; in death, Hunter would bring the organization he had despised the status it had lacked during his lifetime. Clift was appointed its first curator.

Now that the fate of the museum was settled, Home suddenly demanded all Hunter's papers. Clift stalled as long as possible, "having a kind of presentiment that if they were removed to his house some accident might befall them," while feverishly copying as much as he could.[52] When Home finally issued an ultimatum, Clift could resist no longer, and all Hunter's writings—manuscripts, casebooks, lecture notes, catalogs, and letters—were delivered to Home's house in a cart. With a proper salary at last, Clift married in 1801, and five years later he moved into the college's new headquarters in Lincoln's Inn Fields, Hunter's unique collection safe in his capable hands.

For the next twenty years, while Home carved out a glittering reputation within the Royal Society, contributing a colossal ninety-two papers on anatomy and natural history, Clift and the trustees of the museum attempted in vain to recover Hunter's papers. Without the original catalogs, it was impossible to produce a new guide to the museum. And because the trustees were denied his unpublished manuscripts, Hunter's most controversial ideas were kept in the dark. Requests for Home to return them were constantly rebuffed, however. For his voluminous contributions, Home was awarded the Royal Society's Copley Medal, and he became vice president in 1814. Rising fast in his profession, he was appointed sergeant surgeon to George III in 1808, first master of the Royal College of Surgeons in 1813—the same year he was knighted by the Prince Regent—and first president of the college in 1822. But no appeal from the college nor from Clift, now an eminent naturalist in his own right, could persuade Sir Everard to hand over the missing manuscripts.

Finally, on July 26, 1823, when Home was riding in a chaise with Clift to a meeting at Kew, he mentioned there had been a fire at his house.[53] So fierce was the blaze that the flames had leapt out of the chimney and the fire brigade had had to be called. When Clift inquired how the fire had started, Home casually revealed that he had been burning Hunter's manuscripts. The loss was incalculable; Clift broke down in tears. There was, then, only one more thing to do, he lashed back at Home, "and that is to burn the collection itself."[54] Although Clift had presciently copied many of Hunter's original documents, and a few more were later surrendered by Home and his family, these having somehow escaped the blaze, the majority had gone up in smoke. Even the letters from Hunter's favorite pupils, Jenner and Physick, had perished—burned in a fit of jealous rage.

Seeking to justify his actions over the next decade, both to the college and to a parliamentary committee, which in 1834 was inquiring into the destruction, an arrogant Sir Everard insisted he had simply carried out Hunter's dying wishes. It was a lame excuse. Even if Hunter had requested that Home destroy his manuscripts—and Home had not even been present at his death—that did not explain why the act had taken him thirty years. The fact that he later surrendered several unharmed manuscripts, including some of Hunter's casebooks, further belied his plea. The truth was that Home had roundly plagiarized his master's unpublished works, then brutally erased all evidence of his crime. The numerous papers he had read to the Royal Society and the collected lectures he had just finished proof-reading would reveal passages identical to many in Hunter's surviving writings. Home would never atone for his wanton act of destruction. He died, "a regular sot," according to *The Lancet,* in 1832.[55] Three years later, Hunter's surviving writings and lectures derived from pupils' notes were published in four volumes; after the rediscovery of his works on geology and the origins of life, these were published in 1859 and 1861.

Despite its lack of an explanatory catalog, Hunter's museum had become a magnet worldwide. By 1833, Clift had conducted more than 32,000 visitors around the collection, including Archduke Maximilian of Austria, Prince Christian of Denmark, and Prince Louis Napoléon, as well as numerous foreign scientists.[56] Georges Cuvier, the French expert in comparative anatomy, was astounded when he visited, confiding that he had had no idea "that there was such a collection as the Hunterian Museum."[57] And in 1837, a year after returning from his five-year voyage in the *Beagle,* Charles Darwin donated his fossil bones to the museum, asking Richard Owen, the talented assistant curator, who had married Clift's daughter Elizabeth, to study them.[58] Over the next few years, Darwin would become a regular visitor.

More than twenty years later, as Darwin nervously put the finishing touches to *On the Origin of Species,* John Hunter was about to have a dramatic resurrection. The naturalist Francis Buckland, son of William Buckland, who had become one of Darwin's fiercest opponents, was reading his *Times* over breakfast one morning in January 1859, when he spotted a small notice announcing that the vaults of St. Martin's church were about to be cleared.[59] Vaguely recalling that John Hunter had been buried in the church, Buckland launched a frantic search to salvage his remains. There was no time to lose: Any coffins unclaimed by the end of February were to be destroyed. But with no record of where Hunter had been buried, and more than three thousand coffins jumbled in the vaults, it was a daunting challenge. Equipped like a latter-day body snatcher, Buckland examined each of the rotting caskets by lamplight in the foul-smelling vault. After a full seven days of searching, with only two coffins remaining, he had all but given up hope. Approaching the second-to-last coffin, a black lead casket, he faintly discerned the initial "J." Wiping the brass plate clean, at last he read the name: John Hunter. Honored with a second funeral, Hunter was reinterred on March 28 in Westminster Abbey with full ceremony. A plaque placed over his new grave by the Royal College of Surgeons commemorated Hunter as the "Founder of Scientific Surgery."

More than his surviving writings, more even than his unique museum with its astonishing explication of life, Hunter's greatest legacy would remain his revolutionary impact on surgery. Hunter had dedicated his life to applying the principles of science to surgery. By the time of his death, although his backward-looking colleagues at St. George's still resisted, Hunter's scientific approach had already become widely accepted. After he died, his devoted disciples—the thousand or more pupils he had taught—spread his creed of scientific inquiry throughout Britain and the United States. Inspired by Hunter's example and teachings, determined young surgeons such as John Abernethy, Astley Cooper, William Blizard, Henry Cline, and Anthony Carlisle headed the great nineteenth-century British teaching hospitals and led their profession to a respected new status. Across the Atlantic, followers such as William Shippen, John Morgan, Wright Post, and Philip Syng Physick promulgated Hunter's teachings in their new hospitals and established the foundations of American medical education. Progress was inevitably haphazard and Hunter's followers would sometimes forget his conviction in the powers of nature and his conservative approach to surgery in their zeal to ascend in social rank and establish glowing reputations. But through Hunter's pervasive influence, the future practice of surgery would be based largely on the doctrine of observation, experimentation, and application of scientific evidence. When Edward Jenner tested his smallpox vaccine on an eight-year-old boy in 1796, thus establishing the practice of vaccination, which would save millions of lives, he was studiously following his tutor's principles. When Joseph Lister tried out his carbolic-soaked lint on eleven patients in 1867, thus launching antiseptic practices that would prevent countless deaths, he was purposefully adopting his hero's methods. And numberless pioneering surgeons down the years would similarly follow Hunter's scientific principles in helping to render surgery safe and effective.

Yet of all the people who carried Hunter's torch, perhaps the one who understood him best was the poor, orphaned Cornish lad William Clift, who said, "From the beginning, I fancied, without being able to account for it, that nobody about Mr Hunter seemed capable of appreciating him. He seemed to me to have lived before his time and to have died before he was sufficiently understood."[60]

ACKNOWLEDGMENTS

Without the assistance of a great many people who have given their time and expertise freely and generously, it would have been impossible to turn my ambition to write a biography of John Hunter into a reality. In particular, I am indebted to Andrew Cunningham, Wellcome Trust senior research fellow in the history of medicine at Cambridge University, for his patient encouragement and expert advice. He diligently read and commented on my manuscript and offered enthusiastic support throughout, helping me to appreciate "the reasons of things."

Numerous people at museums and libraries in England and Scotland have helped me enormously. Fil Dearie of the Hunter House museum in East Kilbride kept me going with his spirited enthusiasm, John McLeish of East Kilbride public library gave vital help with local history, and Bill Niven, former provost (mayor) of East Kilbride, provided wonderful insights on local history and a fascinating guided tour of Hunter sites. At the Royal College of Surgeons in London, Liz Allen, former curator of the Hunterian Museum, gave me essential early help; Simon Chaplin, the current senior curator, provided invaluable guidance and specialist advice; and head librarian Tina Craig met my seemingly endless demands for manuscripts with unerring patience. At the Wellcome Library for the History and Understanding of Medicine (one of my favorite places to spend a day), staff provided generous assistance and expert knowledge on an almost daily basis. I would also like to thank everybody at the Wellcome Trust Centre for the History of Medicine at University College London, especially Professor Hal Cook, for welcoming a mere journalist into their academic fold. My research was completed while I was an honorary research assistant at the center. Many more staff at various libraries and museums have been instrumental in my research. In particular, I wish to thank everybody who helped me at the Natural History Museum, the library of the Royal Society, the British Library, the Guildhall Library, the City of Westminster Archives Centre, the London Metropolitan Archives, the library of the Royal Academy, the Royal Humane Society, the British Dental Association, Dr. Johnson's House, the Society of Apothecaries, and, at the University of Glasgow, the Special Collections Library, the Hunterian Museum, and the Hunterian Art Gallery.

I have been extraordinarily lucky in enjoying the support of many individual experts who have advised me on specialist areas. Mick Crumplin, honorary consultant surgeon and honorary curator at the Royal College of Surgeons, kindly helped me appreciate eighteenth-century army conditions. Professor Harold Ellis, Professor Sir Peter Bell, and Alan Scott helped elucidate aneurysm surgery today and in the eighteenth century. Malcolm Bishop provided generous advice on dental issues, Dr. Michael Waugh gave me technical assistance on questions concerning sexual diseases, and Peter Baskett and John Zorab both helped me when I was researching resuscitation today and in the past. I am also particularly grateful to all the staff at the dissecting rooms of Guy's, King's, and St. Thomas' School of Biomedical Sciences, London, where I was privileged to be allowed to witness a contemporary dissecting class conducted with respect and dignity. Dr. Colin Stolkin, senior lecturer in anatomy, was encouraging from the beginning. Dr. Alistair Hunter, academic manager of the dissecting rooms, helped me to understand the scientific compulsion that motivated John Hunter to explore the beauty of the human form and answered my numerous and often strange anatomical queries with good humor and generosity throughout. I am grateful, too, to Dr. James Munro, who kindly read my manuscript and provided vital medical insights.

On the publishing front, I was extremely fortunate to find my superb agent, Patrick Walsh, who helped me shape my ideas with a gentle but firm touch. I have also been lucky to find very supportive and enthusiastic editors both in the United Kingdom and in the United States. My special thanks go to Brenda Kimber, my editor at Transworld in the United Kingdom, for her warm help and kindness. In the United States, I am grateful to all of the staff at Broadway, and particularly to Becky Cole, my editor, for her constant encouragement, expert guidance, and enduring patience from beginning to end. Thanks also to Carol Edwards for her skillful and diligent copyediting.

Finally, a big thank-you is due to my family and friends—too many to mention individually, but they know who they are—who have steadfastly given me practical support and spiritual encouragement. My children, Sam and Susannah, deserve special praise for their forbearance and understanding (and my failure to attend numerous school trips). My mum and dad have also given loyal support. Most of all, this book would never have been begun, let alone completed, without the constant enthusiasm and encouragement of my partner, Peter Davies, who believed in this project, and in me, throughout.

National and International Events	John Hunter's Life
1727 George II crowned; Isaac Newton died	
	1728 John Hunter born February 13/14 at Long Calderwood, East Kilbride
1734 St. George's Hospital opened	
1740 War of the Austrian Succession begins	
1745 Jacobite uprising	
1746 Jacobites defeated at Culloden	1746 William Hunter sets up Covent Garden anatomy school
1748 Treaty of Aix-la-Chapelle	1748 John joins William in London
1749 Royal fireworks celebrating peace	1749 John trains with Cheselden at the Royal Hospital Chelsea; becomes instructor in Covent Garden anatomy school; brothers move to house in Great Piazza
	1750 John spends second summer at Chelsea
1751 Gin Act introduced	1751 John trains with Pott at St. Bartholomew's Hospital; mother, Agnes, dies
1752 Murder Act allows courts to direct that murderers be dissected	1752 John treats first recorded patient, a chimney sweep
1753 British Museum founded	
	1754 John becomes a pupil at St. George's Hospital; discovers placental circulation

1755 Samuel Johnson's *Dictionary of the English Language* published

1756 Seven Years War begins

1760 George III crowned

1761 George III marries Princess Charlotte of Mecklenburg-Strelitz

1763 Treaty of Paris ends Seven Years War; Boswell meets Johnson

1768 Royal Academy founded, with William Hunter as its first professor of anatomy; Cook's first voyage, with Joseph Banks and Daniel Solander, 1768–1771

1771 Cook returns and Jenner helps classify specimens; Smollett dies

1755 John at Oxford for a brief spell

1756 John spends five months as house surgeon at St. George's; William moves to Jermyn Street

1757 Hunter brothers quarrel with Pott and the Monros

1760 John falls ill and gives up anatomy; enlists as army surgeon

1761 John takes part in capture of Belle-Ile

1762 John sails for Portugal; William delivers Queen Charlotte's first son, the future George IV; John's first research paper, on the descent of the testes and congenital hernias, published in William's *Medical Commentaries*

1763 John leaves army and returns to London; begins partnership with dentist James Spence

1764 John becomes engaged to Anne Home

1766 John's first paper, on the greater siren, read to the Royal Society

1767 John elected fellow of the Royal Society; begins experiment on venereal disease; Hunter brothers investigate large bones discovered near Ohio River

1768 John moves to William's former house on Jermyn Street; gains diploma from Company of Surgeons; elected surgeon at St. George's

1769 John assists in cesarean operation

1770 Edward Jenner becomes John's first house pupil

1771 John publishes first major work, *The Natural History of the Human Teeth*; marries Anne Home

1773 Boston Tea Party

1774 Humane Society (later Royal
Humane Society) founded

1775 American War of Independence
begins

1776 Adam Smith's *Wealth of Nations*
published; American
Declaration of Independence;
first volume of Gibbon's *History
of the Decline and Fall of the
Roman Empire* published

1778 Banks elected president of the
Royal Society

1779 Cook killed in Hawaii

1780 Gordon riots in London

1772 John Banks Hunter born;
William Lynn becomes assistant; Everard Home becomes
John's apprentice; John gives
private lectures on surgery and
physiology

1773 Inuit visitors meet Hunter;
Jenner returns home; John's first
angina attack; Mary-Ann
Hunter born

1774 John performs a cesarean operation; John dissects electric eels;
James Hunter born; William
Hunter publishes *The Anatomy
of the Human Gravid Uterus*

1775 James dies; William Bell becomes assistant; John advertises
lectures

1776 Agnes Hunter born; Mary-Ann
dies; John appointed surgeon extraordinary to George III; treats
David Hume; gives lectures at
William's Great Windmill Street
school, 1776–1778

1777 John attempts to revive the
Reverend Dr. William Dodd after hanging; second angina attack

1778 Everard Home leaves to join
navy; John publishes part two of
his work on teeth, *A Practical
Treatise on the Diseases of the
Teeth*; the Reverend James
Baillie, John's brother-in-law,
dies and Dorothy moves to Long
Calderwood with son and
daughters

1779 John's "An Account of the Free-
Martin" published in
*Philosophical Transactions of the
Royal Society*

1780 John accuses William of stealing
his discovery of placental circulation; John given remains of

first giraffe brought to Europe; "An Account of an Extraordinary Pheasant" published in *Philosophical Transactions*

1781 British surrender at Yorktown

1782 Charles Byrne, the "Irish giant," arrives in London; Hunter conducts autopsy on prime minister (marquis of Rockingham)

1783 Treaty of Paris signed; William Pitt the Younger becomes prime minister

1783 John moves to Leicester Square; William Hunter dies; John steals body of Irish giant; gives lectures in home on Leicester Square (Castle Street)

1784 Dr. Johnson dies

1784 Everard Home returns to London; John conducts autopsy on father-in-law, Robert Home; Matthew Baillie cedes Long Calderwood to John

1785 *The Times* (originally *Daily Universal Register*) founded

1785 Hunter moves collection to Leicester Square; consulted by Benjamin Franklin; visits Tunbridge Wells and Bath after further angina attacks; performs successful operation on popliteal aneurysm

1786 Robert Burns's *Poems, Chiefly in the Scottish Dialect* published

1786 John appointed deputy surgeon general of the army; awarded Royal Society Copley Medal; publishes *A Treatise on the Venereal Disease*; embalms George III's aunt, Princess Amelia; treats William Pitt the Younger; Reynolds's portrait of Hunter unveiled at Royal Academy

1787 First convicts transported to New South Wales

1787 John treats Adam Smith; Home appointed John's assistant at St. George's; "Observations Tending to Show that the Wolf, Jackal, and Dog Are All of the Same Species" published in *Philosophical Transactions*

1788 George III falls ill during winter

1788 Museum opens twice a year;

1788–1789; riots against surgeons over body snatching in New York

1789 Fall of the Bastille; William Blake's *Songs of Innocence* published

1791 Paine's *Rights of Man,* part one, published; Boswell's *Life of Johnson* published

1792 French republic declared; Paine's *Rights of Man,* part two, published

1793 Louis XVI executed; France declares war on Britain

1794 First volume of Erasmus Darwin's *Zoonomia* published

1796 Edward Jenner successfully tests first smallpox vaccine

John treats Gainsborough and young Byron

1789 John elected to Company of Surgeons' court of assistants; William Bell leaves; Philip Syng Physick becomes house pupil

1790 John appointed surgeon general of the army; John's descriptions of antipodean mammals included in John White's *Journal of a Voyage to New South Wales*; Home takes over delivering lectures

1791 John helps found Veterinary College of London (later Royal Veterinary College); Haydn declines treatment

1792 John writing *Observations and Reflections on Geology*; "Observations on Bees" published in *Philosophical Transactions*; William Clift taken on as apprentice; John performs autopsy on Reynolds; treats Farington; Home defeated in election for surgeon at St. George's; row over hospital pupils' fees begins

1793 John supervises medical provision for overseas expeditions; St. George's adopts stringent rules on surgeons' duties and pupils' enrollment; Hunter dies, October 16, at St. George's

1794 Observations on fossil bones from Germany published posthumously in *Philosophical Transactions*

1799 Pitt's government agrees to buy Hunter's museum and gives custody to Company of Surgeons; Clift appointed first curator

1800 Company of Surgeons renamed
       Royal College of Surgeons

1801 Home demands Hunter's manu-
       scripts

1803 Napoleonic War resumes

1805 Nelson killed at Trafalgar

1806 Museum moves into RCS's new
       headquarters in Lincoln's Inn
       Fields

1815 Napoléon defeated at Waterloo

1823 Home burns Hunter's manu-
       scripts

1859 Darwin's *On the Origin of
Species* published

1859 Hunter reinterred in
       Westminster Abbey;
       *Observations and Reflections on
Geology* published

1861 *Essays and Observations on
Natural History, Anatomy,
Physiology, Psychology and
Geology* published

1871 Darwin's *The Descent of Man*
published

Key Sources

Publications of John Hunter Mentioned in the Notes

The Natural History of the Human Teeth: Explaining Their Structure, Use, Formation, Growth, and Diseases (London: printed for J. Johnson, 1771).

"On the Descent of the Testis," in William Hunter, *Medical Commentaries* (London: A. Hamilton, 1762; and S. Baker and G. Leigh, 1777).

A Practical Treatise on the Diseases of the Teeth; Intended as a Supplement to the Natural History of Those Parts (London: printed for J. Johnson, 1778).

A Treatise on the Venereal Disease (London: privately printed, 1786).

A Treatise on the Blood, Inflammation and Gun-shot Wounds, by the Late John Hunter (London: George Nicol, 1794).

The Works of John Hunter, ed. James Palmer, 4 vols. (London: Longman, Rees, Orme, Brown, Breen, 1835).

Observations and Reflections on Geology (London: Taylor and Francis, 1859).

Essays and Observations on Natural History, Anatomy, Physiology, Psychology and Geology, ed. Richard Owen, 2 vols. (London: Van Voorst, 1861).

Letters from the Past, from John Hunter to Edward Jenner, ed. E. H. Cornelius and A. J. Harding Rains (London: Royal College of Surgeons of England, 1976).

The Case Books of John Hunter FRS, ed. Elizabeth Allen, J. L. Turk, and Sir Reginald Murley (London: Royal Society of Medicine, 1993).

Biographies of Hunter Mentioned in the Notes

Adams, Joseph, *Memoirs of the Life and Doctrines of the Late John Hunter* (London: J. Callow, 1818).

Dobson, Jessie, *John Hunter* (Edinburgh: E. & S. Livingstone, 1969).

Foot, Jessé, *The Life of John Hunter* (London: T. Becket, 1794).

Home, Everard, "A Short Account of the Life of the Author," in *A Treatise on the Blood, Inflammation and Gun-shot Wounds, by the Late John Hunter* (London: George Nicol, 1794).

Kobler, John, *The Reluctant Surgeon: The Life of John Hunter* (London: Heinemann, 1960).

Ottley, Drewry, "The Life of John Hunter, FRS," in *The Works of John Hunter,* ed. James Palmer, vol. 1 (London: Longman, Rees, Orme, Brown, Breen, 1835).

Paget, Stephen, *John Hunter, Man of Science and Surgeon* (London: Fischer Unwin, 1897).

Peachey, George C., *A Memoir of William & John Hunter* (Plymouth: Brendon, 1924).

Qvist, George, *John Hunter 1728–1793* (London: W. Heinemann, 1981).

Abbreviations for Sources Used in the Notes

British Library: BL
The Hunter Album, RCS: HA
The Hunter-Baillie Collection (at RCS): HBC
John Hunter: JH
Loudoun Papers (at RCS): LP
Royal College of Surgeons of England: RCS
The Wellcome Library for the History and Understanding of Medicine: WL

Hunter Sites

Scotland

John Hunter's birthplace at Long Calderwood is preserved as a museum, the Hunter House Museum, which provides information on his life and the development of modern surgery. The Hunter House Museum, 126 Maxwellton Road, East Kilbride, South Lanarkshire, G74 3LR, tel. 441355 261261.

In East Kilbride, the Hunter brothers are remembered in the Hunter Primary School, Hunter High School, Hunter Health Centre, Hunter Street in the old East Kilbride village, and The Hunters pub in Stewartfield.

London

John Hunter's collection of anatomical specimens and naturalist items is housed in the Hunterian Museum at the headquarters of the Royal College of Surgeons of England. Of the original fourteen thousand preparations, more than half were destroyed in an air raid in 1941, while others have not withstood the ravages of time. Today about 3,500 survive, including the specimens displaying his successful popliteal aneurysm operations and the skeleton of the Irish giant Charles Byrne. The museum reopened after refurbishment in 2005. Hunterian Museum, Royal College of Surgeons, 35–43 Lincoln's Inn Fields, London WC2A 3PE, tel. 4420 7405 3474.

A bust of John Hunter stands in Leicester Square, opposite the site of his former house at number 28, which is now a pub, The Moon Under Water. A blue plaque denotes his former house at 31 Golden Square. William Hunter's Great Windmill Street school is now the rear of the Lyric Theatre, Shaftesbury Avenue.

A Note on Money

Making direct financial comparisons between the eighteenth century and modern times is not straightforward. As a rough guide, the Bank of England (Retail Price Index, October 2003) estimates the following:

£1 in 1750 would be worth £81.63 today.
£1 in 1760 would be worth £71.69 today.
£1 in 1770 would be worth £61.05 today.
£1 in 1780 would be worth £60.24 today.
£1 in 1790 would be worth £53.70 today.

A guinea was equivalent to 21 shillings, so, for example, 5 guineas was worth £5 5s.

1. The Coach Driver's Knee

1. Everard Home, "An Account of Mr Hunter's Method of Performing the Operation for the Cure of the Popliteal Aneurysm from Materials Furnished by Mr Hunter," *Transactions of a Society for the Improvement of Medical and Chirurgical Knowledge* 1 (1793): 138–181. Details about the coach driver and his operation are taken from the above unless otherwise indicated.

2. My thanks for their advice to Sir Peter Bell, professor of surgery at Leicester Royal Infirmary; Professor Harold Ellis, clinical anatomist at Guy's, King's, and St. Thomas' School of Biomedical Sciences; and Alan Scott, honorary consultant in vascular surgery and director of the UK multicenter aneurysm screening study. The causes of aneurysms are still unclear, but they occur when the elastin in an artery loses its elasticity, causing the walls to bulge outward. Professor Bell and Professor Ellis believe it possible that the top of a coachman's boots could injure the artery walls at the back of the knee if these had already been weakened by the underlying condition.

3. JH, *Case Books*, pp. 450–451.

4. The figures are cited by John Gunning, William Walker, and Thomas Keate (Hunter's enemies) in a letter to the governors, n.d. (1793), transcribed in George C. Peachey, *A Memoir*, pp. 282–296.

5. Jessé Foot, *The Life*, p. 280.

6. Ibid., p. 242.

7. E. H. Cornelius, "John Hunter As an Expert Witness," *Annals of the Royal College of Surgeons of England* 60 (1978): 412–418. Giving evidence as an expert witness at a murder trial in 1781, Hunter had been asked, "I presume you have dissected more than any man in Europe?" and he agreed, saying, "I have dissected some thousands during these thirty-three years."

8. Percivall Pott, *The Chirurgical Works of Percivall Pott* (London: printed for T. Lowndes, 1779), vol. 3, p. 202.

9. JH, *The Works*, vol. 1, p. 536, citing Bromfield's views.

10. JH to Edward Jenner, August 2 (no year), in JH, *Letters from the Past*, p. 9.

11. Steven G. Friedman, *A History of Vascular Surgery* (New York: Futura, 1989), passim.

12. JH, *The Works*, vol. 1, p. 536.

13. Home, "An Account."

14. Details of the stag experiment were reputedly handed down from William Bell, Hunter's assistant at the time, to a later assistant, William Clift, who retold the story to the nineteenth-century naturalist Richard Owen. See Lloyd G. Stevenson, "The Stag of Richmond Park," *Bulletin of the History of Medicine* 22 (1948): 467–475; Stevenson, "A Further Note on John Hunter and Aneurysm," *Bulletin of the History of Medicine* 26 (1952): 162–167; *British Medical Journal* 1 (1879): 284–285; Jessie Dobson, ed., *Descriptive Catalogue of the Physiological Series in the Hunterian Museum* (Edinburgh and London: E. & S. Livingstone, 1970), part 1, pp. 4–8.

15. JH, *The Works*, vol. 1, p. 536.

16. Paolo Assalini, *Manuale di Chirurgia del Cavaliere Assalini* (Milan: Pirola, 1812), p. 86. My thanks to Robinetta Gaze for her translation from the Italian.

17. Particular thanks to Professor Harold Ellis for explaining Hunter's operation, and to Alan Scott and Sir Peter Bell (see note 2) for describing modern treatment. Today, the problem is commonly tackled with bypass surgery, using either a synthetic vessel or a section of vein taken from the thigh, although smaller aneurysms are still occasionally treated by tying the artery above and below and relying on collateral circulation, as in Hunter's day.

18. David C. Schechter and John J. Bergan, "Popliteal Aneurysm: A Celebration of the Bicentennial of John Hunter's Operation," *Annals of Vascular Surgery* 1 (1986): 118–126; Assalini, *Manuale di Chirurgia*, p. 86.

19. The leg of the first coach driver can still be seen in the Hunterian Museum, specimen P 275, alongside the limb of the fourth patient, specimen P 279, which was obtained by a Hunter disciple, Thomas Wormald, from the man's widow in 1837.

2. The Dead Man's Arm

1. Everard Home, "A Short Account," p. xv. Here Home describes Hunter's first human dissection.

2. In William Hunter to James Hunter, September 17, 1743, HBC, vol. 2, p. 5, William refers to his "darling London." There are several studies detailing William's life. The most useful include C. Helen Brock, "The Happiness of Riches," in *William Hunter and the Eighteenth-Century Medical World*, ed. William Bynum and Roy Porter (Cambridge: Cambridge University Press, 1985), pp. 35–54; C. Helen Brock, ed., *William Hunter 1718–1783: A Memoir by Samuel Foart Simmons and John Hunter* (Glasgow: University of Glasgow Press, 1983), in which JH annotated an early biography of his brother; C. Helen Brock, "The Early Years of James, William and John Hunter" (Ph.D. thesis, 1977), East Kilbride public li-

brary; Sir Charles Illingworth, *The Story of William Hunter* (Edinburgh and London: E. & S. Livingstone, 1967).

3. Controversy surrounds the date of Hunter's birth. Although the parish register (old parish register of East Kilbride, copy in East Kilbride public library) records his birth on February 13, 1728, Hunter always celebrated his birthday on February 14. Most likely, the exact time of his birth in a dimly lighted room in the middle of the night went unrecorded. Even today the Hunterian Society celebrates Hunter's birthday on February 13, while the RCS honors him on February 14.

4. Hunter's parents' date of marriage and his siblings' dates of birth and death are recorded in the Hunter family Bible, RCS. The children's dates of birth are also given in the old parish register of East Kilbride.

5. For details of the history of East Kilbride and the Hunter brothers' early lives, see David Ure, *The History of Rutherglen and East Kilbride* (Glasgow: David Niven, 1793); and T. E. Niven, *East Kilbride: The History of Parish and Village* (Glasgow: Gavin Watson, 1965). Many thanks for his advice and clarification to Bill Niven, local historian and former provost (mayor) of East Kilbride, whose father, Eric (T. E.) Niven, previously charted the history of the village.

6. The house where John Hunter was born at Long Calderwood is preserved as a museum, the Hunter House Museum (see Hunter sites on pages 287–288). Contrary to popular belief, William Hunter was born in the village, before the family moved to Long Calderwood, according to the old parish register. Thanks to Bill Niven for pointing this out.

7. Stephen Paget, *John Hunter*, p. 27.

8. Reminiscences of Dorothea Baillie, told to her daughters Agnes and Joanna Baillie, HBC, vol. 2, p. 1, and vol. 6, p. 18.

9. Hunter's pupils Astley Cooper and John Abernethy both said that Hunter had men read for him: See R. C. Brock, *The Life and Work of Astley Cooper* (Edinburgh and London: E. & S. Livingstone, 1952), p. 47; and John Abernethy, *Physiological Lectures* (London: Longman, 1825), p. 202. Hunter displayed classic symptoms of dyslexia, according to the British Dyslexia Association (personal communication to the author, August 2003).

10. Article from *European Magazine* in 1782, reprinted in Abernethy, *Physiological Lectures,* pp. 341–352.

11. Jessé Foot, *The Life,* p. 60.

12. W. R. Le Fanu, *A Bibliography of Edward Jenner* (Winchester: St. Paul's Bibliographies, 1985), p. 101. Hunter's comment is recorded by Benjamin Waterhouse in a letter to Jenner, April 24, 1801.

13. JH, *The Works,* vol. 1, p. 217; Richard Owen, ed., *Descriptive and Illustrated Catalogue of the Physiological Series of Comparative Anatomy* (London: RCS, 1840), vol. 5, p. xiii.

14. Peter Laslett, *The World We Have Lost—Further Explored,* 3d ed. (London: Methuen, 1983), p. 108. The figure for life expectancy at birth in England in 1751 is given as 36.6 years.

15. M. Dorothy George, *London Life in the Eighteenth Century* (London: Peregrine, 1966), p. 399.

16. Roy Porter, *The Greatest Benefit to Mankind* (London: HarperCollins, 1997), pp. 245–303. For more discussion of medicine in the age of enlightenment, see Roy Porter, *Enlightenment: Britain and the Creation of the Modern World* (London: Penguin, 2000).

17. *Gentleman's Magazine*, 18 (1748), pp. 348–350.

18. T. Jock Murray, "The Medical History of Dr Samuel Johnson," *The Nova Scotia Medical Bulletin* (June–August 1982): 71–78.

19. *The New Dispensatory of the Royal College of Physicians of London, with Copious and Accurate Indexes. Faithfully Translated from the Latin of the Pharmacopoeia Londinensis* (London: W. Owen, 1746).

20. Reminiscences of Dorothea Baillie, told to her daughters Agnes and Joanna Baillie, HBC, vol. 6, p. 18.

21. George, *London Life in the Eighteenth Century*, p. 37.

22. Hogarth published his etching *Gin Lane* on February 1, 1753.

23. Susan C. Lawrence, *Charitable Knowledge: Hospital Pupils and Practitioners in Eighteenth-Century London* (Cambridge: Cambridge University Press, 1996), p. 85.

24. John P. Blandy and John S. P. Lumley, eds., *The Royal College of Surgeons of England: 200 Years of History at the Millennium* (London and Oxford: Royal College of Surgeons of England, Blackwell Science, 2000), pp. 6–7.

25. William Hunter, *Two Introductory Lectures* (London: printed by order of the trustees for J. Johnson, 1784), pp. 88–89.

26. Ernest Finch, "The Influence of the Hunters on Medical Education," *Annals of the Royal College of Surgeons of England* 20 (1957): 205–248.

27. Home, "A Short Account," p. xv.

28. William Hunter, *Two Introductory Lectures*, p. 102.

3. The Stout Man's Muscles

1. Middlesex Sessions Rolls, December 1747 (ms. 2889), London Metropolitan Archives. The case was also reported in *Gentleman's Magazine*, 17 (1747), p. 591; and *Westminster Journal*, December 26, 1747, BL.

2. William Hunter, *Two Introductory Lectures* (London: printed by order of the trustees for J. Johnson, 1784), p. 87 (William's italics).

3. Christopher Lawrence, "Alexander Monro Primus and the Edinburgh Manner of Anatomy," *Bulletin of the History of Medicine* 62 (1988): 193–214.

4. William Hunter, *Two Introductory Lectures*, p. 87.

5. William Hunter, *Medical Commentaries* (London: A. Hamilton, 1762), p. 8.

6. Alan F. Guttmacher, "Bootlegging Bodies: A History of Body-Snatching," *Bulletin of the Society of Medical History of Chicago* 4 (1935): 352–402.

7. Samuel Pepys, *The Shorter Pepys*, ed. Robert Latham (London: Penguin, 1985), p. 261.

8. Toby Gelfand, "The 'Paris Manner' of Dissection: Student Anatomical Dissection in Early Eighteenth-Century Paris," *Bulletin of the History of Medicine* 16 (1972): 99–130; and Guttmacher, "Bootlegging Bodies." The legal situation would remain unchanged until the Anatomy Act of 1832 provided a source of unclaimed bodies from workhouses.

9. Ralph Hyde, ed., *The A to Z of Georgian London* (Lympne Castle, Kent: Harry Margary in association with Guildhall Library, London, 1981). The site of Tyburn Tree is shown on John Rocque's map of 1747 at the junctions of Tiburn Lane and Tiburn Road (today Park Lane and Oxford Street), near to where Marble Arch now stands on the northeast corner of Hyde Park. The ten convicts hanged on October 28, 1748, are named and their crimes are listed on "The Proceedings of the Old Bailey" Web site (www.oldbaileyonline.org). General information on hanging and the anatomists' battles for bodies can be found in Andrew Langley, *Georgian Britain 1714 to 1837* (London: Hamlyn, 1994); Peter Linebaugh, *The London Hanged: Crime and Civil Society in the Eighteenth Century* (London: Allen Lane, 1991); and Linebaugh, "The Tyburn Riot Against the Surgeons," in *Albion's Fatal Tree: Crime and Society in Eighteenth-Century England*, ed. Douglas Hay (London: Allen Lane, 1975), pp. 65–117.

10. Samuel Richardson, *Familiar Letters on Important Occasions*, cited in Linebaugh, "The Tyburn Riot." He was describing a scene in 1740.

11. Linebaugh, *The London Hanged*, pp. 38–39.

12. Martin Fido, *Bodysnatchers: A History of the Resurrectionists, 1742–1832* (London: Weidenfeld and Nicolson, 1988), pp. 4–5. The skeleton of Jonathan Wild remains at RCS headquarters in Lincoln's Inn Fields.

13. Ruth Richardson, *Death, Dissection and the Destitute* (London: Phoenix Press, 2001). Richardson's book provides a detailed analysis of public opinion toward dissection and body snatching.

14. Cecil Howard Turner, *The Inhumanists* (London: A. Ouseley, 1932), p. 50.

15. JH, *The Works*, vol. 2, p. 159.

16. Richardson, *Death, Dissection and the Destitute*, p. 31.

17. Report from the Select Committee of the House of Commons on Anatomy, July 22, 1828, BL. Evidence to the select committee was given by two of John Hunter's pupils, Sir Astley Cooper and John Abernethy, and by Abernethy's former assistant, Dr. James Macartney, as well as by two anonymous body snatchers, A. B. and C. D.

18. Turner, *The Inhumanists*, p. 107 (Turner reports the grave robbing in Scotland); George C. Peachey, *A Memoir*, p. 42 (Peachey records the situation in Oxford).

19. JH, *Case Books*, pp. 308–309.

20. General information on body snatching can be found in Turner, *The Inhumanists*; Richardson, *Death, Dissection and the Destitute*; Fido, *Bodysnatchers*; James Blake Bailey, ed., *The Diary of a Resurrectionist 1811–1812* (London: Swan Sonnenschein, 1896); and in the Report from the Select Committee of the House of Commons on Anatomy.

21. Linebaugh, "The Tyburn Riot."

22. Bransby Blake Cooper, *The Life of Sir Astley Cooper* (London: J. W. Parker, 1843), passim. Astley Cooper's nephew, Bransby Cooper, outlined the body snatchers' methods in his biography of the surgeon. Rigor mortis—the process in which the dead body stiffens—usually begins a few hours after death and is complete in about twelve hours. It then wears off and the body becomes limp again within twenty-four to thirty-six hours of death, although the process takes longer in colder temperatures. My thanks to Dr. Alistair Hunter, academic manager of the dissecting rooms at Guy's, King's, and St. Thomas' School of Biomedical Sciences, for advice.

23. Liza Picard, *Dr Johnson's London* (London: Weidenfeld and Nicolson, 2000), p. 297. Picard gives the monthly wage of a seaman with the East India Company as one pound, fifteen shillings in 1762.

24. Report from the Select Committee of the House of Commons on Anatomy, evidence from Sir Astley Cooper.

25. Bailey, ed., *The Diary of a Resurrectionist*, p. 145. The incident described relates to an evening in 1812. The writer has been identified as the notorious gang leader Joshua Naples.

26. JH, *Case Books*, pp. 313–314.

27. Drewry Ottley, "The Life," p. 10.

28. JH, *The Works*, vol. 2, p. 158.

29. JH, *The Works*, Atlas, pp. 5–6.

30. William Hunter, *The Anatomy of the Human Gravid Uterus Exhibited in Figures* (Birmingham: J. Baskerville, S. Baker and G. Leigh, 1774).

31. Article from *European Magazine* in 1782, reprinted in John Abernethy, *Physiological Lectures* (London: Longman, 1825), pp. 341–352.

32. In *The Dissecting Room*, c. 1770, William is shown standing just left of center, with John at his left elbow.

33. William Hunter, *Two Introductory Lectures*, p. 113.

34. C. Helen Brock, ed., *William Hunter 1718–1783: A Memoir by Samuel Foart Simmons and John Hunter* (Glasgow: University of Glasgow Press, 1983), p. 5. The description is given in John Hunter's annotations.

35. Jane M. Oppenheimer, "John and William Hunter and Some Contemporaries in Literature and Art," *Bulletin of the History of Medicine* 23 (1949): 21–47.

36. Joseph Adams, *Memoirs*, p. 37; John Abernethy, Hunterian Oration (1819), in Abernethy, *Physiological Lectures*, p. 54.

37. Ottley, "The Life," p. 10.

38. Jessé Foot, *The Life*, pp. 81–82.

4. The Pregnant Woman's Womb

1. C. G. T. Dean, *The Royal Hospital Chelsea* (London: Hutchinson, 1950), passim. The Royal Hospital Chelsea still provides a home for army pensioners today, while the grounds are the venue for the annual Chelsea Flower Show.

2. Biographical detail about William Cheselden can be found in Sir Zachary

Cope, *William Cheselden 1688–1752* (Edinburgh and London: E. & S. Livingstone, 1953); Knut Haeger, *The Illustrated History of Surgery.* Revised and updated ed., ed. Sir Roy Calne (London: Harold Starke; Chicago: Fitzroy Dearborn, 2000); Sidney Lee and Leslie Stephen, eds., *Dictionary of National Biography* (London: Smith, Elder and Co., 1908–1909); John P. Blandy and John S. P. Lumley, eds., *The Royal College of Surgeons of England: 200 Years of History at the Millennium* (London and Oxford: Royal College of Surgeons of England and Blackwell Science, 2000); Sir Zachary Cope, *The Royal College of Surgeons of England: A History* (London: Anthony Blond, 1959).

3. Details of the lithotomy operation and ancient surgery in general can be found in Ghislaine Lawrence, "Surgery (Traditional)," in *Companion Encyclopaedia of the History of Medicine,* ed. William Bynum and Roy Porter (London and New York: Routledge, 1993), pp. 961–983. Today, bladder stones are much less common, although doctors are uncertain why; different diets may be the reason. Surgeons today normally remove bladder stones in an operation that involves opening the bladder via the abdomen, or by using a special instrument, a lithotrope, which is inserted through the urethra into the bladder to crush a stone. In either case, the patient would be under a general anesthetic. Occasionally, lasers or ultrasound is used to destroy a stone. The lithotomy position, as used by Cheselden, survives in operations for hemorrhoids and similar conditions (personal communication to the author from Professor David Kirk, consultant urologist, Gartnavel General Hospital, Glasgow, November 2003).

4. Roy Porter, *The Greatest Benefit to Mankind* (London: HarperCollins, 1997), p. 235.

5. Cheselden's lithotomy operations are described in detail in Cope, *William Cheselden 1688–1752;* and Cope, *The Royal College of Surgeons of England.* Other aspects are discussed in Ira M. Rutkow, *Surgery: An Illustrated History* (St. Louis: Mosby-Year Book, Inc., in collaboration with Norman Pub, 1993), pp. 263–267; Toby Gelfand, "The "Paris Manner" of Dissection: Student Anatomical Dissection in Early Eighteenth-Century Paris," *Bulletin of the History of Medicine* 16 (1972): 99–130. Dr. James Douglas, who witnessed many of Cheselden's lithotomies, recorded that "he seldom exceeds half a Minute" (quoted in Cope, *William Cheselden 1688–1752,* p. 29).

6. William Hunter, *Two Introductory Lectures* (London: printed by order of the trustees for J. Johnson, 1784), p. 73.

7. Everard Home, "A Short Account," p. xvi.

8. Joan Lane, "The Role of Apprenticeship in Eighteenth-Century Medical Education in England," and Toby Gelfand, " 'Invite the Philosopher, As Well As the Charitable': Hospital Training as Private Enterprise in Hunterian London," both in *William Hunter and the Eighteenth-Century Medical World,* ed. William Bynum and Roy Porter (Cambridge: Cambridge University Press, 1985), pp. 57–103 and 129–151; Susan C. Lawrence, *Charitable Knowledge* (Cambridge: Cambridge University Press, 1996), passim.

9. There are no records of Hunter's work with Cheselden. General information

on eighteenth-century operations can be found in Guy Williams, *The Age of Agony* (London: Constable, 1975), while the surgical instruments used are discussed in David J. Warren, *Old Medical and Dental Instruments* (Princes Risborough, Buckinghamshire: Shire Publications, Ltd., 1994).

10. Rutkow, *Surgery,* pp. 266–267.

11. Le Dran, *The Operations in Surgery of Mons. Le Dran, Translated by Thomas Gataker with Remarks, Plates of the Operations, and a Sett of Instruments by William Cheselden* (London: C. Hitch and R. Dodsley, 1749), passim. Details of Cheselden's methods of amputation, trepanning, and the operation for a harelip are all taken from his comments on Le Dran's surgery. General details are also taken from Lorenz Heister, *A General System of Surgery,* 2d ed. (London: W. Innys, 1745).

12. Cope, *William Cheselden 1688–1752,* pp. 75–79. Today, cataract surgery entails a similar technique in removing the hard, opaque part of the lens, although nowadays this is replaced with an artificial implant and, naturally, the operation is performed with the use of an anesthetic.

13. Home, "A Short Account," p. xvi.

14. Agnes Baillie to Matthew Baillie, RCS. Relating the memories of her mother, Dorothea, William and John's sister, Agnes described the dissecting room as being behind their house in Covent Garden (HBC, vol. 2, p. 1). Detail on Covent Garden can be found in *Survey of London* (London Survey Committee, 1970), vol. 36; and in Sheila O'Connell, *London 1753* (London: British Museum Press, 2003), catalog to accompany an exhibition of the same name at the British Museum, May 23 to November 23, 2003.

15. For help in understanding the process of dissection in Hunter's day, as well as for the privilege of witnessing modern medical students undertaking a dissection class, I am indebted to Dr. Alistair Hunter, academic manager of the dissecting rooms, and Professor Harold Ellis, clinical anatomist, at Guy's, King's, and St. Thomas' School of Biomedical Sciences, as well as all the staff in the dissecting rooms at Guy's.

16. JH, *Case Books,* pp. 401–402, 405.

17. John M. T. Ford, ed., *A Medical Student at St Thomas's Hospital, 1801–1802: The Weekes Family Letters* (London: Wellcome Institute for the History of Medicine, 1987), p. 78.

18. JH, *The Works,* vol. 4, "Some Observations on Digestion," pp. 81–116, for description of the gastric juices; JH, *Essays and Observations,* vol. 1, p. 189, for description of the taste of semen.

19. William Hunter, *Two Introductory Lectures,* pp. 110, 91.

20. Hunter described his methods in JH, *Essays and Observations,* vol. 1, "On Making Anatomical Preparations by Injections, etc.," pp. 385–398. General information on preparations can be found in F. J. Cole, *A History of Comparative Anatomy from Aristotle to the Eighteenth Century* (London: Macmillan, 1944), pp. 445–450.

21. JH, *Essays and Observations,* vol. 1, p. 124.

22. Drewry Ottley, "The Life," pp. 75–76.

23. C. Helen Brock, *Calendar of the Correspondence of Dr William Hunter, 1740–83* (Cambridge: Wellcome Unit for the History of Medicine, 1996), p. i. Maggie Riley, at the Hunterian Museum, Glasgow, has estimated that William's original collection numbered 3,755 (personal communication to the author, 2002). Only a handful can now be positively identified as JH's handiwork, based on his descriptions of his research, but hundreds more must have sprung from his knife.

24. William Hunter, *The Anatomy of the Human Gravid Uterus Exhibited in Figures* (Birmingham: J. Baskerville, S. Baker and G. Leigh, 1774). William Hunter gives the date of the arrival of the body as 1751 in the preface of his *Gravid Uterus,* but in the text he refers to the winter of 1750. In fact, the drawings by van Rymsdyk are dated 1750, as confirmed by the staff at Glasgow University Library, Special Collections (personal communication to the author, 2003). All quotes from William are from the preface unless otherwise stated.

25. John H. Teacher, *Catalogue of the Anatomical and Pathological Preparations of Dr William Hunter in the Hunterian Museum, University of Glasgow* (Glasgow: James MacLehose and Sons, 1900), vol. 1, pp. xlix–l. Teacher has shown that twelve plates are dated—I to X in 1750, XV and XXVI in 1754—and that four undated plates—XIII, XXI, XXII, and XXXII—are probably from the years 1750–1754. The plates depict six women. William refers to the second and third "subjects" arriving before the first ten plates were completed.

26. JH, *The Works,* vol. 4, p. 62.

27. John H. Teacher, *Catalogue of the Anatomical and Pathological Preparations of Dr William Hunter in the Hunterian Museum,* pp. 326, 396–404, 509, 571–573, 589–590. Teacher details which of the specimens in William's collection (at Glasgow University) can be identified as John's work. Just four of the items recorded in JH's first catalog, created in about 1764, can categorically be dated to his time in Covent Garden. JH, "A Copy of the Oldest Portion of Catalogue, in Mr Hunter's Own Handwriting," n.d., ms. 49 e 53, RCS.

28. J H, *The Works,* vol. 1, p. 210.

29. JH, *A Treatise on the Venereal Disease* (London: Gand W Nicol, 3rd ed., 1810), pp. 128–129. Hunter said he treated the sweep in "about the year 1752," dating it to his spell at St. Bartholomew's.

5. The Professor's Testicle

1. Biographical details of Alexander Monro, Jr., and the Monro dynasty in general can be found in T. V. N. Persaud, *A History of Anatomy: The Post-Vesalian Era* (Springfield, IL: Charles C. Thomas, c. 1997); and Malcolm Nicolson, "Medicine," in *Scotland: A Concise Cultural History,* ed. P. H. Scott (Edinburgh: Mainstream, 1993), pp. 327–342.

2. William Hunter, *Medical Commentaries* (London: A. Hamilton, 1762), p. 19. The dates, details, and comments regarding the arguments between the Hunters and the Monros, and the Hunters with Pott, are all taken from William's *Medical Commentaries* unless otherwise stated.

3. C. Helen Brock, *Calendar of the Correspondence of Dr William Hunter, 1740–83* (Cambridge: Wellcome Unit for the History of Medicine, 1996), p. 7. Since his degree came from a Scottish university instead of from Oxford or Cambridge, William was allowed to become only a licentiate of the college, rather than a full fellow, much to his chagrin.

4. Roy Porter, *The Greatest Benefit to Mankind* (London: HarperCollins, 1997), pp. 182, 183, 242.

5. William refused to allow John to visit their dying mother in 1751, despite her appeals, arguing, "I cannot consent this season to her request, for my brother's sake, for my own sake, and even for my mother's sake." John visited Long Calderwood in 1752 for the last time and took Dorothy back to London on his return. See John Thomson, *An Account of the Life, Lectures and Writings of William Cullen, MD* (Edinburgh: W. Blackwood and Sons, 1859), vol. 1, appendix, correspondence with William Hunter, p. 540–543.

6. C. Helen Brock, ed., *William Hunter 1718–1783: A Memoir by Samuel Foart Simmons and John Hunter* (Glasgow: University of Glasgow Press, 1983), p. 41. Observation made in John Hunter's annotations.

7. "Reviews and Notices," *British Medical Journal* (1861): 303–305. The statement quoted is that of Alexander Carlyle.

8. Brock, ed., *William Hunter 1718–1783,* p. 8. The remark is found in John Hunter's annotations.

9. Copy of John's signature, written as Johannes Hunter, in the enrollment register at Oxford University, June 5, 1755, preserved in HA.

10. Drewry Ottley, "The Life," p. 14, quoting Hunter speaking to his pupil Sir Anthony Carlisle.

11. Letter from Minson Hales to John Hunter, April 11, 1753, quoted in *Calendar of the Correspondence of Dr William Hunter 1740–83,* ed. Brock. Glasgow University Library, Special Collections, 1993; refers to William having been "dangerously ill." Ottley, "The Life," p. 16. Ottley states that John took over a portion of the lectures in 1754.

12. Useful sources on the history of anatomy include Persaud, *A History of Anatomy*; Roger French, "The Anatomical Tradition," in *Companion Encyclopaedia of the History of Medicine,* ed. William Bynum and Roy Porter (London and New York: Routledge, 1993), pp. 81–101; and Porter, *The Greatest Benefit to Mankind.* William Hunter provided his students with a summary in the first lecture of each course. See William Hunter, *Two Introductory Lectures* (London: printed by order of the trustees for J. Johnson, 1784).

13. The history of discoveries of the lymphatic system is explained in Nellie B. Eales, "The History of the Lymphatic System, with Special Reference to the Hunter-Monro Controversy," *Journal of the History of Medicine* 29 (1974): 280–294, as well as in sources mentioned heretofore on the general history of anatomy.

14. Everard Home, "A Short Account," p. xviii. The animals mentioned are listed in JH, "A Copy of the Oldest Portion of Catalogue, in Mr Hunter's Own Handwriting," n.d., ms. 49 e 53 RCS. Hunter is understood to have made his

arrangement with the menagerie at some point in the 1750s. The specimens listed in this first catalog were all collected before 1764, and since Hunter spent most of the latter period in the army, most of these animals were probably obtained before 1760, while he worked with William.

15. JH, *The Works*, vol. 3, pp. 76–77; Stephen Inwood, *The Man Who Knew Too Much: The Strange and Inventive Life of Robert Hooke 1635–1703* (London: Macmillan, 2002), p. 47.

16. Andrew Cunningham, "The Pen and the Sword: Recovering the Disciplinary Identity of Physiology and Anatomy Before 1800, I: Old Physiology—the Pen," *Studies in History and Philosophy of Biological and Biomedical Sciences* 33 (2002): 631–665. Details of eighteenth-century embryology research can also be found in Brian Cook, *Contributions of the Hunter Brothers to Our Understanding of Reproduction: An Exhibition from the University Library's Collections* (leaflet, Glasgow University Library, Special Collections Department, 1992).

17. Hunter's experiments on embryology were published in JH, *Essays and Observations*. He described the methods of investigating chicken embryos in JH, "Of the Different Methods to Be Taken to Examine the Progress of the Chick in Incubated Eggs," n.d., ms. 49 d 11, RCS; also reprinted in Richard Owen, ed., *Descriptive and Illustrated Catalogue of the Physiological Series of Comparative Anatomy*, Vol. 5 (London: RCS, 1840).

18. For general background on contemporary microscopes, see James B. McCormick, *Eighteenth Century Microscopes: Synopsis of History and Workbooks* (Lincolnwood, IL: Science Heritage, Ltd., 1987).

19. JH, *The Works*, vol. 4, pp. 187–192.

20. JH, "A Copy of the Oldest Portion of Catalogue," p. 16. Specimen C4, showing the first pair of cranial nerves, is dated 1754 and was therefore in Hunter's original collection. The specimen can no longer be located.

21. William Hunter, *Medical Commentaries*, pp. 1–4.

22. Nellie B. Eales, "The History of the Lymphatic System"; the quote is from Alexander Monro, Jr., *De testibus et de semine in variis animalibus* (Edinburgh, 1755), p. 55. Other details of the Hunters' row with the Monros are taken from William Hunter, *Medical Commentaries*.

23. William Hunter, *Medical Commentaries*, p. 7 (William's italics).

24. Ibid., p. 34. The injections were made at some point in 1753 or 1754— William was later unsure of the exact year. William reported John's ambition to trace the entire system.

25. The pupils' register of St. George's Hospital, 3 vols., transcript at RCS. Vol. 1, p. 1, records that John Hunter was appointed a house surgeon on May 5, 1756.

26. Information on the hospital's history is taken from George C. Peachey, *History of St George's Hospital* (London: J. Bale, 1910–1914).

27. Jessé Foot, *The Life*, pp. 75–76. Foot describes the duties of a house surgeon as well as that person's responsibility for the keys of the "dead house."

28. C. Helen Brock, *Calendar of the Correspondence of Dr William Hunter*

1740–83, p. 10. According to Brock, William took a lease on the Jermyn Street house in the summer of 1756.

29. William describes the interest in John's new research in William Hunter, *Medical Commentaries*, p. 89. John himself describes the investigations in JH, "Observations on the State of the Testis in the Foetus, and on the Hernia Congenita," ibid., pp. 75–89.

30. The argument with Pott is described fully in William Hunter, *Medical Commentaries*, supplement to the first part, pp. 12–27; it is also discussed in Fenwick Beekman, "The 'Hernia Congenita' and an Account of the Controversy It Provoked Between William Hunter and Percivall Pott," *Bulletin of the New York Academy of Medicine* 22 (1946): 486–500.

31. Beekman, "The 'Hernia Congenita,' " quoting Pott's second edition of *A Treatise on Ruptures* (London: L. Hawes, W. Clarke, and R. Collins, 1763), p. 139.

32. Article from *European Magazine* in 1782, reprinted in John Abernethy, *Physiological Lectures* (London: Longman, 1825), pp. 341–352.

33. William Hunter, *Medical Commentaries*, supplement to the first part, p. iii.

34. Nellie B. Eales, "The History of the Lymphatic System."

35. The details of the experiments on the five animals are described by John Hunter in William Hunter, *Medical Commentaries*, pp. 42–48. Technically, certain fats can enter the veins, but only in minute quantities.

36. Joseph Wright, *An Experiment on a Bird in the Air Pump*, exhibited 1768, National Gallery, London.

37. JH, "The Modern History of the Absorbing System," n.d., ms. 49 e 5, RCS. These notes, in the handwriting of Hunter's later assistant and brother-in-law Everard Home, were apparently part of an original catalog to Hunter's museum.

38. John Wiltshire, *Samuel Johnson in the Medical World, the Doctor and the Patient* (Cambridge: Cambridge University Press, 1991), pp. 128–138, quoting Samuel Johnson, *The Idler* 17 (1758).

39. Wiltshire argues persuasively that Haller, not Hunter, was the target of Johnson's attack. Haller's experiments are discussed in Andrew Cunningham, "The Pen and the Sword," p. 653.

40. JH, *The Works*, vol. 4, "Some Observations on Digestion," pp. 81–116.

41. Betsy Copping Corner, ed., *William Shippen, Jr., Pioneer in American Medical Education*, With Notes, and the Original Text of Shippen's Student Diary, London, 1759–60, Together with a Translation of His Edinburgh Dissertation, 1761 (Philadelphia: American Philosophical Society, 1951), p. 7; quoting a letter from William Shippen, Sr., to his brother, Edward Shippen, September 1, 1758.

42. Ibid., Shippen's Student Diary, October 2 and 9, 1759, p. 25.

43. *Dictionary of American Biography* (Oxford: Oxford University Press; New York: Charles Scribner's Sons, 1928–1958), vol. 42, pp. 117–118.

44. The period from 1760 to March 1761 in Hunter's life is fairly obscure. Various biographers refer to his falling ill in 1759 or 1760, but the latter seems most likely, since he was clearly in fine form while Shippen lodged in his house (from fall of 1759 until January 1760). Foot and Home, early biographers, both say

his health suffered in 1760, while Ottley states he suffered from an inflammation of the lungs. William refers to his brother's ill health in his *Medical Commentaries* without being specific about the date. In his "The Animal Oeconomy," John says unequivocally that he "quitted" anatomy in 1760, and in an unpublished ms., he says this was in the "early summer" of 1760. Yet he records two postmortems and one case of treatment in his casebooks that summer. An article on his lectures in the *European Magazine* in 1780, apparently approved by Hunter, attributes his decision to quit to his desire "for a more enlarged field of observation." He was living in Covent Garden again in the winter of 1760–1761, when Morgan was staying there, according to Bell's biography of Morgan. Foot, *The Life*, p. 74; Home, "A Short Account," p. xviii; Ottley, "The Life," p. 20; William Hunter, *Medical Commentaries*, p. 35; JH, *The Works*, vol. 4, p. 292; JH, "The Modern History of the Absorbing System," n.d., ms. 49 e 5, RCS; JH, *Case Books*, p. 317 (two postmortems) and pp. 111–112 (surgical care); article in the *European Magazine* (1782); Whitfield J. Bell, *John Morgan, Continental Doctor* (Philadelphia: University of Pennsylvania Press, 1965), p. 48.

6. The Lizard's Tails

1. T. Keppel, *The Life of Augustus Viscount Keppel* (London: Henry Colburn, 1842), pp. 293–320. Keppel's concerns about the troops are related in letters from Maj.-Gen. Studholme Hodgson to the earl of Albermarle. His "secret instructions" are also detailed here.

2. Details of the Seven Years War and the Battle for Belle-Ile are related in Tom Pocock, *Battle for Empire: The Very First World War 1756–63* (London: Michael O'Mara, Ltd., 1998); J. Fortescue, *A History of the British Army* (London: Macmillan and Co., 1899), vol. 2, pp. 521–538; and W. Clowes, *The Royal Navy: A History* (New York: AMS Press, 1966), vol. 3, pp. 234–236.

3. Roy Porter, *Blood and Guts: A Short History of Medicine* (London: Penguin, 2003), p. 109.

4. A. Peterkin, William Johnston, and R. Drew, *Commissioned Officers in the Medical Services of the British Army 1660–1960* (London: Wellcome Historical Medical Library, 1968), vol. 1, p. 33.

5. Jeremy Lewis, *Tobias Smollett* (London: Jonathan Cape, 2003), p. 28. The excerpt from Smollett's novel is from Tobias Smollett, *The Adventures of Roderick Random* (Oxford: Oxford University Press, 1748), p. 187.

6. Keppel, *The Life of Augustus Viscount Keppel*, p. 309, quoting a letter from Hodgson to the earl of Albemarle, April 12, 1761.

7. *Gentleman's Magazine*, 31 (1761), p. 229, quoting a letter from Hodgson reporting the victory, April 23.

8. JH, *The Works*, vol. 3, p. 559.

9. JH, *Case Books*, p. 65; Jessie Dobson, *John Hunter*, p. 55.

10. Sir Neil Cantlie, *A History of the Army Medical Department* (Edinburgh and London: Churchill Livingstone, 1974), vol. 1, p. 133.

11. Ibid., p. 107.

12. Dobson, *John Hunter,* p. 51.

13. JH to William Hunter, May 28, 1762, published in the *Medical Times and Gazette* (1867), pp. 515–516. Four of John's letters to William from Belle-Ile are preserved in the RCS Library, HBC, vol. 2, while two more—cited above and March 23, 1762—were discovered later and published in the *Medical Times and Gazette.*

14. G. E. Gask, "John Hunter in the Campaign in Portugal, 1762–3," *British Journal of Surgery* 24 (1936–1937): 640–667. Page 664 cites pay rates of different medical staff.

15. JH to William Hunter, September 28, 1761, HBC, vol. 2, p. 18.

16. Many thanks to Mick Crumplin, honorary consultant surgeon and honorary curator at the Royal College of Surgeons, for advice on Hunter's treatment of gunshot wounds compared to that of his contemporaries, as well as for general advice on army surgery. Debridement—cutting away deadened tissue—is the accepted practice today.

17. JH, *The Works,* vol. 3, p. 549.

18. Sir Robert Drew, "John Hunter and the Army," *Journal of the Royal Army Medical Corps* 113, no. 1 (1967): 11.

19. JH, *Case Books,* pp. 274–275.

20. JH, *The Works,* vol. 3, pp. 549–550; JH, *Case Books,* p. 275.

21. JH, *The Works,* vol. 3, p. 555.

22. Ibid., pp. 574–575.

23. JH to William Hunter, March 23, 1762, published in the *Medical Times and Gazette* (1867), p. 516. This letter refers to his appointment as director of the hospital, the titles he was awarded, and his need to keep a horse for the job.

24. JH to William Hunter, July 11, 1761, HBC, vol. 2, p. 14.

25. JH to William Hunter, July 11, 1761, HBC, vol. 2, p. 14; and September 14 and 28, 1761, HBC, vol. 2, p. 18.

26. C. Helen Brock, *Calendar of the Correspondence of Dr William Hunter 1740–83* (Cambridge: Wellcome Unit for the History of Medicine, 1996), p. 23.

27. Alexandre Dumas, *The Man in the Iron Mask* (Oxford: Oxford University Press, 1991), p. 517. Porthos dies in Belle-Ile after the island is attacked by Louis XIV's forces.

28. JH, *The Works,* vol. 3, p. 21; Richard Owen, ed., *Descriptive and Illustrated Catalogue of the Physiological Series of Comparative Anatomy,* Vol. 5 (London: RCS, 1840), p. 61.

29. JH, *Essays and Observations,* vol. 1, p. 245.

30. Jessie Dobson, ed., *Descriptive Catalogue of the Physiological Series in the Hunterian Museum* (Edinburgh and London: E. & S. Livingstone, 1970), part 1, p. 265.

31. Details of Hunter's time in Portugal are fully described in the Loudoun Papers (LP), a set of letters, records, and sick returns archived at the RCS Library. His spell in Portugal is also described in Gask, "John Hunter in the Campaign in

Portugal," pp. 640–667; Fenwick Beekman, "John Hunter in Portugal," *Annals of Medical History* 8 (1936): 288–296; and Sir Robert Drew, "John Hunter and the Army."

32. W. Young to Lord Loudoun, September 8, 1762, LP, ms. 86.

33. JH, *The Works*, vol. 4, p. 293.

34. JH, *Observations and Reflections on Geology*, p. xvi.

35. R. Craig, "Gunshot Wounds Then and Now: How Did John Hunter Get Away with It?" in *Papers Presented at the Hunterian Bicentenary Commemorative Meeting* (London: Royal College of Surgeons of England, 1995), pp. 15–19. Craig describes the three gunshot-wound specimens with impressions of French musket balls. JH, "A Copy of the Oldest Portion of Catalogue, in Mr Hunter's Own Handwriting," n.d., ms. 49 e 53, RCS. Hunter lists fifty specimens of lizards and the soldier's intestine.

7. The Chimney Sweep's Teeth

1. Anne S. Hargreaves, *White as Whales Bone: Dental Services in Early Modern England* (Leeds: W. S. Maney and Son, Ltd., 1998), p. 9 (citing the duchess of Northumberland describing George III) and p. 11 (citing Horace Walpole describing the duke of Newcastle).

2. Details about the history of dentistry can be found in Hargreaves, *White as Whales Bone*; Malvin E. Ring, *Dentistry: An Illustrated History* (New York: Abrams, 1985); and John Woodforde, *The Strange Story of False Teeth* (London: Routledge and Kegan Paul, 1968).

3. James Woodforde, *The Diary of a Country Parson*, ed. J. Beresford (Oxford: Oxford University Press, 1981), vol. 1, p. 183.

4. Smollett mentions the disagreement in a letter to William: Tobias Smollett to William Hunter, June 14, 1763, HBC, vol. 1, p. 91.

5. C. Helen Brock, "The Happiness of Riches," in *William Hunter and the Eighteenth-Century Medical World*, ed. William Bynum and Roy Porter (Cambridge: Cambridge University Press, 1985), p. 41. Brock says William's bank account shows several payments to John.

6. Jessie Dobson, *John Hunter*, p. 108. Hunter gave his address as Covent Garden in June 1765, when purchasing land at Earls Court, so it is reasonable to assume he had settled there on returning from Portugal. Foot states that Hunter embarked on dentistry in an alliance with James Spence at this time, and Hunter himself refers to working with Spence in his treatise on teeth, although he does not give the dates. Jessé Foot, *The Life*, p. 132.

7. Foot, *The Life*, p. 131.

8. Sidney Lee and Leslie Stephen, eds., *Dictionary of National Biography* (London: Smith, Elder and Co., 1908–1909), vol. 7, pp. 367–368; Foot, *The Life*, pp. 133–134.

9. JH, *The Works*, vol. 2, p. 108.

10. Ibid., p. 104.

11. The specimen showing a human tooth transplanted into a cock's comb is P 56 in the Hunterian Museum. The cockerel's testicle in the hen's abdomen is P 53.

12. Sir Roy Calne, "Replacement Surgery and Transplantation," in *Papers Presented at the Hunterian Bicentenary Commemorative Meeting* (London: Royal College of Surgeons of England, 1995), pp. 12–14.

13. Dr. William Irvine to Professor Thomas Hamilton, June 17, 1771, cited in "The Hunters and the Hamiltons: Some Unpublished Letters," *The Lancet,* 214 (1928): 354–360.

14. John Woodforde, *The Strange Story of False Teeth,* p. 24.

15. Henry W. Noble, "Tooth Transplantation: A Controversial Story," shortened version of a lecture given to the Scottish Society for the History of Medicine, June 15, 2002, available on the History of Dentistry Research Group's Web site: www.rcpsglasg.ac.uk/hdrg/home.html. The examples of earlier tooth transplants are related here.

16. Thomas Rowlandson, *Transplanting of Teeth,* published 1790. For a discussion of Rowlandson's cartoon, see Fiona Haslam, *From Hogarth to Rowlandson: Medicine in Art in Eighteenth-Century Britain* (Liverpool: Liverpool University Press, 1996), pp. 252–253; Ruth Richardson, "Transplanting Teeth: Reflections on Thomas Rowlandson's *Transplanting Teeth,*" *The Lancet* 354 (1999): 1740. The dentist depicted was Bartholomew Ruspini, a disciple of Hunter's who practiced in England from the 1750s onward; see J. Menzies Campbell, *Dentistry Then and Now* (Glasgow: Pickering and Inglis, 1963), p. 64.

17. JH, *The Works,* vol. 2, p. 100.

18. Thomas Bell, who edited Hunter's treatise on teeth in *The Works,* called tooth transplanting Hunter's favorite operation. Ibid., p. 104, note by Bell, lecturer in comparative anatomy at Guy's Hospital, writing in 1835. Hunter only referred to the practice on a handful of occasions, although examples are also mentioned in case studies by others.

19. Noble, "Tooth Transplantation."

20. John Woodforde, *The Strange Story of False Teeth,* p. 84.

21. Ibid., p. 81.

22. Thomas Berdmore, *A Treatise on the Disorders and Deformities of the Teeth and Gums* (London: privately printed, 1768), p. 102.

23. William Rae, "Lectures on the Teeth," *British Journal of Dental Science* (1857): 517–521, taken from a verbatim manuscript note of lectures beginning April 12, 1782, by Mr. Tomes. See also Christine Hillam, "New Notes on the Lectures of William Rae," *Dental Historian* 33 (1998): 50–72.

24. JH to William Cullen, September 24, 1777, transcript of letter, ms. Cullen 204, Glasgow University Library, Special Collections.

25. Thanks for his advice to Bob Corfield of the UK Transplant Authority.

26. Hunter's contribution to dentistry is discussed in Jerry J. Herschfeld, "John Hunter and His Practical Treatise on Diseases of the Teeth," *Bulletin of the History*

of Dentistry 29, no. 1: 32–36; Campbell, Dentistry Then and Now, pp. 88–106; Irwin D. Mandel, "Revisiting John Hunter," Journal of the History of Dentistry 48, no. 2: 57–60; and David E. Poswillo, "John Hunter's Contribution to Dentistry," Dental Historian 38 (2001): 13–17. Many thanks to Malcolm Bishop, dental surgeon, for his help in explaining Hunter's contribution to dentistry and its significance today. My thanks also to the American Dental Association.

8. The Debutante's Spots

1. John Hadley, "An Account of a Mummy, Inspected at London 1763, in a Letter to William Heberden, MD FRS, from John Hadley, MD FRS," Philosophical Transactions of the Royal Society 54 (1764): 1–14 (New York: Johnson Reprint Co., 1965).

2. Everard Home, "A Short Account," p. xviii. Home says Hunter taught practical anatomy and operational surgery for several winters at this time. His spell as an army surgeon gave him an automatic right to practice on his return to civilian life.

3. The casebooks detail numerous dissections in the winter of 1763–1764.

4. Roy Porter, The Greatest Benefit to Mankind (London: HarperCollins, 1997), pp. 263–264. The gradual acceptance of postmortems in Georgian society and Hunter's role in this trend have been described by Simon Chaplin in a lecture, " 'An Excellent Hand for the Business': John Hunter and the Art of Dissection in Johnson's London," given at Dr. Johnson's house, London, November 20, 2003.

5. John Abernethy, Hunterian Oration (1819), in Abernethy, Physiological Lectures (London: Longman, 1825), pp. 40–41.

6. Sidney Lee and Leslie Stephen, eds., Dictionary of National Biography, vol. 3, p. 584; Jessie Dobson, "John Hunter and the Byron Family," Journal of the History of Medicine 10 (1955): 333–335. The case is recorded in JH, Case Books, pp. 213, 347.

7. JH, Case Books, p. 332.

8. Ibid., pp. 356–357.

9. Biographical details about Anne Home Hunter can be found in Aileen K. Adams, " 'I Am Happy in a Wife': A Study of Mrs John Hunter (1742–1821)," in Papers Presented at the Hunterian Bicentenary Commemorative Meeting (London: Royal College of Surgeons of England, 1995), pp. 32–37; Jane M. Oppenheimer, "Anne Home Hunter and Her Friends," Journal of the History of Medicine 1 (1946): 434–445; and Sir Arthur Porritt, "John Hunter's Women," Transactions of the Hunterian Society 17 (1958–1959): 81–111. Anne had at least two sisters.

10. Reminiscences of Mrs. John Hunter by her great-niece, Mrs. E. Milligan, in 1866, HBC, vol. 2, p. 58; Professor William Hamilton to his father, December 25, 1777, cited in "The Hunters and the Hamiltons: Some Unpublished Letters," The Lancet, 214 (1928): 354–360.

11. Dr William Hunter by Allan Ramsay, c. 1765, Hunterian Art Gallery, University of Glasgow.

12. John Hunter FRS by Robert Home, c. 1770, Royal Society.

13. JH, *The Works,* vol. 4, p. 320.

14. *John Hunter* by Robert Home, c. 1775–1778, Royal College of Surgeons. An engraving taken from this painting by H. Cook (at the Royal Society) dates the painting c. 1765 and also shows the mummy more clearly.

15. Jessie Dobson, "Some of John Hunter's Patients," *Annals of the Royal College of Surgeons of England* 42 (1968): 124–133.

16. Jessé Foot, *The Life,* p. 240.

17. Dobson, *John Hunter,* pp. 114–115.

18. George C. Peachey, *A Memoir,* pp. 144–145. Peachy cites the details of Hunter's purchase of land at Earls Court. W. W. Hutchings, *London Town—Past and Present* (London: Cassell, 1909), p. 674. Hutchings refers to Hunter living at 31 Golden Square. A blue plaque denotes the house.

19. JH, *The Works,* vol. 4, pp. 131–155. The experiments were first related in two papers to the Royal Society, "Experiments on Animals and Vegetables with Respect of the Power of Producing Heat," read June 22, 1775, and "On the Heat etc of Animals and Vegetables," read June 19, and November 13, 1777.

20. JH, *The Works,* vol. 1, p. 284.

21. JH, *Case Books,* p. 233.

22. JH, *The Works,* vol. 1, p. 512.

23. JH, *Case Books,* p. 233. Two specimens of dogs' tendons survive as P 109 and P 110, Hunterian Museum, RCS.

24. JH, "A Copy of the Oldest Portion of Catalogue, in Mr Hunter's Own Handwriting," n.d., ms. 49 e 53, RCS.

25. Museum notes, Hunterian Museum, RCS.

26. JH, "The Modern History of the Absorbing System," n.d., ms. 49 e 5, RCS. These notes, in the handwriting of Hunter's later assistant and brother-in-law Everard Home, were apparently part of an original catalog to Hunter's collection.

27. JH, *The Works,* vol. 4, pp. 394–397, paper first published in the *Philosophical Transactions of the Royal Society* 56 (1767): 307–310; JH, "General Observations on the Pnumobrankes," miscellaneous papers, n.d., ms. 49 d 11, RCS.

28. *Royal Society Journal Book Copy,* vol. 26, 1767–1770, February 5, 1767 (no page numbers).

29. William was elected a fellow of the Royal Society on April 30, 1767.

30. There are two specimens showing the heart of the greater siren in the Hunterian Museum, 912 and 913.

9. The Surgeon's Penis

1. JH, *The Works,* vol. 2, p. 417. *A Treatise on the Venereal Disease* was first published in 1786.

2. Scholars and disciples have argued for decades over whether Hunter inoculated himself or an anonymous victim. Hunter obscured the identity of the subject throughout, but evidence shows quite plainly that it was himself. Two sets of

pupils' lecture notes quote Hunter stating that "I have produced in myself a Chancre." See JH, "Lectures on the Principles of Surgery by John Hunter 1787," transcription of notes by Mr. Twigge, n.d., ms. 49 e 28, p. 390, RCS; and Philip J. Weimerskirch and Goetz W. Richter, "Hunter and Venereal Disease," *The Lancet* 313 (1979): 503–504, citing notes in the Edward G. Miner Library of the University of Rochester Medical Center, New York: JH, "Lectures on Venereal Diseases," ms. c. 1800. When Hunter's *Works* were first published in 1835, the editor of the treatise on venereal disease, George Babington, stated categorically that Hunter had performed the experiment on himself. Hunter's biographer in *The Works,* Drewry Ottley, even recorded Hunter joking about his self-experiment during his lectures. See JH, *The Works,* vol. 2, pp. 146–147; Ottley, "The Life," p. 47. It is highly unlikely, despite what Hunter's devotees have suggested, that he performed the experiment on a hospital or private patient or managed to procure a volunteer who lived in his house. In May 1767, he had no hospital patients on which to practice, and he would have been extremely foolish to risk such a reckless experiment on a private patient during his early career. It is implausible that he procured a volunteer he could monitor daily for the three-year duration of the experiment. The counterargument—that Hunter did not experiment on himself—can be found in W. J. Dempster, "Towards a New Understanding of John Hunter," *The Lancet* 1 (1978): 316–318; George Qvist, "John Hunter's Alleged Syphilis," *Annals of the Royal College of Surgeons of England* 59 (1977): 205–209; George Qvist, "Some Controversial Aspects of John Hunter's Life and Work," *Annals of the Royal College of Surgeons of England* 61 (1979): 138–141; and George Qvist, *John Hunter 1728–1793,* pp. 47–50.

3. James Boswell, *Boswell's London Journal, 1762–1763,* ed. Frederick A. Pottle (New Haven: Yale University Press, 2000), pp. 83–84 (entry for December 14, 1762).

4. *Harris's List of Covent-Garden Ladies, or Man of Pleasure's Kalendar for the Year 1779,* from excerpts at Dr. Johnson's House, Gough Square, London.

5. Derek Parker, *Casanova* (Stroud, England: Sutton Publishing, 2003), pp. 156–164.

6. William B. Ober, *Boswell's Clap and Other Essays: Medical Analyses of Literary Men's Afflictions* (Carbondale: Southern Illinois University Press, 1979), pp. 1–42.

7. Boswell, *Boswell's London Journal,* p. 139 (entry for January 12, 1763).

8. Ibid., p. 156 (entry for January 20, 1763).

9. Ibid., p. 227 (entry for March 25, 1763). Contemporary condoms are described in Sheila O'Connell, *London 1753* (London: British Museum Press, 2003), p. 144, catalog to accompany an exhibition of the same name at the British Museum, May 23 to November 23, 2003.

10. *Descriptive Catalogue of the Pathological Series in the Hunterian Museum,* vol. 1 (London: E. & S. Livingstone, 1966), p. 15 (describes the penis as specimen P 30); *A Guide to the Hunterian Museum,* bicentenary ed. (London: Royal College of Surgeons, 1993), p. 21 (details the bone and skull specimens, series P 714 to 746).

11. JH, *Case Books*. The cases cited below are on pp. 267, 266, and 269, respectively. Hunter gives no dates for his consultations with his venereal patients.

12. JH, *The Works*, vol. 2, p. 387. The case referred to happened in 1782.

13. Boswell, *Boswell's London Journal*, p. 156 (entry for January 20, 1763).

14. Information on the history of venereal disease is given in Claude Quétel, *History of Syphilis* (Cambridge: Polity Press, 1990); William Bynum, "Treating the Wages of Sin: Venereal Disease and Specialism in Eighteenth-Century Britain," in *Medical Fringe and Medical Orthodoxy 1750–1850*, ed. William Bynum and Roy Porter (London, Sydney, and Wolfeboro, NH: Croom Helm, c. 1987), pp. 5–28; and Allan M. Brandt, "Sexually Transmitted Diseases," in *Companion Encyclopedia of the History of Medicine*, ed. William Bynum and Roy Porter (London: Routledge, 1993), vol. 1, pp. 562–584. I am grateful for the advice of Dr. Michael Waugh, consultant genito-urinary physician at Leeds General Infirmary, on the nature and treatment of venereal diseases.

15. Bynum, "Treating the Wages of Sin," pp. 8–9.

16. JH, *The Works*, vol. 2, p. 187.

17. Ibid., p. 190.

18. Ibid., pp. 163–164.

19. Ibid., p. 193. This excerpt also describes the bread-pills test.

20. Lord Holland [Henry Richard Vassall Fox], *Further Memoirs of the Whig Party 1807–1821* (London: John Murray, 1905), pp. 343–344.

21. JH, *The Works*, vol. 2, p. 425.

22. JH, *Case Books*, p. 259.

23. JH, *The Works*, vol. 2, pp. 417–419. As Hunter's comments show, it is not true, despite what some of Hunter's defenders have argued, that he described the subject in the third person. Neither did he always describe his own health in the first person. Writing about his own angina in *The Works*, vol. 3, p. 150, he begins by saying, "A gentleman was attacked with a pain in the situation of the pylorus," then continues in the same vein.

24. Deborah Hayden, *Pox: Genius, Madness and the Mysteries of Syphilis* (New York: Basic Books, 2003), pp. 29–31; Diane Beyer Perett, *Ethics and Error: The Dispute Between Ricord and Auzias-Turenne over Syphilization 1845–70* (Ph.D. diss., Stanford University, 1977), pp. 13–29.

25. John Sheldon, *The History of the Absorbent System* (London: privately printed, 1784), p. 31. Sheldon states that Hunter "informed me that he had fed himself with madder" and that it had turned his urine red.

26. Richard Lovell Edgeworth, *Memoirs of Richard Lovell Edgeworth Esq* (London: R. Hunter, 1820), vol. 1, pp. 190–191. The circumstances are described more fully in Chapter 12.

27. Benjamin Bell, *A Treatise on Gonorrhoea Virulenta, and Lues Venerea* (Edinburgh: James Watson and Co., 1793).

28. Joseph Adams, *Memoirs*, p. 235.

29. JH, *The Works*, vol. 2, p. 123.

30. H. Clutterbuck to Joseph Adams, 1799, ms. 27 c 5, p. 18, RCS. Clutterbuck complained that Hunter's views led to reduced use of mercury.

31. Ober, *Boswell's Clap and Other Essays*, p. 23.

32. JH, *The Works*, vol. 2, pp. 304–307.

33. Ibid., pp. 307–308.

34. The explanation is given by the artist Joseph Farington: Joseph Farington, *The Farington Diary*, ed. J. Greig (London: Hutchinson and Co., 1922–1928), vol. 3, p. 660 (entry for September 13, 1796).

35. Foot referred to an instance where he treated a patient in consultation with Hunter. See Jessé Foot, *Observations Upon the New Opinions of John Hunter in His Late Treatise on the Venereal Disease* (London: T. Becket, 1786), in three parts, p. 28, RCS. He said he called for a syllabus for Hunter's lectures in 1773; see Foot, *The Life,* p. 243.

36. Foot, *Observations*, pp. 90, 110.

37. Charles Brandon Trye, *A Review of Jesse Foote's Observations on the New Opinions of John Hunter in His Late Treatise on the Venereal Disease* (London: John Murray, 1788), pp. 55, 1, 57.

38. JH to Charles Brandon Trye, February 24, (no year), copy of letter, Hunterian Society Catalogue, ms. 5610, 29/5, WL.

39. Ottley, "The Life," p. 22. The remark was found on a scrap of paper among Hunter's manuscripts after his death.

40. Everard Home, "A Short Account," p. xx. Home suggests Hunter set up the group shortly after joining the RS, although Joseph Adams later argued it was unlikely such a novice member could have organized the club so soon. However, Richard Lovell Edgeworth (see next note) confirmed that the club did begin in the late 1760s. Edgeworth said that the club met at Young Slaughter's.

41. Richard Lovell Edgeworth, *Memoirs of Richard Lovell Edgeworth Esq,* vol. 1, pp. 188–189; Jenny Uglow, *The Lunar Men* (London: Faber and Faber, 2002), pp. 124–125.

42. JH, *The Works,* vol. 4, pp. 461–463.

43. Everard Home, "An Account of a Hermaphrodite Dog," first read to the Royal Society on March 7, 1799, tracts RCS. In this paper, Home referred to Hunter's successful experiment in artificial insemination, although he did not give a date. The experiment is discussed in Brian Cook, *Contributions of the Hunter Brothers to Our Understanding of Reproduction: An Exhibition from the University Library's Collections* (leaflet, Glasgow University Library, Special Collections Department, 1992).

44. C. Helen Brock, *Calendar of the Correspondence of Dr William Hunter 1740–83* (Cambridge: Wellcome Unit for the History of Medicine, 1996), p. 39.

45. C. Helen Brock, ed., *William Hunter 1718–1783: A Memoir by Samuel Foart Simmons and John Hunter* (Glasgow: University of Glasgow Press, 1983), p. 59.

46. William Wadd, *Mems. Maxims and Memoirs* (London: Callow and Wilson, 1827), p. 283.

47. William Hunter, "Observations on the Bones, Commonly Supposed to Be Elephants Bones, Which Have Been Found Near the River Ohio in America," *Philosophical Transactions of the Royal Society* 58 (1768): 34–45 (New York: Johnson Reprint Co., 1965).

48. Company of Surgeons Examination Book, 1745–1800 (entry for July 7, 1768), facsimile at RCS.

10. The Kangaroo's Skull

1. Details of the *Endeavour*'s voyage are taken from a variety of sources, principally Ernest Rhys, ed., *The Voyages of Captain Cook* (Ware, Hertfordshire: Wordsworth Editions Ltd., 1999), which is based on Cook's journals; J. C. Beaglehole, *The Life of Captain James Cook* (London: Adam and Charles Black, 1974); and Patrick O'Brian, *Joseph Banks* (London: Harvill Press, 1997).

2. Ray Desmond, *Kew: The History of the Royal Botanic Gardens* (London: Harvill Press, 1995), p. 87. Desmond cites the letter from J. Ellis to C. Linnaeus, August 19, 1768.

3. Information about London's animal collections is taken from Julia Allen, *Samuel Johnson's Menagerie* (Norwich: Erskine Press, 2002); Geoffrey Parnell, *The Royal Menagerie at the Tower of London* (pamphlet, Royal Armouries Museum, Leeds, 1999); and Daniel Hahn, *The Tower Menagerie* (London: Simon and Schuster, 2003). My thanks for further information to Geoffrey Parnell of the Royal Armouries Library.

4. *An Historical Account of the Curiosities of London and Westminster* (London: J. Newbury, 1767), pp. 12–27, copy at the BL.

5. Jessie Dobson, "John Hunter's Animals," *Annals of the Royal College of Surgeons of England* 17 (1962): 379–486.

6. Richard D. Altick, *The Shows of London* (Cambridge, MA: Belknap Press, 1978), p. 35.

7. Drewry Ottley, "The Life," p. 30. Castle Street was later renamed Charing Cross Road.

8. Julia Allen, *Samuel Johnson's Menagerie,* p. 15.

9. Everard Home, "A Short Account," p. xxxi.

10. Details of Hunter's Earls Court home, known as Earl's Court House, are taken from contemporary news cuttings in the HA. It was demolished in 1886.

11. "John Hunter at Earl's Court, Kensington 1764–93," leaflet reprinted from the *Atheneum* of 1870, p. 5, HA.

12. Laszlo A. Magyar, *John Hunter and John Dolittle* at www.geocities.com/tapir 32hu/hunter.html. This theory is supported, I believe, by a letter from Hunter when asked for his view by Edward Jenner on a patient. In typical conservative style, his reply was, "I believe the best thing you can do is to do little," see JH, *Letters from the Past,* p. 10.

13. Ottley, "The Life," p. 125.

14. JH, miscellaneous notes and extracts, n.d., ms. 49 e 19, RCS.

15. JH, *The Works,* vol. 4, pp. 422–466.

16. Home, "A Short Account," p. xix.

17. Jessie Dobson, "The Hunter Specimens at Kew Observatory," *Annals of the Royal College of Surgeons of England* 8 (1951): pp. 457–462.

18. Home, "A Short Account," p. xxxviii.

19. Andrew Cunningham, "The Pen and the Sword: Recovering the Disciplinary Identity of Physiology and Anatomy Before 1800, II: Old Anatomy—the Sword," *Studies in History and Philosophy of Biological and Biomedical Sciences* 34 (2003): pp. 51–76.

20. The work of Buffon, Daubenton, and others is described in Charles Coulston Gillispie, ed., *Dictionary of Scientific Biography* (New York: Charles Scribner's Sons, 1970), various entries; John Gribben, *Science: A History 1543–2001* (London: Allen Lane, 2002), pp. 221–229; F. J. Cole, *A History of Comparative Anatomy from Aristotle to the Eighteenth Century* (London: Macmillan, 1944), p. 20 and appendix. For a comprehensive discussion of the pursuit of comparative anatomy, see Andrew Cunningham, *The Anatomist Anatomis'd: an Experimental Discipline in Eighteenth-Century Europe* (due to be published in 2005).

21. Stephen J. Cross, "John Hunter, the Animal Oeconomy, and Late Eighteenth-Century Physiological Discourse," *Studies in the History of Biology* 5 (1981): 1–110. Cross provides a good explanation of Hunter's purpose.

22. JH, miscellaneous notes and extracts, n.d., ms. 49 e 19, RCS.

23. JH, *The Works,* vol. 4, pp. 292–298.

24. JH, miscellaneous notes and extracts, n.d., ms. 49 e 19, RCS.

25. Hunter's arrangement can be discerned from his earliest catalog: JH, "A Copy of the Oldest Portion of Catalogue, in Mr Hunter's Own Handwriting," n.d., ms. 49 e 53, RCS.

26. JH, *Essays and Observations,* vol. 1, p. 107.

27. JH, *The Works,* vol. 4, pp. 331–392. This paper, entitled "Observations on the Structure and Oeconomy of Whales," was first published in *Philosophical Transactions of the Royal Society* in 1787. Hunter's research is cited in Herman Melville, *Moby-Dick* (London: Penguin, 1972), p. 85. Melville's novel was first published in England as *The Whale* in 1851.

28. JH, *Essays and Observations,* vol. 1, p. 52.

29. JH, *The Works,* vol. 4, pp. 131–155.

30. Ibid., pp. 315–318; Jessie Dobson, ed., *Descriptive Catalogue of the Physiological Series in the Hunterian Museum* (Edinburgh and London: E. & S. Livingstone, 1970), pp. 8–12. The bone-growth experiments are also discussed in G. Bentley, "John Hunter's Studies of the Musculoskeletal System," in *Papers Presented at the Hunterian Bicentenary Commemorative Meeting* (London: Royal College of Surgeons of England, 1995), pp. 20–25.

31. JH, *Essays and Observations,* vol. 1, p. 194. Hunter's observations were cited in Charles Darwin, *The Descent of Man,* ed. Richard Dawkins (London: Gibson Square Books, 2003), p. 540.

32. JH, *The Works,* Atlas, p. 18. The six sparrows are still in a perfect state in the Hunterian Museum, specimens 2457–2462.

33. Ottley, "The Life," p. 28.

34. David Morris, "John Hunter—Myth or Legend?," *Hunterian Society Transactions* (1974–1976): 43–55.

35. Jessie Dobson, *John Hunter,* p. 113.

36. Jane M. Oppenheimer, "John and William Hunter and Some Contemporaries," *Bulletin of the History of Medicine* 23 (1949): 41–42.

37. The cesarean operation is described by William Cooper, the physician initially called in by the midwife, and by Henry Thomson, the surgeon who performed the operation, in *Medical Observations and Inquiries* 4 (1771): 261–271 and 272–279.

38. J. H. Young, *The History of the Caesarean Section* (London: H. K. Lewis and Co. Ltd., 1944), pp. 1–54.

39. William Hunter, ms. notes H 56, August 13, 1774, Glasgow University Library, Special Collections; William Cooper, "An Account of the Caesarean Operation," *Medical Observations and Inquiries* 5 (1776): 217–232. The second operation, in which Hunter was assisted by James Patch, a former pupil from the Covent Garden school, took place on August 13, 1774.

40. Otto Sonntag, ed., *John Pringle's Correspondence with Albrecht von Haller* (Basel: Schwabe, 1999), pp. 135–136 (letter from Pringle to Haller, April 3, 1770); JH, *The Works,* vol. 4, pp. 81–121; JH, *Case Books,* pp. 374–375.

41. JH, "On the Digestion of the Stomach After Death," *Philosophical Transactions of the Royal Society* 62 (1772): 447–454 (New York: Johnson Reprint Co., 1965). The youth's stomach is specimen 592 in the Hunterian Museum; a similar preparation of a human stomach destroyed by gastric acid after death, dating from about 1755, is specimen 591.

42. JH, *Case Books,* pp. 370–371.

43. Martin Myrone, *George Stubbs* (London: Tate Publishing, 2002), p. 46.

44. William Hunter, "An Account of the Nyl-ghau, an Indian Animal, Not Hitherto Described," *Philosophical Transactions of the Royal Society* 61 (1771): 170–181 (New York: Johnson Reprint Co., 1965). John's observations are also recorded in JH, "Note on Teeth and Colon of Nyl-ghau and Goat," n.d., ms. H 147, Glasgow University Library, Special Collections. The nilgai skeleton is specimen RC-SHC/CO 1347 in the Hunterian Museum.

45. Sonntag, ed., *John Pringle's Correspondence with Albrecht von Haller,* p. 165 (letter from Pringle to Haller, June 13, 1771).

46. For details on Jenner's life, see Richard B. Fisher, *Edward Jenner 1749–1823* (London: André Deutsch, 1991); and John Baron, *The Life of Edward Jenner* (London: Henry Colburn, 1827).

47. Jenner's vaccine, using a small dose of cowpox to protect against the deadly and disfiguring smallpox, eventually led to the complete eradication of the disease, as declared by the World Health Organization in 1979.

48. Harold B. Carter, *Sir Joseph Banks 1743–1820* (London: British Museum [Natural History], 1988), pp. 95–96.

49. It is difficult to identify the animals Hunter obtained from Cook's first voyage, since records are incomplete. There are numerous antipodean animals in the Hunterian Museum, but it is not always clear whether they were brought back on Cook's first voyage or by later explorers, such as John White, who donated many animals to Hunter after his journey to New South Wales in 1788. For animals named in the text see the following. Sea pen: (museum preparation 2925); sharks' eggs and eels: Richard Owen, ed., *Descriptive and Illustrated Catalogue of the Physiological Series of Comparative Anatomy*, vol. 5 (London: RCS, 1840), p. 61; mole rat and zorilla: JH, *Essays and Observations*, vol. 2, pp. 235–236, 69–70; giant squid: museum preparation 308.

50. Carter, *Sir Joseph Banks 1743–1820*, pp. 89–91; Joseph Banks, *The Endeavour Journal of Joseph Banks 1768–1771*, ed. J. C. Beaglehole (Sydney: Trustees of the Public Library of New South Wales in association with Angus and Robertson, 1962), vol. 2, pp. 93–94.

51. Much debate has centered on the identity of the kangaroos seen by Banks and Cook. T. C. S. Morrison-Scott and F. C. Sawyer argue convincingly that the skull given to Hunter came from a great gray kangaroo, most probably the second one shot by Gore, based on a photograph of the skull; see Morrison-Scott and Sawyer, "The Identity of Captain Cook's Kangaroo," *Bulletin of British Museum (Natural History)* 1 (1950): 45–50. Hunter says Banks gave him a skull from Cook's first voyage; see JH, *The Works*, vol. 4, p. 485. Hunter's description of the teeth is from the same source. The skull was destroyed in 1941, during World War II. Many thanks to the staff of the Natural History Museum for help.

52. John White, *Journal of a Voyage to New South Wales*, ed. A. Chisholm (Sydney and London: Angus and Robertson, 1962; first published 1790).

53. Dr. William Irvine to Professor Thomas Hamilton, June 17, 1771, cited in "The Hunters and the Hamiltons: Some Unpublished Letters," *The Lancet* 214 (1928): 354–360. Irvine states that Hunter received two hundred pounds for his treatise on teeth.

54. JH to William Hunter, Saturday [sic] evening, n.d. (July 21, 1771), HBC, vol. 2, p. 10. The day was, in fact, Sunday.

55. Copy of Register of Marriage at St. James's Westminster, 1771, p. 21, HA.

56. John Kobler, *The Reluctant Surgeon*, p. 157. Kobler states that Cook and Banks attended the wedding, but without any reference. The story of the travelers presenting hickory-wood logs to the couple is contained in a curious pamphlet by DRAGM (believed to be a pseudonym for Robert Anstruther Goodsir), *Only an Old Chair* (Edinburgh: David Douglas, 1884).

57. Tobias Smollett, *The Letters of Tobias Smollett*, ed. Lewis M. Knapp (Oxford: Clarendon Press, 1970), p. 140 (letter from Smollett to Hunter, January 9, 1771); Jeremy Lewis, *Tobias Smollett* (London: Jonathan Cape, 2003), pp. 270, 278–279. Other writers have suggested Smollett promised his corpse to William Hunter, but

Knapp, who is regarded as the authority on the writer, and Lewis, his most recent biographer, both state his offer was made to John. The letter extract has no addressee.

11. The Electric Eel's Peculiar Organs

1. Everard Home, "A Short Account," p. xxii.

2. John Baron, *The Life of Edward Jenner* (London: Henry Colburn, 1827), p. 10 (Baron says Jenner called Hunter "the dear man"); Jessie Dobson, *William Clift FRS* (London: William Heinemann, 1954), p. 109 (Clift says Lynn called Hunter "Glorious John").

3. For details about Anne Hunter and her literary contributions, see Janet Todd, ed., *A Dictionary of British and American Women Writers* (London: Methuen and Co., 1987), pp. 169–170; Aileen K. Adams, " 'I Am Happy in a Wife': A Study of Mrs John Hunter (1742–1821)," in *Papers presented at the Hunterian Bicentenary Commemorative Meeting* (London: Royal College of Surgeons of England, 1995), pp. 32–37.

4. J Dobson, *William Clift FRS*, p. 109.

5. Drewry Ottley, "The Life," p. 41. William Clift, Hunter's last assistant, threw doubt on this tale, but a remark in a 1787 newspaper declaring that "Mrs Hunter's poetry is banished" from Hunter's house may well be a reference to the incident; see *Morning Herald*, April 13, 1787, p. 14, HA.

6. Hester Thrale, *Thraliana: The Diary of Mrs Hester Lynch Thrale 1776–1809*, ed. Katherine C. Balderston (Oxford: Clarendon Press, 1951), vol. 1, p. 67.

7. Lord Holland [Henry Richard Vassall Fox], *Further Memoirs of the Whig Party 1807–1821* (London: John Murray, 1905), p. 345.

8. *Kensington Express*, February 20, 1886, cutting appears on page 9 in the HA.

9. James Beattie, *James Beattie's London Diary*, ed. Ralph C. Walker (Aberdeen: Aberdeen University Press, 1946), p. 40 (entry for May 25, 1773). Beattie visited in May 1773.

10. George Cartwright, *A Journal of Transactions and Events, during a Residence of Nearly Sixteen Years on the Coast of Labrador* (Newark, Nottinghamshire: Allin and Ridge, 1792), vol. 1, p. 271; Anthony A. Pearson, "John Hunter and the Woman from Labrador," *Annals of the Royal College of Surgeons of England* 60 (1978): 7–13. The portraits are in the Hunter drawing books at the RCS. The Inuit family visited in early 1773.

11. These figures are cited by John Gunning, William Walker, and Thomas Keate (Hunter's enemies at St. George's) in a letter to the governors, n.d. (1793), printed verbatim in George C. Peachey, *A Memoir*, pp. 282–296. Figures for the fees are quoted in JH, letter to his colleagues, July 9, 1792, cited in Peachey, *A Memoir*, p. 272.

12. JH, letter to the governors, February 28, 1793, cited in Peachey, *A Memoir*, pp. 275–282.

13. Hunter states that he began lecturing privately, inviting pupils from St.

George's free of charge, in 1772 both in his lecture notes and in an article, which he sanctioned, in the *European Magazine*. JH, *The Works,* vol. 1, p. 210; article from *European Magazine* in 1782, reprinted in John Abernethy, *Physiological Lectures* (London: Longman, 1825), pp. 341–352. There has been debate about when Hunter opened his lectures to all comers, with different biographers giving 1773 or 1774 as the year, but Peachey makes a convincing case for 1775, when the earliest-known advertisement was placed; see Peachey, *A Memoir,* p. 162. The *European Magazine* article refers to Hunter lecturing privately in 1772, 1773, and 1774. Further evidence for the lectures not being given publicly until after 1774 is given in a paper signed by Hunter in May 1774, certifying that a pupil attended lectures "which I gave several of the pupils of St George's hospital," ms. 49 e 59, RCS.

14. Jessé Foot, *The Life,* pp. 243–244.

15. Letter from John Gunning, William Walker, and Thomas Keate to the governors, n.d. (1793), in Peachey, *A Memoir,* p. 289.

16. Henry Cline, Hunterian Oration at the RCS, 1824.

17. Article from *European Magazine* in 1782, reprinted in Abernethy, *Physiological Lectures,* pp. 341–352.

18. JH, *The Works,* vol. 1, p. 208. Hunter's idea of educating medical students so that they can educate themselves has been revived only recently by the United Kingdom's General Medical Council.

19. Joseph Adams, *Memoirs,* p. 75.

20. George Macilwain, *Memoirs of John Abernethy* (London: Hurst and Blackett, 1854), vol. 1, p. 253.

21. Foot, *The Life,* p. 245.

22. Ottley, "The Life," p. 48.

23. Ibid.

24. Abernethy, *Physiological Lectures,* p. 6; Adams, *Memoirs,* p. 73.

25. JH, *The Works,* vol. 1, p. 208.

26. The contents of Hunter's lectures are taken from JH, "Lectures on the Principles of Surgery," in *The Works,* vol. 1, unless otherwise specified. These are transcribed from shorthand notes taken by Nathaniel Rumsey in 1786 and 1787, although they follow very much the pattern of notes taken by other pupils at various times.

27. JH, *The Works,* vol. 1, p. 20.

28. Ibid., pp. 625–628.

29. The two quotes regarding Hunter's errors are from Ibid., p. 495, and from Stephen Jacyna, "Physiological Principles in the Surgical Writings of John Hunter," in *Medical Theory, Surgical Practice,* ed. Christopher Lawrence (London and New York: Routledge, 1992), p. 145, citing notes of Hunter's lectures, ms. 49 e 23, RCS.

30. JH, *The Works,* vol. 1, p. 406.

31. Ibid., p. 405.

32. Mary Coke, *The Letters and Journals of Lady Mary Coke,* ed. J. A. Home (Edinburgh: David Douglas, 1889–1896), vol. 4, p. 102 (entry for July 23, 1772).

33. Jacyna, "Physiological Principles in the Surgical Writings of John Hunter," citing JH, "Surgical Lectures," notes at the Royal College of Physicians of Edinburgh, ms. M8 47, pp. 208–209.

34. Sir D'Arcy Power, *Hunterian Oration 1925* (Bristol: John Wright and Sons Ltd., 1925), p. 9.

35. William Clift, relating details told to him by Henry Cline, in a note on the inside cover of JH, "Lectures on the Principles of Surgery," notes taken by Hopkinson, probably between 1781 and 1785, ms. RCS. Cline attended in 1774 or 1775.

36. R. C. Brock, *The Life and Work of Astley Cooper* (Edinburgh and London: E. & S. Livingstone, 1952), p. 145.

37. Abernethy, *Physiological Lectures,* p. 126 (introductory lecture, 1815).

38. Sir Arthur Porritt, "John Hunter: Distant Echoes," *Annals of the Royal College of Surgeons of England* 41 (1967): 10.

39. Foot, *The Life,* p. 280.

40. Abernethy, *Physiological Lectures,* p. 199.

41. JH, *Letters from the Past,* p. 10. The booklet reprints thirty-three of Hunter's letters. There are known to be fifty-one in all, of which thirty-two are kept at the library of the RCS. Others are also given in Baron, *The Life of Edward Jenner.* Many are undated. The letters quoted here are cited in the RCS booklet.

42. Ottley, "The Life," p. 36.

43. Home, "A Short Account," p. lxv.

44. Foot, *The Life,* p. 250; Lord Holland [Henry Richard Vassall Fox], *Further Memoirs of the Whig Party 1807–1821,* pp. 341–342.

45. Dobson, *William Clift FRS,* p. 11.

46. JH, *The Works,* vol. 1, p. 244.

47. F. Dudley Hart, "William Heberden, Edward Jenner, John Hunter and Angina Pectoris," *Journal of Medical Biography* 3 (1995): 56–58.

48. The poem, entitled "To the Memory of a Lovely Infant," was published in Anne Hunter, *Poems* (London: T. Payne, 1802).

49. Benjamin Franklin, *The Autobiography and Other Writings* (New York, Harmondsworth: Penguin Classics, 1986), pp. 214–215.

50. Fiona Haslam, *From Hogarth to Rowlandson: Medicine in Art in Eighteenth-Century Britain* (Liverpool: Liverpool University Press, 1996), pp. 196–198.

51. Edward Duyker and Per Tingbrand, *Daniel Solander, Collected Correspondence 1753–82* (Melbourne: Melbourne University Press, 1995), pp. 340–341 (letter from Solander to Ellis, November 7, 1774). More details on Walsh and his electric fish experiments can be found in Marco Piccolino and Marco Bresadola, "Drawing a Spark from Darkness: John Walsh and Electric Fish," *Trends in Neurosciences* 25 (2002): 51–57.

52. JH, "Anatomical Observations on the Torpedo," *Philosophical Transactions of the Royal Society* 63 (1773): 481–489 (New York: Johnson Reprint Co., 1965). Several preparations of the electric organs of the torpedo fish survive in Hunter's museum as specimens 2168–2179.

53. Duyker and Tingbrand, *Daniel Solander, Collected Correspondence 1753–82,* pp. 342–343 (letter from Solander to Banks, November 10, 1774).

54. JH, "An Account of the Gymnotus Electricus," *Philosophical Transactions of the Royal Society* 65 (1775): 395–407 (New York: Johnson Reprint Co., 1965). The specimens are 2185 and 2186 in the Hunterian Museum.

55. Otto Sonntag, ed., *John Pringle's Correspondence with Albrecht von Haller* (Basel: Schwabe, 1999), pp. 348–349 (letter from Pringle to Haller, December 13, 1776).

56. Duyker and Tingbrand, *Daniel Solander, Collected Correspondence 1753–82,* pp. 354–355 (letter from Solander to Banks, August 14, 1775).

57. JH to the Reverend James Baillie, November 23, 1775, HBC, vol. 7, p. 17.

58. Dobson, *William Clift FRS,* p. 13, citing Clift's description of Hunter's new coach.

59. David Hume, *The Letters of David Hume,* ed. J. Greig (Oxford: Clarendon Press, 1932), vol. 2, pp. 324–325 (letter from Hume to his brother John Home [Hume], June 10, 1776).

60. William Hickey, *Memoirs of William Hickey,* ed. A. Spencer (London: Hurst and Blackett, 1913), vol. 2, pp. 86–87.

61. James Perry, *The Torpedo: A Poem to the Electric Eel,* 1777, RCS.

12. The Chaplain's Neck

1. Details of Dodd's life and even more famous death are taken principally from Jessie Dobson, "John Hunter and the Unfortunate Doctor Dodd," *Journal of the History of Medicine* 10 (1955): 369–378; Rev. W. Foster, *Samuel Johnson and the Dodd Affair* (Lichfield: Johnson Society, 1951); and accounts in *Gentleman's Magazine,* as cited hereafter.

2. *Gentleman's Magazine,* 47 (1777), pp. 293–294.

3. Early nineteenth-century records reveal that out of thirty-six-bodies dissected after hanging between 1812 and 1830, the heart was still beating in ten, although the dissection still went ahead. See Jessie Dobson, "Cardiac Action after 'Death' by Hanging," *Lancet* 261 (1951): 1222–1224. The Greene and Duell cases are from Peter Linebaugh, "The Tyburn Riot Against the Surgeons," in *Albion's Fatal Tree: Crime and Society in Eighteenth-Century England,* ed. Douglas Hay (London: Allen Lane, 1975), p. 103.

4. JH, *The Works,* vol. 4, p. 153. The rabbit's ear experiment is on p. 152.

5. P. J. Bishop, *A Short History of the Royal Humane Society* (London: RHS, 1974). I am grateful for the help of Janet Smith at the RHS.

6. JH, "Proposals for the Recovery of Persons Apparently Drowned," read to the Royal Society on March 21, 1776; in *The Works,* vol. 4, pp. 165–175.

7. D. Duda, L. Brandt, and M. El Gindi, "The History of Defibrillation," in *The History of Anaesthesia: Proceedings of the Second International Symposium on the History of Anaesthesia Held in London 20–23 April 1987,* ed. R. Atkinson and T. Boulton (London and New York: Royal Society of Medicine, 1987), pp. 464–468.

I am grateful to Peter Baskett, retired anesthetist and editor of the journal *Resuscitation,* and John Zorab, retired consultant anesthetist, for advice on this topic.

8. *Gentleman's Magazine,* 47 (1777), p. 346.

9. Dobson, "John Hunter and the Unfortunate Doctor Dodd," citing the *London Review of English and Foreign Literature* of September 1777.

10. *Gentleman's Magazine,* 60 (1790), pp. 1010, 1066, 1077–1078.

11. Dobson, "John Hunter and the Unfortunate Doctor Dodd," citing the *Aberdeen Journal,* August 19, 1794.

12. Ibid., citing a letter from Charles Hutton in *Newcastle Magazine,* March 1822. Hutton related the tale only after all the participants in the story were dead.

13. JH, *Hunterian Reminiscences, Being the Substance of a Course of Lectures in the Principles and Practices of Surgery Delivered by Mr John Hunter in the Year 1785, Taken in Shorthand and Afterwards Fairly Transcribed by the Late Mr James Parkinson,* ed. J. W. K. Parkinson (London: Sherwood, Gilbert and Piper, 1833), p. 149.

14. Everard Home, "A Short Account," p. xxvii. Home refers to the episode in 1776, but it is understood to have been 1777.

15. "The Hunters and the Hamiltons, Some Unpublished Letters" *The Lancet* 214 (1928): 354–360. The article reprints several letters sent by William Hamilton to his father, Thomas Hamilton, in 1777–1778.

16. Sidney Lee and Leslie Stephen, eds., *Dictionary of National Biography* (London: Smith, Elder and Co., 1908–1909), vol. 9, pp. 1121–1122. The experiments on venom are described by Home in miscellaneous manuscripts; see ms. 49 e 5, RCS, with reference to 1782.

17. Cutting from *St James's Chronicle,* n.d. (1776), p. 10, HA.

18. Matthew Baillie, "A Short Memoir of My Life," mss. Baillie, p. 7, RCS.

19. Jessé Foot, *The Life,* p. 250.

20. *Royal Society Journal Book Copy,* 1777–1780, vol. 29, pp. 573–584 (entry for January 27, 1780).

21. JH, "On the Structure of the Placenta," *Royal Society Letters and Papers* 65 (1780): 138. The paper was later published in revised form in JH, *The Works,* vol. 4, pp. 60–71.

22. William Hunter to RS, February 3, 1780, *Royal Society Letters and Papers* 65 (1780): 138.

23. JH to Joseph Banks, February 17, 1780, *Royal Society Letters and Papers* 65 (1780): 140.

24. William Hunter, *Two Introductory Lectures* (London: printed by order of the trustees for J. Johnson, 1784), p. 64.

25. JH, *The Works,* vol. 1, p. 214.

26. JH, miscellaneous notes and extracts, n.d., ms. 49 e 19, RCS.

27. C. Helen Brock, *Calendar of the Correspondence of Dr William Hunter 1740–83* (Cambridge: Wellcome Unit for the History of Medicine, 1996), pp. 75–76.

28. Ibid., p. 89.

29. C. Helen Brock, ed., *William Hunter 1718–1783: A Memoir by Samuel Foart Simmons and John Hunter* (Glasgow: Glasgow University Press, 1983), p. 28.

30. Jessie Dobson, "John Hunter's Giraffe," *Annals of the Royal College of Surgeons of England* 24 (1959): 124–128; Lee and Stephen, eds., *Dictionary of National Biography*, vol. 15, pp. 471–473.

31. John Hunter's drawing books, RCS.

32. Foot, *The Life*, p. 246.

33. Edward Duyker and Per Tingbrand, *Daniel Solander, Collected Correspondence 1753–82* (Melbourne: Melbourne University Press, 1995), pp. 394–395 (letter from Solander to Banks, September 6, 1781). Hunter says he obtained the bottle-nosed whale in 1783 in the atlas accompanying "The Animal Oeconomy," but he must be mistaken, for Solander refers to the same animal in his letter of 1781, while a newspaper cutting of the same year is plainly the same whale. JH, *The Works*, vol. 4, pp. 331–392; newspaper cutting (untitled and undated, but handwritten note says 1781), facing p. 14 in HA.

34. JH, *The Works*, vol. 4, pp. 331–392.

35. Home, "A Short Account," p. xxii.

13. The Giant's Bones

1. Sidney Lee and Leslie Stephen, eds., *Dictionary of National Biography* (London: Smith, Elder and Co., 1908–1909), vol. 3, p. 579.

2. Jessie Dobson, *William Clift FRS* (London: Heinemann, 1954), pp. 118–119, citing research in the 1840s by a retired naval surgeon, Mr. Gough, in Ireland.

3. Sylas Neville, *The Diary of Sylas Neville 1767–1788*, ed. Basil Cozens-Hardy (Oxford: Oxford University Press, 1950), p. 290 (entry for February 22, 1782).

4. C. J. S. Thompson, *The Mystery and Lore of Monsters* (London: Williams and Norgate, 1930), pp. 63, 66.

5. Liza Picard, *Dr Johnson's London* (London: Weidenfeld and Nicolson, 2000), p. 251. Thanks to Dr. James Munro for suggesting that the man covered with scales suffered from ichthyosis.

6. William Le Fanu, "Hunter's Dwarfs," *Annals of the Royal College of Surgeons of England* 6 (1950): 446–449. The portrait *Teresa, the Corsican Fairy* was painted by William Hincks in 1774. Details of her subsequent death, after trying to deliver a normal-size child, were later added to Hunter's casebooks; see JH, *Case Books*, pp. 477–478.

7. Details about the giants exhibited in London are from Edward J. Wood, *Giants and Dwarfs* (London: Bentley, 1868), passim.

8. Jan Bondeson, *A Cabinet of Medical Curiosities* (Ithaca: Cornell University Press, 1997), p. 74.

9. *Morning Herald,* April 24, 1782, BL.

10. Ibid., April 30, 1782, BL.

11. *London Chronicle,* August 17–20, 1782, Guildhall Library.

12. Joseph Boruwlaski, *Memoirs of the Celebrated Dwarf Joseph Boruwlaski, a Polish Gentleman* (London, 1788), pp. 199–200; Wood, *Giants and Dwarfs,* pp. 330–343.

13. Dr. William Blackburne to Professor William Hamilton, 1789, Glasgow University Library, Special Collections MS Gen 1356/78. Blackburne said Boruwlaski had been "much countenanced by John Hunter." Hunter commissioned a painting by Philip Reinagle: *Joseph Boruwlaski* (1782), Royal College of Surgeons of England.

14. Boruwlaski, *Memoirs of the Celebrated Dwarf,* pp. 199–201.

15. *Morning Herald,* July 22, 1782, BL.

16. Bondeson, *A Cabinet of Medical Curiosities,* pp. 85–86.

17. JH to William Petty (earl of Shelburne), July 29, 1782, cited in S. Wood, "Two Further Letters of John Hunter and Notes on Rockingham's Last Illness from Hunter's Case Book," *Annals of the Royal College of Surgeons of England* 5 (1949): 347–350.

18. JH, *The Works,* vol. 4, pp. 34–43. The paper "An Account of the Free-Martin" was first presented to the Royal Society in 1779.

19. JH, *The Works,* vol. 4, pp. 44–49. The paper "An Account of an Extraordinary Pheasant" was first presented to the Royal Society in 1780. See also Brian Cook, *Contributions of the Hunter Brothers to Our Understanding of Reproduction, an Exhibition from the University Library's Collections* (leaflet, Glasgow: Glasgow University Library, Special Collections Department, 1992). As well as discussing Hunter's discoveries and Darwin's comments, this paper also notes that one of the three freemartins was not, in fact, a true example—although this made little difference to Hunter's conclusions.

20. JH, *The Works,* vol. 4, pp. 277–278.

21. JH, n.d., ms. 49 e 19, RCS.

22. Everard Home, "A Short Account," p. lxvi.

23. Article from *European Magazine* in 1782, reprinted in John Abernethy, *Physiological Lectures* (London: Longman, 1825), pp. 341–352. The £10,000 expenditure would be equivalent to about £600,000 today.

24. A set of notes of William's lectures bears an inscription stating that they were purchased from the executor of the late Mr. Howison, who had long been an assistant, probably "in a manual capacity," to William Hunter and Matthew Baillie, and records that Baillie made provision for Howison (personal communication to the author from Alan Callender, special collections assistant). John Howison, "Lectures Anatomical and Chirurgical by William Hunter," ms., 2 vols., 1775, Newcastle University Library, Pybus Collection. Howison also took notes of John's lectures. JH, *The Works,* vol. 1, p. 202. The preface refers to notes of Hunter's lectures owned by Benjamin Brodie, which were ascribed to Mr. Howison.

25. Drewry Ottley, "The Life," pp. 106–107. Ottley refers to Howison as "his [Hunter's] man" and describes how he was set to follow Byrne.

26. A. M. Landolt and M. Zachmann, "The Irish Giant: New Observations

Concerning the Nature of His Ailment," *The Lancet* (1980), no. 1: 1311–1312. Thanks to Professor John Wass, professor of endocrinology at the Radcliffe Infirmary, Oxford, and to the Pituitary Foundation for medical advice. Childhood-onset acromegaly, or gigantism, is the term for overproduction of a growth hormone in childhood. Acromegaly is the same condition, after normal growth has stopped, in adults. Both usually have the same cause: a benign tumor on the pituitary gland. Today the tumor would be removed by surgery.

27. The three entertainments were advertised in the *Morning Herald,* April 11, 1782 (Mr. Astley's show), August 1, 1782 (Mr. Katterfelto's show), and October 29, 1782 (Mr. Breslaw's show), as well as on various other dates, Guildhall Library.

28. Jessie Dobson, ed., *Descriptive Catalogue of the Physiological Series in the Hunterian Museum* (Edinburgh and London: E. & S. Livingstone, 1970), part 2, p. 201.

29. *Morning Herald,* November 18, 1782, Guildhall Library.

30. G. Frankcom and J. H. Musgrave, *The Irish Giant* (London: Duckworth, 1976), pp. 17, 26. Cotter eventually arrived in London in 1785. He was variously described as being between seven eight and eight seven.

31. *Morning Herald,* April 23, 1783, Guildhall Library; John Howison is listed as a house occupier in the Poor Rate Collector Books 1783, St Mary's in the Fields Parish Records, Guildhall Library.

32. C. Helen Brock, ed., *William Hunter 1718–1783: A Memoir by Samuel Foart Simmons and John Hunter* (Glasgow: University of Glasgow Press, 1983), p. 27.

33. JH, *Case Books,* p. 98.

34. Brock, ed., *William Hunter 1718–1783,* p. 71.

35. Obituary of William Hunter in *Gentleman's Magazine,* 53 (1783), p. 364.

36. Brock (ed.), *William Hunter 1718–1783,* p. 27.

37. Joseph Adams, *Memoirs,* pp. 133–134.

38. Matthew Baillie to R. Barclay, cited in Stephen Paget, *John Hunter,* p. 238.

39. Brock, ed., *William Hunter 1718–1783,* p. 27. The remark is found in John Hunter's annotations.

40. Adams, *Memoirs,* p. 92.

41. Tom Taylor, *Leicester Square: Its Associates and Its Worthies* (London: Bickers and Son, 1874), pp. 281, 341.

42. The house at 28 Leicester Square was later demolished and a pub currently stands on its site. A tatty bust of John Hunter stands in the square opposite. Details of the house's interior are shown in a plan sketched by Hunter's last assistant, William Clift in HA, p. 39.

43. Home, "A Short Account," p. xxix. The £6,000 expenditure would be equivalent to £180,000 today.

44. *Gentleman's Magazine,* 53 (1783), p. 541. Although the article states Byrne lost £700, a later court case attests to the theft of £770 in two banknotes.

45. *Morning Herald,* June 5, 1783, BL.

46. Ibid., June 16, 1783, BL.

47. *Parker's General Advertiser,* June 5, 1783, BL.

48. *Gentleman's Magazine,* 53 (1783), p. 541; *British Magazine,* n.d. (1783), in HA; *Annual Reporter Chronicle,* June 1783, cited in "The Demolition of Earl's Court House," *West London Observer,* February 6, 1886, facing p. 10 in HA.

49. Details of Hunter's theft of the giant's body are given in Ottley, "The Life," pp. 106–107; and Taylor, *Leicester Square,* pp. 403–407. Both relate essentially the same story. The latter description, which includes details of the barn swap, is told by Richard Owen, based on the story handed down from Hunter's last assistant, William Clift. Hunter would later claim he had paid 130 guineas.

50. Hunter's outlay would be the equivalent of about thirty thousand pounds today.

51. JH to Edward Jenner, n.d. (1783), in John Baron, *The Life of Edward Jenner* (London: Henry Colburn, 1827), p. 65.

52. JH to Joseph Banks, n.d. (1787), transcribed in a leaflet, "John Hunter at Earl's Court Kensington 1764–93," reprinted in *The Atheneum* (1869–1870), HA. The original letter is said to have been destroyed when a bomb fell on the Hunterian Museum in 1941.

53. JH, *Case Books,* p. 382 (the Reverend Mr. Vivian, specimen P 205), p. 9 (Lady Beauchamp, specimen P 389), and p. 393 (Lieutenant General Desaguliers, specimen P 292); Jessie Dobson, "Some of John Hunter's Patients," *Annals of the Royal College of Surgeons* 42 (1968): 124–133 (Hon. Frederick Cornwallis, specimens P 378 and 379).

54. JH, *Case Books,* pp. 548–550.

14. The Poet's Foot

1. JH to Edward Jenner, April 22, 1785, in JH, *Letters from the Past,* p. 36.

2. Joseph Adams, *Memoirs,* p. 93.

3. *The Sketch,* February 24, 1897, excerpt facing p. 39 in HA. This article, reporting the planned demolition of Hunter's home at 28 Leicester Square in 1897, relates that Stevenson "is said to have chosen" the house as the scene for the activities of Dr. Jekyll and Mr. Hyde. Stevenson quotations are from Robert Louis Stevenson, *The Strange Case of Dr Jekyll and Mr Hyde* (London and Glasgow: Collins, 1953; first published 1886), pp. 37–38.

4. William Clift, "Ground Plan of John Hunter's House Based on the Plan Drawn from Memory and Annotated by William Clift in 1832," p. 39, in HA. Everard Home gave the dimensions of the extension in Home, "A Short Account," p. xxx.

5. James Williams to Mary Williams, October 8, 1793, cited in G. Edwards, "John Hunter's Last Pupil," *Annals of the Royal College of Surgeons of England* 42 (1968): 68–70.

6. Home, "A Short Account," p. l–li.

7. Lord Holland [Henry Richard Vassall Fox], *Further Memoirs of the Whig Party 1807–1821* (London: John Murray, 1905), p. 344; Drewry Ottley, "The Life," p. 121.

8. JH to Edward Jenner, May 1788, cited in Ottley, "The Life," p. 110.

9. Home, "A Short Account," p. l–li. Home describes Hunter's various episodes of illness and treatment in detail.

10. Benjamin Franklin to Benjamin Vaughan, July 1785, cited in George Corner and Willard E. Goodwin, "Benjamin Franklin's Bladder Stone," *Journal of the History of Medicine* 8 (1953): 359–377.

11. JH et al. to Benjamin Vaughan, n.d. (1785), cited in Corner and Goodwin, "Benjamin Franklin's Bladder Stone," p. 366. One of the physicians was also named John Hunter, here Latinized as Ionnes Hunter, who is occasionally confused with the surgeon.

12. H. Leigh Thomas, Hunterian Oration 1827, WL.

13. Adams, *Memoirs*, p. 199.

14. "John Hunter," newspaper clipping, no title, n.d., p. 13, in HA.

15. George C. Peachey, *A Memoir*, pp. 164–165, citing an advertisement in the *Gazetteer and New Daily Advertiser*, October 1, 1785. Clift's plan shows the hat pegs and pupils' register.

16. JH, *Hunterian Reminiscences, Being the Substance of a Course of Lectures in the Principles and Practices of Surgery Delivered by Mr John Hunter in the Year 1785, Taken in Shorthand and Afterwards Fairly Transcribed by the Late Mr James Parkinson*, ed. J. W. K. Parkinson (London: Sherwood, Gilbert and Piper, 1833).

17. John Abernethy, *Physiological Lectures* (London: Longman, 1825), p. 134 (introductory lecture, 1815).

18. Bransby Blake Cooper, *The Life of Sir Astley Cooper* (London: J. W. Parker, 1843), vol. 1, p. 142.

19. The pupil's register of St. George's, vol. 1 (with reference to May 1789), transcript at RCS. Physick's career is also discussed in Sir Ernest Finch, "The Influence of the Hunters on Medical Education," *Annals of the Royal College of Surgeons of England* 20 (1957): 205–248.

20. Cuthbert E. Dukes, "London Medical Societies in the Eighteenth Century," *Proceedings of the Royal Society of Medicine Section of the History of Medicine* 53 (1960): 699–706.

21. Jessie Dobson, *John Hunter*, pp. 241–242.

22. Thomas Chevalier, Hunterian Oration 1821, WL. Chevalier was one of Hunter's pupils.

23. Peachey, *A Memoir*, pp. 164–165, citing an advertisement in the *Gazetteer and New Daily Advertiser*, October 1, 1785.

24. Letter from John Gunning, William Walker, and Thomas Keate to the governors of St. George's, n.d. (1793), printed verbatim in Peachey, *A Memoir*, pp. 282–296.

25. Cooper, *The Life of Sir Astley Cooper*, p. 232.

26. JH, *Treatise on the Venereal Disease*; Adams, *Memoirs*, p. 101.

27. Home, "A Short Account," p. lxvi; Clift, "Ground Plan of John Hunter's House." The plan shows Hunter's name by the door.

28. David Mannings, *Sir Joshua Reynolds: A Complete Catalogue of His Paintings* (New Haven: Yale University Press, 2000), pp. 271–272. Hunter's portrait was

no. 223, entitled *Portrait of a gentleman, half length,* in the customary style of many portrait titles, in the original catalog. Thanks to Elizabeth King, research assistant at the Royal Academy Library, for information on the portrait and the exhibition. Thanks also to Andrew Cunningham for explaining the archetypal philosopher's pose.

29. Selwyn Taylor, *John Hunter and His Painters* (London: Royal College of Surgeons of England, 1993), p. 1. Reynolds's first portrait now hangs in the Court Room at the headquarters of the Society of Apothecaries, Apothecaries' Hall. It is understood to have been donated by Weatherall's nephew, Thomas Knight. Many thanks to Dee Cook, archivist of the Society of Apothecaries. The life mask is preserved at the RCS.

30. The items displayed in the Reynolds portrait are discussed in Taylor, *John Hunter and His Painters;* Sir Arthur Keith, "The Portraits and Personality of John Hunter," *British Medical Journal* (1928): 205–209; Lord Brock, "Background Details in Reynolds's Portrait of John Hunter," *Annals of the Royal College of Surgeons of England* 48 (1971): 219–226; and George Qvist, *John Hunter,* pp. 188–189.

31. The sketch of angled lines has been variously considered to show facial angles and branching arteries. My thanks to Dr. Alistair Hunter, academic manager of the dissecting rooms at Guy's, King's and St Thomas' School of Biomedical Sciences, who convincingly suggests they depict muscle fibers. This is probably a reference to the Croonian lectures on the muscles, which Hunter delivered to the RS.

32. Ottley, "The Life," p. 121; Ernest E. Irons, "The Last Illness of Sir Joshua Reynolds," *Bulletin of the Society of Medical History of Chicago* 5 (1939): 119–142.

33. John Ehrman, *The Younger Pitt* (London: Constable, 1969), vol. 1, p. 594. My thanks to Cyrus Kerawala, maxillofacial surgeon at the Royal Surrey County Hospital, Guildford, for advice on the likely form of the cyst.

34. Home, "A Short Account," pp. lvii–lviii.

35. John Richardson, *The Annals of London* (London: Cassell and Co., 2000), p. 218.

36. JH, *Essays and Observations,* vol. 1, pp. 398–400.

37. W. S. Lewis et al., eds., *The Yale Edition of Horace Walpole's Correspondence* (New Haven: Yale University Press, 1937–1961), vol. 33, p. 535 (letter from Walpole to Lady Ossory, November 4, 1786).

38. Adam Sisman, *Boswell's Presumptuous Task* (London: Hamish Hamilton, 2000), pp. 156–157.

39. James Boswell, *Private Papers of James Boswell from Malahide Castle,* ed. G. Scott and F. Pottle (New York: W. E. Rudge, 1933), vol. 17, p. 17 (entry for March 21, 1787), BL.

40. Ibid., pp. 74–75, 77, 83 (entries for March 7, 12, and 18, 1788).

41. R. H. Campbell and A. S. Skinner, *Adam Smith* (London and Canberra: Croom Helm, 1982), p. 202; John Rae, *Life of Adam Smith* (London: Macmillan and Co., 1895), p. 402.

42. Adam Smith to William Strahan, December 20, 1777, in Adam Smith, *The Correspondence of Adam Smith,* ed. Ernest Campbell Mossner and Ian Simpson Ross (Oxford: Clarendon Press, 1977), pp. 229–230.

43. Rae, *Life of Adam Smith,* p. 406.

44. Adam Smith to Henry Dundas, July 18, 1787, in Smith, *The Correspondence of Adam Smith,* pp. 306–307.

45. Benita Eisler, *Byron: Child of Passion, Fool of Fame* (London: Hamish Hamilton, 1999), pp. 12–13; A. B. Morrison, "Byron's Lameness," *The Byron Journal* (1975): 24–31.

46. JH, *Case Books,* p. 148.

47. Eisler, *Byron,* pp. 42–43.

48. Thomas Gainsborough to unknown recipient, n.d. (but assumed to be April 1788), ms. 5610, Hunterian Society collection at WL.

49. The relationship between Haydn and Anne Hunter is discussed in Aileen K. Adams, " 'I Am Happy in a Wife': A Study of Mrs John Hunter (1742–1821)," in *Papers Presented at the Hunterian Bicentenary Commemorative Meeting* (London: Royal College of Surgeons of England, 1995), pp. 32–37. Hadyn's experience as Hunter's patient is described in B. Bugyi, "J. Haydn and the Hunters," in *Proceedings of the XXIII International Congress of the History of Medicine, London September 2–9, 1972* (London: Wellcome Institute of the History of Medicine, 1974), vol. 2, pp. 904–907. In JH, *The Works,* vol. 1, pp. 568–569, Hunter says of nasal polyps that "the best mode of removing them is with a forceps."

50. James Joseph Walsh, *History of Medicine in New York: Three Centuries of Medical Progress* (New York: National Americana Society, 1919), vol. 2, pp. 382–389.

51. Dobson, *John Hunter,* p. 179.

52. Jane M. Oppenheimer, "A Note on William Blake and John Hunter," *Journal of the History of Medicine* 1 (1946): 41–45; William Blake, *Poetry and Prose of William Blake,* ed. G. Keynes (London: Nonesuch Press, 1927), pp. 865–887; Peter Ackroyd, *Blake* (London: Sinclair-Stevenson, 1995), pp. 30, 81.

15. The Monkey's Skull

1. Newspaper cutting (no title), n.d. (penciled 1788), p. 13 in HA.

2. Richard Owen, ed., *Descriptive and Illustrated Catalogue of the Physiological Series of Comparative Anatomy* (London: RCS, 1840), vol. 5, pp. 177–178.

3. *A Guide to the Hunterian Museum* (London: RCS, 1993), p. 21.

4. Newspaper cutting (no title), n.d. (penciled 1788), p. 13 in HA.

5. Ibid.

6. Jessie Dobson, *John Hunter,* p. 190.

7. Benjamin Hutchinson, *Biographia Medica, or Historical and Critical Memoirs of the Lives and Writings of the Most Eminent Medical Characters That Have Existed from the Earliest Account of Time to the Present Period* (London: J. Johnson, 1799), vol. 1, pp. 495–496. The total number of fossils is given as 2,773 by Richard Owen

in JH, *Essays and Observations,* vol. 1, p. 293. Frederic Wood Jones, in "John Hunter as a Geologist," *Annals of the Royal College of Surgeons of England* 12 (1953): 219–245, especially p. 233, calculated the collection at 2,957. Editors of the 1859 publication of *Observations and Reflections on Geology* erroneously gave the total as 415.

8. Drewry Ottley, "The Life," p. 72; *Gentleman's Magazine* 63 (1793), MS 5610, Hunterian Society Collection at WL.

9. Ottley, "The Life," p. 116.

10. Jessie Dobson, "John Hunter's Animals," *Annals of the Royal College of Surgeons of England* 17 (1962): 379–486.

11. JH to an unknown recipient, January 15, 1793, Hunterian Letters 49 b 18a, Grey-Turner Bequest, RCS.

12. Owen, ed., *Descriptive and Illustrated Catalogue of the Physiological Series of Comparative Anatomy,* p. 41.

13. W. S. Lewis et al., eds., *The Yale Edition of Horace Walpole's Correspondence* (New Haven: Yale University Press, 1937–1961), vol. 15, p. 241 (letter from Walpole to the Reverend Robert Nares, October 5, 1793).

14. "Penny Cyclopaedia," cited in Tom Taylor, *Leicester Square: Its Associates and Its Worthies* (London: Bickers and Son, 1874), pp. 418–419.

15. Jessie Dobson, *A Guide to the Hunterian Museum (Physiological Series)* (London: E & S Livingstone, 1958), passim.

16. Newspaper cutting (no title), n.d. (penciled 1788), p. 13 in HA.

17. Everard Home, "A Short Account," p. xxxv.

18. Richard Owen, preface to "The Animal Oeconomy" in JH, *The Works,* vol. 4, p. xxxviii. The Italian anatomist Antonio Scarpa visited the collection in 1781 and the Dutch anatomist Peter Camper in 1785. Johann Friedrich Blumenbach, professor of medicine at Göttingen, visited in the early 1790s. My thanks to Simon Chaplin for additional information.

19. Stephen Paget, *John Hunter,* p. 230.

20. For a comprehensive overview of the development of ideas on evolution, see Peter J. Bowler, *Evolution: The History of an Idea,* 3d ed. (Berkeley: Los Angeles, London: University of California Press, 2003). Other useful summaries include John C. Greene, *The Death of Adam: Evolution and Its impact on Western Thought,* 5th ed. (Ames: Iowa State University Press, 1981); and Roy Porter, *The Making of Geology: Earth Sciences in Britain 1660–1815* (Cambridge: Cambridge University Press, 1977). Very many thanks to Andrew Cunningham for helping me to understand Hunter's contribution to this field.

21. Stephen Inwood, *The Man Who Knew Too Much: The Strange and Inventive Life of Robert Hooke 1635–1703* (London: Macmillan, 2002), pp. 125–126.

22. Bowler, *Evolution,* p. 51

23. Bowler, *Evolution,* pp. 61–62; Greene, *The Death of Adam,* p. 76; Porter, *The Making of Geology,* p. 159.

24. Bowler, *Evolution,* pp. 69–70; John Gribben, *Science: A History 1543–2001* (London: Allen Lane, 2002), pp. 218–219.

25. Greene, *The Death of Adam,* pp. 189–191; Bowler, *Evolution,* p. 52. Camper's conclusions were published in 1791, two years after his death.

26. Bowler, *Evolution,* pp. 52–53.

27. Gribben, *Science,* pp. 221–226; Charles Coulston Gillispie, ed., *Dictionary of Scientific Biography* (New York: Charles Scribner's Sons, 1970), vol. 2, pp. 576–582.

28. Other contemporaries of Hunter were similarly investigating pre-Darwinian evolutionary ideas, but they published their conclusions after his death. Erasmus Darwin, who had attended the Hunter brothers' school back in 1753, put his views on the origins of life into verse. His first poetic efforts had been published in 1789, although his theory that all life developed from a single ancestor—"one living filament"—would only clearly be outlined in his two-volume work *Zoonomia* (1794 and 1796). See Charles Coulston Gillispie, ed., *Dictionary of Scientific Biography,* vol. 3, pp. 577–580. The French naturalist Jean Baptiste Pierre Antoine de Monet de Lamarck published his belief in a primitive common ancestor in 1801 and then more fully in 1815. See Gillispie, ed., *Dictionary of Scientific Biography,* vol. 7, pp. 584–593; Gribben, *Science,* pp. 335–338; Bowler, *Evolution,* pp. 86–95.

29. For details on Charles Darwin, see Adrian Desmond and James Moore, *Darwin* (London: Penguin, 1991). A useful short summary of Darwin and his ideas can be found in Patrick Tort, *Charles Darwin: The Scholar Who Changed Human History* (London: Thames & Hudson, 2001).

30. JH to Edward Jenner, July 6, 1777, and March 29, 1778, in JH, *Letters from the Past,* pp. 17, 20. Hunter told Jenner, "I am matching my Fossill [sic] as far as I can with the resent [sic]."

31. JH, *The Works,* vol. 4, p. 36.

32. JH, *The Works,* vol. 4, pp. 277–285.

33. JH, *Essays and Observations,* vol. 1, p. 228.

34. JH, *The Works,* vol. 4, pp. 319–330, The paper "Observations Tending to Show That the Wolf, Jackal, and Dog Are All of the Same Species" was first published in *Philosophical Transactions of the Royal Society* in 1787.

35. Edward Jenner to JH (n.d.), cited in JH, "Observations Tending to Show That the Wolf, Jackal, and Dog Are All of the Same Species."

36. JH, *The Works,* vol. 4, pp. 470–480. The paper "Observations on the Fossil Bones Presented to the Royal Society by His Most Serene Highness the Margrave of Anspach by the Late John Hunter," was read to the Royal Society by Everard Home on May 8, 1794.

37. JH, *Observations and Reflections on Geology.* Two copies of the ms. version of this treatise exist. One is believed to have been taken in dictation by William Bell and William Clift (ms. 49 c 1), the other to have been copied by Clift between 1793 and 1800 (ms. 49 c 2). Both are preserved at the RCS Library.

38. JH, *Observations and Reflections on Geology,* p. x. Hunter's contribution to geology is discussed in Jones, "John Hunter as a Geologist," pp. 219–245; and in George Qvist, "Some Controversial Aspects of John Hunter's Life and Work: Part 5, Geology and Palaeontology," *Annals of the Royal College of Surgeons of England* 61 (1979): 381–384.

39. JH, *Observations and Reflections on Geology,* pp. xlvi.

40. J. Rennell to JH (n.d.), transcript in RCS, ms. 49 c 2, pp. 102–105. JH, ms. 49 c 1 and ms. 49 c 2, RCS. It has always been assumed that Rennell was commenting on the *Observations and Reflections on Geology* treatise, partly because later editors changed "thousands of years" to "thousands of centuries" in this, and partly because Rennell's letter is attached to the two mss. from which it was published. It is conceivable that he was really commenting on Hunter's "Observations on the Fossil Bones," which also contains the phrase "many thousands of years." In some ways, this would make more sense, as the paper was intended for the RS, while it is unclear why Hunter would seek Rennell's views on his other treatise. Home read the fossils paper to the RS in 1794, after Hunter's death, and either he or Hunter might have made the amendment.

41. JH to Edward Jenner, August 17 (no year, but identified as 1789), in JH, *Letters from the Past,* p. 39; JH, miscellaneous notes and extracts, n.d., ms. 49 e 19, RCS. Joseph Adams, a former pupil, stated that Hunter had been described as a materialist; see Adams, *Memoirs,* p. 233.

42. JH, *Essays and Observations,* vol. 1, p. 3. It was Richard Owen who compiled Hunter's various notes into the two volumes.

43. Ibid., p. 4.

44. Ibid., p. 246.

45. Ibid., p. 9.

46. Ibid., p. 203.

47. Charles Darwin, *The Descent of Man,* ed. Richard Dawkins (London: Gibson Square Books Ltd., 2003; first published 1871), pp. 9–10.

48. JH, *Essays and Observations,* vol. 1, p. 43.

49. Ibid., p. 37. This section is headed "On the Origin of Species," although the title was probably added by Richard Owen while editing the original manuscript. My thanks to Simon Chaplin for elucidating this point and other advice on these notes.

50. George Qvist, "Some Controversial Aspects of John Hunter's Life and Work: Part 6, Evolution," *Annals of the Royal College of Surgeons of England* 61 (1979): 478–483. Qvist called the collection "a museum of evolution." Thanks to Andrew Cunningham for explaining that it was not.

51. JH, *Essays and Observations,* vol. 1, preface by Owen.

16. The Anatomist's Heart

1. Details of Clift's early life are taken from William Clift, "On His Condition as Hunter's Clerk," ms. 49 e 45, RCS; William Clift, "A Short Account of My Life" (1840), ms. 49 e 45, RCS; Richard Owen, biography of William Clift, n.d., ms. 49 e 45, RCS; Jessie Dobson, *William Clift FRS* (London: Heinemann, 1954); Frances Austin, ed., *The Clift Family Correspondence 1792–1846* (Sheffield: University of Sheffield, 1991).

2. William Clift, "List of John Hunter's Household" (1792), ms. 49 e 68, RCS.

3. William Clift to Elizabeth Clift, March 5, 1792, in Austin, *The Clift Family Correspondence 1792–1846,* p. 30.

4. Dobson, *William Clift FRS,* pp. 10–11.

5. John Abernethy, *Physiological Lectures* (London: Longman, 1825), p. 209, quoting William Clift.

6. Benjamin Hutchinson, *Biographia Medica, or Historical and Critical Memoirs of the Lives and Writings of the Most Eminent Medical Characters That Have Existed from the Earliest Account of Time to the Present Period* (London: J. Johnson, 1799), vol. 1, p. 483.

7. William Clift to Elizabeth Clift, December 24, 1792, in Austin, *The Clift Family Correspondence 1792–1846,* pp. 49–50.

8. William Clift to John Clift, October 11, 1792, in Austin, *The Clift Family Correspondence 1792–1846,* pp. 47–48; Jessie Dobson, "A Note on John Hunter," *Annals of the Royal College of Surgeons of England* 15 (1954): 345–346.

9. Dobson, *William Clift FRS,* p. 109.

10. R. H. Franklin, "John Hunter and His Relevance in 1977," *Annals of the Royal College of Surgeons of England* 60 (1978): 266–273.

11. William Clift, "A Short Account of My Life," n.d., ms. 49 e 45, RCS.

12. Richard Owen, ed., *Descriptive and Illustrated Catalogue of the Physiological Series of Comparative Anatomy* (London: RCS, 1840), vol. 5, pp. 359, 11–12.

13. JH, *The Works,* vol. 4, pp. 422–466. The paper "Observations on Bees" was first read to the Royal Society on February 23, 1792.

14. Dobson, *William Clift FRS,* p. 10.

15. Jessé Foot, *The Life,* p. 242.

16. W. R. Le Fanu, "John Hunter's Buffaloes," *British Medical Journal* 2 (1931): 574.

17. Dobson, *William Clift FRS,* p. 8.

18. Details of Home's life can be found in A. W. Beasley, *Home Away from Home* (Wellington: Central Institute of Technology, 2000).

19. Everard Home, "A Short Account," p. lxv.

20. JH to his colleagues, July 9, 1792, in George C. Peachey, *A Memoir,* pp. 272–273. The complete correspondence between Hunter and his colleagues in their row of 1792–1793 is printed verbatim in Peachey's biography.

21. John Gunning, William Walker, and Thomas Keate to JH, October 4, 1792, in Peachey, *A Memoir,* pp. 274–275.

22. Owen, ed., *Descriptive and Illustrated Catalogue of the Physiological Series of Comparative Anatomy,* vol. 5, pp. 120–126.

23. A. Peterkin, William Johnston, and R. Drew, *Commissioned Officers in the Medical Services of the British Army 1660–1960* (London: Wellcome Historical Medical Library, 1968), vol. 1, p. 33. Details of Hunter's later army career can be found in Lloyd G. Stevenson, "John Hunter, Surgeon-General 1790–1793," *Journal of the History of Medicine* 19 (1964): 239–266. Stevenson cites Hunter's correspondence collected at the Public Record Office. Thanks to Andrew Cunningham for helping put Hunter's approach into an eighteenth-century context.

24. JH to the governors of St. George's Hospital, February 28, 1793, in Peachey, *A Memoir,* pp. 275–282.

25. The surgeons' reply (n.d.), in Peachey, *A Memoir,* pp. 282–296.

26. Surgeons' letter to the committee appointed to examine the laws relative to the surgeons' pupils and to consider the best method of improving their education, May 27, 1793, in Peachey, *A Memoir,* pp. 297–303.

27. James Williams to Mary Williams, October 8, 1793, in G. Edwards, "John Hunter's Last Pupil," *Annals of the Royal College of Surgeons of England* 42 (1968): 68–70.

28. James Williams to Mary Williams, October 16, 1793, in Edwards, "John Hunter's Last Pupil."

29. William Clift, note added in JH, n.d., ms. 49 e 19, RCS.

30. William Clift to Elizabeth Clift, October 18, 1793, in Austin, *The Clift Family Correspondence 1792–1846,* p. 79.

31. Jessie Dobson, "John Hunter's Animals," *Annals of the Royal College of Surgeons of England* 17 (1962): 379–486.

32. James Williams to Mary Williams, October 16, 1793, in Edwards, "John Hunter's Last Pupil."

33. William Clift to Elizabeth Clift, October 18, 1793, in Austin, *The Clift Family Correspondence 1792–1846,* p. 79.

34. Jessie Dobson, *John Hunter,* p. 344.

35. Peachey, *A Memoir,* pp. 219–221.

36. Drewry Ottley, "The Life," pp. 131–132. Details of Hunter's last minutes are also related in James Williams to Mary Williams, October 16, 1793, in Edwards, "John Hunter's Last Pupil"; William Clift to Elizabeth Clift, October 18, 1793, in Austin, *The Clift Family Correspondence 1792–1846,* p. 79; and Home, "A Short Account," p. lxi.

37. William Clift to Elizabeth Clift, October 18, 1793, in Austin, *The Clift Family Correspondence 1792–1846,* p. 79; James Williams to Mary Williams, October 16, 1793, in Edwards, "John Hunter's Last Pupil."

38. William Clift to Elizabeth Clift, October 18, 1793, in Austin, *The Clift Family Correspondence 1792–1846,* p. 79.

39. W. S. Lewis et al., eds., *The Yale Edition of Horace Walpole's Correspondence* (New Haven: Yale University Press, 1937–1961), vol. 12, p. 38 (letter from Walpole to M. Berry, October 19, 1793), and vol. 15, p. 244 (letter from Walpole to the Reverend Robert Nares, October 20, 1793).

40. Joseph Farington, *The Farington Diary,* ed. J. Greig (London: Hutchinson and Co., 1922–1928), vol. 1, pp. 6–7 (entry for October 17, 1793).

41. *European Magazine,* November 1793, clipping on p. 11 of the HA; *Gentleman's Magazine,* 63 (1793), pp. 964–965; The *Sun,* October 1793, cutting on p. 16 of the HA.

42. Lloyd Allan Wells, "Aneurysm and Physiological Surgery," *Bulletin of the History of Medicine* 44 (1970): 422.

43. Home, "A Short Account," pp. lxii–lxv.

44. Dobson, *John Hunter,* p. 349.

45. William Clift to Elizabeth Clift, November 20, 1793, in Austin, *The Clift Family Correspondence 1792–1846,* p. 81.

46. William Clift to Elizabeth Clift, October 18, 1793, in Austin, *The Clift Family Correspondence 1792–1846,* p. 79.

47. Clift, "List of John Hunter's Household."

48. Ottley, "The Life," p. 137.

49. Sir Arthur Porritt, "John Hunter's Women," *Transactions of the Hunterian Society* 17 (1958–1959): 81–111.

50. William Clift to Elizabeth Clift, October 18, 1793, in Austin *The Clift Family Correspondence 1792–1846,* p. 79.

51. Ottley, "The Life," p. 142.

52. William Clift, "Note on the Preservation of the Hunterian Observations Before Their Destruction by Sir Everard Home," n.d., ms. 49 e 45, RCS.

53. William Clift, evidence to the parliamentary committee on medical education (1834), in JH, *Essays and Observations,* vol. 2, pp. 493–500; and in Stephen Paget, *John Hunter,* pp. 252–256.

54. Paget, *John Hunter,* p. 253, quoting Clift.

55. Beasley, *Home Away from Home,* p. 83, quoting article in *The Lancet* (1832).

56. George Qvist, *John Hunter,* p. 72.

57. Abernethy, *Physiological Lectures,* p. 5.

58. Sir Arthur Keith, "Memorable Visits of Charles Darwin to the Museum of the Royal College of Surgeons," *Annals of the Royal College of Surgeons of England* 11 (1952): 362–363.

59. Newspaper clipping on p. 24 in the HA; *Illustrated London News of the World,* April 2, 1859, on p. 27 in the HA; newspaper clipping (n.d.), Hunterian Society Catalogue, MS 5616, 29/2, WL.

60. Richard Owen, biography of William Clift, ms. 49 e 45, RCS.

NOTE: The initials JH refer to John Hunter; WH refers to William Hunter.